# STUDENT INTERACTIVE WORKBOOK

for Starr, Evers, and Starr's

# BIOLOGY TODAY AND TOMORROW

*with Physiology*
SECOND EDITION

### RICHARD W. CHENEY, JR.
*Christopher Newport University*

### JOHN D. JACKSON
*North Hennepin Community College*

Australia • Brazil • Canada • Mexico • Singapore • Spain • United Kingdom • United States

© 2007 Thomson Brooks/Cole, a part of The Thomson Corporation. Thomson, the Star logo, and Brooks/Cole are trademarks used herein under license.

ALL RIGHTS RESERVED. No part of this work covered by the copyright hereon may be reproduced or used in any form or by any means—graphic, electronic, or mechanical, including photocopying, recording, taping, Web distribution, information storage and retrieval systems, or in any other manner—without the written permission of the publisher.

Printed in the United States of America
1 2 3 4 5 6 7 10 09 08 07 06

Printer: Thomson/West

0-495-10885-5
Cover image: Jay Barnes

**Thomson Higher Education**
**10 Davis Drive**
**Belmont, CA 94002-3098**
**USA**

> For more information about our products, contact us at:
> **Thomson Learning Academic Resource Center**
> **1-800-423-0563**
>
> For permission to use material from this text or product, submit a request online at
> **http://www.thomsonrights.com.**
> Any additional questions about permissions can be submitted by email to **thomsonrights@thomson.com.**

# CONTENTS

1 Invitation to Biology   1

## I CELLS
2 Molecules of Life   10
3 How Cells Are Put Together   23
4 How Cells Work   33
5 Where It Starts—Photosynthesis   42
6 How Cells Release Chemical Energy   50

## II GENETICS
7 How Cells Reproduce   60
8 Observing Patterns in Inherited Traits   71
9 DNA Structure and Function   82
10 Gene Expression and Control   89
11 Studying and Manipulating Genomes   98

## III EVOLUTION AND DIVERSITY
12 Processes of Evolution   106
13 Evolutionary Patterns, Rates, and Trends   115
14 Early Life   126
15 Plant Evolution   139
16 Animal Evolution   150
17 Plants and Animals: Common Challenges   167

## IV HOW PLANTS WORK
18 Plant Form and Function   173
19 Plant Reproduction and Development   188

## V HOW ANIMALS WORK
20 Animal Tissues and Organ Systems   199
21 How Animals Move   209
22 Circulation and Respiration   216
23 Immunity   228
24 Digestion, Nutrition, and Excretion   237
25 Neural Control and the Senses   248
26 Endocrine Controls   262
27 Reproduction and Development   271

## VI ECOLOGY
28 Population Ecology   285
29 Community Structure and Biodiversity   296
30 Ecosystems   307
31 The Biosphere   317
32 Behavioral Ecology   327

**ANSWERS   336**

## CREDITS

This page constitutes an extension of the copyright page. We have made every effort to trace the ownership of all copyrighted material and to secure permission from copyright holders. In the event of any question arising as to the use of any material, we will be pleased to make the necessary corrections in future printings. Thanks are due to the following authors, publishers, and agents for permission to use the material indicated.

### Chapter 2.
**17:** PDB file courtesy of Dr. Christina A. Bailey.
**18:** PDB ID: 1BBB; Silva, M.M., Rogers, P.H., Arnone, A.; A third quarternary structure of human hemoglobin A at 1.7-A resolution; *J Biol Chem* 267 pp. 17248 (1992).

### Chapter 9.
**84:** PDB ID: 1BBB; Silva, M.M., Rogers, P.H., Arnone, A.; A third quarternary structure of human hemoglobin A at 1.7-A resolution; *J Biol Chem* 267 pp. 17248 (1992).

### Chapter 13.
**118:** (right) Frans Lanting/Minden Pictures; computer enhanced by Lisa Starr
**118:** (center) J. Scott Altenbach, University of New Mexico, computer enhanced by Lisa Starr
**118:** (left) © Stephen Dalton/Photo Researchers, Inc.

### Chapter 14.
**135:** © Science Photo Library/Photo Researchers, Inc.
**138:** (1 top) © Andrew Syred/SPL/Photo Researchers, Inc.
**138:** (1 bottom) Courtesy © James Evarts
**138:** (2) © Astrid Hanns-Frieder Michler/SPL/Photo Researchers, Inc.
**138:** (3) © Michael Wood/mykob.com
**138:** (4) © Wim van Egmond
**138:** (6) © Lewis Trusty/Animals Animals
**138:** (7) © Chris Worden
**138:** (9 top) CNRI/SPL/Photo Researchers, Inc.
**138:** (9 bottom) © P. Hawtin, University of Southampton/SPL/Photo Researchers, Inc.
**138:** (10) © Lawson Wood/CORBIS
**138:** (11) Edward S. Ross
**138:** (12) © Oliver Meckes/Photo Researchers, Inc.

### Chapter 16.
**162:** (1) Monterey Bay Aquarium/Enlightened Images
**162:** (2) © L. Jensen/Visuals Unlimited
**162:** (3) © Frans Lemmens/The Image Bank/Getty Images
**162:** (4) © Cabisco/Visuals Unlimited
**163:** (5) © Peter Parks/Imagequestmarine.com
**163:** (7) © Joe McDonald/CORBIS
**163:** (8) © Ivor Fulcher/Corbis
**164:** (9) Chris Anderson/Darklight Imagery
**164:** (10) Herve Chaumeton/Agence Nature
**164:** (12) © Andrew Syred/SPL/Photo Researchers, Inc.
**165:** (13) Alex Kirstitch
**165:** (14) Jon Kenfield/Bruce Coleman Ltd.
**165:** (15) © Jonathan Bird/Oceanic Research Group, Inc.
**165:** (16) © CORBIS

### Chapter 18.
**176:** (top and bottom left) © Andrew Syred/Photo Researchers, Inc.
**176:** (left center) Dr. Dale M. Benham, Nebraska Wesleyan University
**176:** (right center) D. E. Akin and I. L. Risgby, Richard B. Russel Agricultural Research Center, Agricultural Research Service, U.S. Dept. Agriculture, Athens, GA
**176:** (right) Kingsley R. Stern
**177:** (right) Photo by Mike Clayton/University of Wisconsin Botany Department
**177:** (left) Carolina Biological Supply Company
**179:** (top center) Mike Clayton/University of Wisconsin Department of Botany
**179:** (bottom center) Mike Clayton/University of Wisconsin Department of Botany
**179:** (bottom right) © Omikron/Photo Researchers, Inc.
**181:** (center) H. A. Core, W. A. Cote, and A. C. Day, *Wood Structure and Identification*, 2nd Ed., Syracuse University Press, 1979.

### Chapter 19.
**190:** (top left) © John McAnulty/Corbis
**190:** (top right) © Robert Essel NYC/Corbis

### Chapter 20.
**202:** (18) © Biophoto Associates/Photo Researchers, Inc.
**202:** (19) © University of Cincinnati, Raymond Walters College, Biology
**202:** (20) © Science Photo Library/Photo Researchers, Inc.
**202:** (21) Ed Reschke
**202:** (22) Ed Reschke
**202:** (23) Ed Reschke
**202:** (24) © Science Photo Library/Photo Researchers, Inc.
**202:** (25) Ed Reschke
**202:** (26) © Triarch/Visuals Unlimited
**202:** (27) © John Cunningham/Visuals Unlimited
**202:** (28) © Michael Abbey/Photo Researchers, Inc.

### Chapter 25.
**255:** C. Yokochi and J. Rohen, *Photographic Anatomy of the Human Body*, 2nd Ed., Igaku-Shoin, Ltd., 1979.

# PREFACE

*Tell me and I will forget, show me and I might remember, involve me and I will understand.*
—Chinese Proverb

The proverb outlines three levels of learning, each successively more effective than the method preceding it. The writer of the proverb understood that humans learn most efficiently when they *involve* themselves in the material to be learned. This interactive workbook is like a tutor; when properly used it increases the efficiency of your study periods. The interactive exercises actively involve you in the most important terms and central ideas of your text. Specific tasks ask you to recall key concepts and terms and apply them to life; they test your understanding of the facts and indicate items to reexamine or clarify. Your performance on these tasks provides an estimate of your next test score based on specific material. Most important, though, this biology study guide and text together help you make informed decisions about matters that affect your own well-being and that of your environment. In the years to come, human survival on planet Earth will demand administrative and managerial decisions based on an informed biological background.

## HOW TO USE THE STUDENT INTERACTIVE WORKBOOK

Following this preface, you will find an outline that will show you how this workbook is organized and will help you use it efficiently. Each chapter begins with a title, a brief introduction to the chapter, and a list of chapter focal points, which often refer to figures in the text that cover main points of the chapter. The Interactive Exercises follow, wherein each chapter is divided into sections of one or more of the main (1-level) headings that are labeled 1.1., 1.2, and so on. *For easy reference to an answer or definition, each question and term in this unique study guide is accompanied by the appropriate text page(s), and appears in the form:* [p.352]. The Interactive Exercises begin with a list of Selected Words (other than boldfaced terms) chosen by the authors as those that are most likely to enhance understanding. In the text chapters, the selected words appear in italics, quotation marks, or roman type. This is followed by a list of Boldfaced Terms that appear in the text. These terms are essential to understanding each study guide section of a particular chapter. Space is provided by each term for you to formulate a definition in your own words. Next is a series of different types of exercises that may include completion, short answer, true–false, fill in the blank, matching, choice, dichotomous choice, identification, problems, labeling, sequencing, multiple choice, and completion of tables.

A Self-Quiz immediately follows the Interactive Exercises. This quiz is composed primarily of multiple-choice questions, although sometimes we present another examination device or some combination of devices. Any wrong answers in the Self-Quiz indicate portions of the text you need to reexamine. A series of Chapter Objectives/Review Questions follows each Self-Quiz. These are tasks that you should be able to accomplish if you have understood the assigned reading in the text. Some objectives require you to compose a short answer or long essay while others may require a sketch or supplying correct words. The answers for these questions are not provided except for the identification of organisms in two chapters.

The next part of each chapter is named Integrating and Applying Key Concepts. It invites you to try your hand at applying major concepts to situations in which there is not necessarily a single pat answer and so none is provided in the chapter answer section (except for two genetics problems). Your text generally will provide enough clues to get you started on an answer, but this part is intended to stimulate your thought and provoke group discussions.

*A person's mind, once stretched by a new idea, can never return to its original dimension.*
—Oliver Wendell Holmes

## STRUCTURE OF THE STUDENT INTERACTIVE WORKBOOK

The outline below shows how each chapter in this workbook is organized.

Chapter Number ⟶ **4**

Chapter Title ⟶ **HOW CELLS WORK**

Chapter Introduction ⟶ Brief description of the topics covered in the chapter.

Chapter Focal Points ⟶ A list of figures and other sections of the chapter that provide insight into the key concepts and most important topics discussed in the chapter.

Interactive Exercises ⟶ The interactive exercises are divided into numbered sections by titles of main headings and page references. *This study guide is unique in that each question or term is accompanied by a reference to the text page(s) where that answer or definition may be found in the text.* Each section begins with a list of author-selected words that appear in the text chapter in italics, quotation marks, or roman type. This is followed by a list of important boldfaced, page-referenced terms from each section of the chapter. Each section ends with interactive exercises that vary in type and require constant *interaction* with the important chapter information.

Self-Quiz ⟶ Usually a set of multiple-choice questions or other testing devices that *sample* important blocks of text information.

Chapter Objectives/ Review Questions ⟶ Combinations of relative objectives to be met and questions to be answered.

Integrating and Applying Key Concepts ⟶ Applications of text material to questions for which there may be more than one correct answer.

Answers to Interactive Exercises and Self-Quiz ⟶ Answers for all interactive exercises can be found at the end of this study guide by chapter and title, and the main headings with their page references, followed by answers for the Self-Quiz.

# 1
# INVITATION TO BIOLOGY

## INTRODUCTION

Chapter 1 lays the groundwork for the rest of the book. It also lays out the process of biological inquiry, describing what can and what cannot be studied by this method.

## FOCAL POINTS

- Figure 1.1 [pp.2–3] shows the various levels of organization from atoms to the biosphere.
- Figure 1.6 [p.6] gives examples of life's diversity.
- Sections 1.5 and 1.6 [pp.9, 10] outline biological inquiry and discuss experimentation.
- Section 1.7 [p.12] discusses the limits of science.

## Interactive Exercises

*Note:* In the answer sections of this book, a specific molecule is most often indicated by its abbreviation. For example, adenosine triphosphate is ATP.

***What Am I Doing Here?*** [p.1]

### 1.1. LIFE'S LEVELS OF ORGANIZATION [pp.2–3]

***Selected Words:*** In addition to the boldfaced terms, the text features other important terms essential to understanding the assigned material. "Selected Words" is a list of these terms, which appear in the text in italics, in quotation marks, and occasionally in roman type.

atom [p.2], molecule [p.2], cell [p.2], tissue [p.2], organ [p.2], organ system [p.2], "molecules of life" [p.2], multicelled organism [p.3]

***Boldfaced Terms***

The page-referenced terms are important; they were in boldface type in the chapter. Write a definition for each term in your own words without looking at the text. Next, compare your definition with that given in the chapter or in the text glossary. If your definition seems inaccurate, allow some time to pass and repeat this procedure until you can define each term rather quickly (how fast you answer is a gauge of your learning effectiveness).

[p.2] cell _____

[p.3] population _____

[p.3] community _____

[p.3] ecosystem _____

[p.3] biosphere _____

## Short Answer

1. Can you state some reasons why the study of biology will be beneficial to you? [p.1] _____

## Matching

Choose the most appropriate answer for each term.

2. __d__ organ system [p.2]
3. __c__ cell [p.2]
4. __f__ community [p.3]
5. __h__ ecosystem [p.3]
6. __e__ molecule [p.2]
7. __j__ population [p.3]
8. __k__ tissue [p.2]
9. __b__ biosphere [p.3]
10. __i__ multicelled organism [p.3]
11. __a__ organ [p.2]
12. __g__ atom [p.2]

a. Structural unit of two or more tissues interacting in some task
b. All regions of the Earth's waters, crust, and atmosphere that hold organisms
c. Smallest unit that can live and reproduce on its own or as part of a multicelled organism
d. Organs interacting physically, chemically, or both in some task
e. Two or more joined atoms of the same or different elements
f. All populations of all species occupying a specified area
g. Smallest unit of an element that still retains the element's properties
h. A community that is interacting with its physical environment
i. Individual made of different types of cells
j. Group of single-celled or multicelled individuals of the same species occupying a specified area
k. Structural unit of certain types and proportions of cells interacting in some task

## Sequence

Arrange the levels of organization listed in the matching exercise above in the correct hierarchical order. Write the term representing the most inclusive level next to 13. The least inclusive level is written next to 23. [pp.2–3]

13. _biosphere_; 14. _ecosystem_; 15. _community_; 16. _population_; 17. _multicelled organism_; 18. _organ system_; 19. _organ_; 20. _tissue_; 21. _cell_; 22. _molecule_; 23. _atom_.

## 1.2. OVERVIEW OF LIFE'S UNITY [pp.4–5]

**Selected Words:** *inheritance* [p.4], *reproduction* [p.4], *development* [p.4], *energy* [p.4], *decomposers* [p.4], *internal* environment [p.5]

## Boldfaced Terms

[p.4] DNA _____

_____

[p.4] metabolism _____

_____

[p.4] producers _____

_____

[p.4] consumers _____

_____

[p.5] homeostasis _____

_____

[p.5] receptors _____

_____

[p.5] stimulus _____

_____

## Fill in the Blanks

The signature molecule of life is the nucleic acid called (1) _____ [p.4]. This molecule also holds information for building (2) _____ [p.4] from smaller molecules, the amino acids. Many kinds of (2) serve as (3) _____ [p.4] materials, and others are (4) _____ [p.4]. When (4) receive a boost of energy, they build, split, and rearrange the (5) _____ [p.4] of life in ways that keep cells alive. Each organism inherits its DNA—and its traits—from (6) _____ [p.4].

(7) _____ is the acquisition of traits after parents transmit their DNA to offspring [p.4].

(8) _____ refers to actual mechanisms by which parents transmit DNA to offspring. For large organisms, the information in DNA guides (9) _____ [p.4]. Each cell uses energy to maintain itself, grow, and make more cells; this capacity is known as (10) _____ [p.4].

## Choice

For questions 11–15, choose from the following: [p.4]

a. producers   b. consumers   c. decomposers

11. _____ A lion stalking a zebra on the African savannah
12. _____ Fungi growing on a tree that appears to be dying
13. _____ Ferns, oak trees, apple trees, and mosses
14. __b__ A snake swallowing an egg
15. _____ A bacterium on a rotting animal on the side of the road

Invitation to Biology  3

*Fill in the Blanks*

The great one-way flow of energy into the world of life also flows right out of it—from the environment, through (16) _____ [p.4], then consumers and decomposers, then back to the (17) _____ [pp.4–5]. The one-way energy-flow through organisms and the cycling of materials are very extensive and are responsible for organizing life in the (18) _____ [p.5] in an interconnected way. Living things sense (19) _____ [p.5] in their surroundings and they make compensatory, controlled (20) _____ [p.5] to them. The responses of organisms are made possible by (21) _____ [p.5]. These are molecules and structures that detect (22) _____ [p.5] such as sunlight energy, chemical energy, and mechanical energy.

Sugars leave your gut and then enter your (23) _____ [p.5], part of the (24) _____ [p.5] environment. When there is too much sugar, your (25) _____ [p.5] begins secreting more insulin that stimulates receptors on the living cells of your body to take up more (26) _____ [p.5]. When enough cells respond to insulin, the blood sugar level returns to (27) _____ [p.5]. Organisms keep their internal operating conditions within a range that cells can tolerate; this state is called (28) _____ [p.5], and is a defining feature of life.

## 1.3. IF SO MUCH UNITY, WHY SO MANY SPECIES? [pp.6–7]
## 1.4. AN EVOLUTIONARY VIEW OF DIVERSITY [p.8]

**Selected Words:** Carolus Linnaeus [p.6], genus [p.6], Eukarya [p.6], *prokaryotic* [p.6], eubacteria [p.6], protist [p.6], *eukaryotic* [p.6], *adaptive* traits [p.8], "diversity" [p.8], *artificial* selection [p.8]

**Boldfaced Terms**

[p.6] species _____

[p.6] genus _____

[p.6] bacteria _____

[p.6] archaea _____

[p.6] plants _____

[p.7] fungi _____

[p.7] animals _____

[p.8] mutations _____

[p.8] evolution _____

[p.8] natural selection _____

## *Matching*

Match each of the major organism groups to its best description.

| Major Organism Group | Description |
|---|---|
| 1. animals [p.7] | a. eukaryotic, multicelled, photosynthetic producers |
| 2. archaea [p.6] | b. most common prokaryote |
| 3. bacteria [p.6] | c. eukaryotic, multicelled decomposers and consumers; secrete enzymes to digest food externally and then absorb nutrients |
| 4. fungi [p.7] | d. prokaryotes that live in harsh environments like hot springs |
| 5. plants [pp.6–7] | e. diverse eukaryotes; unicellular to multicelled; diverse lineages |
| 6. protists [pp.6–7] | f. eukaryotic, multicelled consumers; ingest tissues of other organisms |

## *Choice*

Choose from the following: [p.8]

    a. evolution through artificial selection    b. evolution through natural selection

7. _____ Pigeon breeding
8. _____ A favoring of adaptive traits in nature
9. _____ The selection of one form of a trait over another taking place under contrived, manipulated conditions
10. _____ Refers to heritable change that is occurring in a line of descent over time
11. _____ When individuals differ in their ability to survive and reproduce, the traits that help them do so tend to become more common in a population over time
12. _____ Breeders are the "selective agents"
13. _____ Swifter or better-camouflaged pigeons are more likely to avoid peregrine falcons and live long enough to reproduce, compared to the not-so-swift or too-conspicuous pigeons among them

## 1.5. THE NATURE OF BIOLOGICAL INQUIRY [p.9]
## 1.6. THE POWER OF EXPERIMENTAL TESTS [pp.10–11]

## 1.7. THE SCOPE AND LIMITS OF SCIENCE [p.12]

**Selected Words:** if–then process [p.9], scientific method [p.9], "theory" [p.9], *quantitative* results [p.11], *subjective* answers [p.12]

### Boldfaced Terms

[p.9] hypothesis _____

_____

[p.9] prediction _____

_____

[p.9] test _____

_____

[p.9] models _____

_____

[p.9] scientific theory _____

_____

[p.10] variable _____

_____

[p.10] control group _____

_____

[p.10] experimental groups _____

_____

[p.10] mimicry _____

_____

[p.11] sampling error _____

_____

### Sequence

Arrange the following common steps used in the scientific method in correct chronological sequence. Write the letter of the first step next to 1, the letter of the second step next to 2, and so on. [p.9]

1. _____ a. Develop hypotheses about possible answers to questions or solutions to problems.
2. _____ b. Devise ways to test the accuracy of predictions.
3. _____ c. Repeat or devise new tests.
4. _____ d. Make a prediction, using hypotheses as a guide.
5. _____ e. If the tests do not provide the expected results, check to see what might have gone wrong.
6. _____ f. Objectively analyze and report the results from tests and the conclusions drawn from them.
7. _____ g. Identify a problem or ask a question about nature.

*Dichotomous Choice*

Circle one of two possible answers given between parentheses in each statement; questions 8–12 deal with spontaneous generation.

An Italian physician, Francisco Redi, published a paper in 1688 in which he challenged the doctrine of spontaneous generation, the proposition that living things could arise from dead material. Although many examples of spontaneous generation were described in his day, Redi's work dealt particularly with disproving the notion that decaying meat could be transformed into flies. He tested his ideas in a laboratory.

8. "Two sets of jars are filled with meat or fish. One set is sealed; the other is left open so that flies can enter the jars." This description deals with a(n) (hypothesis/experiment). [pp.9–10]
9. The description in the previous question also includes a (prediction/control). [p.10]
10. Prior to his test, Redi suggested that "worms are derived directly from the droppings of flies." This statement represents a (theory/hypothesis). [pp.9–10]
11. "Worms (maggots) will appear only in the second set of jars" represents a (prediction/hypothesis). [p.9]
12. The statement "Mice arise from a dirty shirt and a few grains of wheat placed in a dark corner" is best called a (belief/test). [p.12]
13. The work by Charles Darwin established the concept of organic evolution. After more than a century of many thousands of tests providing evidence, Darwin's work has a high probability of being a good (scientific hypothesis/scientific theory). [p.9]
14. The control group is identical to the experimental group except for the (hypothesis/variable) being studied. [p.10]
15. Systematic observations, model development, and conducting experiments are all methods employed to (make predictions/test predictions). [p.9]
16. Science is distinguished from faith in the supernatural by (cause and effect/experimental design). [p.12]
17. Through the use of samples that are, by chance, not representative of a population in their experiments, scientists encounter (bias in reporting results/sampling error). [p.10]
18. Science emphasizes reporting test results in (quantitative/qualitative) form. [p.11]

*Completion*

19. Questions whose answers are _____ in nature do not readily lend themselves to scientific analysis and experiments. [p.12]
20. Scientists often stir up controversy when they explain a part of the world that was considered beyond natural explanation—that is, belonging to the "_____." [p.12]
21. The external world, not internal _____, must be the testing ground for scientific hypotheses. [p.12]

## Self-Quiz

____ 1. All populations of all species occupying a specified area such as a coral reef best describe a(n) _____. [p.3]
    a. community
    b. ecosystem
    c. population
    d. biosphere
    e. domain

____ 2. A cell that uses energy to maintain itself, grow, and make more cells is exhibiting _____. [p.4]
    a. adaptation
    b. metabolism
    c. responsiveness
    d. homeostasis
    e. development

___ 3. The acquisition of traits after parents transmit their DNA to offspring is _____. [p.4]
   a. adaptation
   b. inheritance
   c. metabolism
   d. homeostasis
   e. reproduction

For questions 4–5, choose from the following answers: a. producers; b. consumers; c. decomposers.

___ 4. _____ are mostly bacteria and fungi that break down complex materials in the environment to simpler molecules and then absorb them. [p.5]

___ 5. _____ cannot make their own food; they must eat producers and other organisms. [p.5]

___ 6. Which match is incorrect? [pp.6–7]
   a. Animals—multicelled consumers, most move about
   b. Plants—mostly multicelled producers
   c. Eubacteria and Archaebacteria—relatively simple, multicelled organisms
   d. Fungi—mostly multicelled decomposers
   e. Protists—many complex single cells, some multicellular

___ 7. A favoring of some forms of a given trait over others in nature is best termed _____. [p.8]
   a. artificial selection
   b. diversity
   c. mutation
   d. natural selection
   e. selected traits

___ 8. The principal point of evolution by natural selection is that _____. [p.8]
   a. it is simply a difference in which individuals of a population survive and reproduce in a given generation; some individuals possess more adaptive forms of traits
   b. even bad mutations can improve survival and reproduction of organisms in a population
   c. evolution does not occur when some forms of traits increase in frequency and others decrease or disappear with time
   d. individuals lacking adaptive traits make up more of the reproductive base for each new generation

___ 9. A hypothesis should *not* be accepted as valid if _____. [p.9]
   a. the sample studied is determined to be representative of the entire group
   b. a variety of different tools and experimental designs yield similar observations and results
   c. other investigators can obtain similar results when they conduct the experiment under similar conditions
   d. several different experiments, each without a control group, systematically eliminate each of the variables except one

___ 10. The experimental group and the control group are identical except for _____. [pp.10–11]
   a. the number of variables studied
   b. the variable under study
   c. the two variables under study
   d. the number of experiments performed on each group

# Chapter Objectives/Review Questions

This section lists general and detailed chapter objectives that can be used as review questions. You can make maximum use of these items by writing answers on a separate sheet of paper. Fill in answers where blanks are provided. To check for accuracy, compare your answers with information given in the chapter or glossary.

1. A _____ is the smallest unit of organization with the capacity to survive and reproduce on its own, given raw materials, energy inputs, information encoded in DNA, and suitable conditions in the environment. [pp.2–3]

2. Arrange in order, from least inclusive to most inclusive, the levels of organization that occur in nature. Define each level as you list it. [pp.2–3]
3. _____ holds information for building proteins from smaller molecules: the amino acids. [p.4]
4. By definition, distinguish between *inheritance, reproduction,* and *development.* [p.4]
5. Explain how the actions of producers, consumers, and decomposers demonstrate the one-way flow of energy and the cycling of materials in nature. [p.5]
6. _____ refers to the maintenance of internal operating conditions of organisms remaining within a range their cells can tolerate. [p.5]
7. List the six major groups of life's organisms; briefly describe the general characteristics of the organisms placed in each. [pp.6–7]
8. Explain the origin of trait variations that function in evolution. [p.8]
9. Explain what is meant by the term *diversity* and speculate about what caused the great diversity of life forms on Earth. [p.8]
10. Be able to list the three main mechanisms that bring about evolution as discovered by Charles Darwin. [p.8]
11. Be able to list the general steps used in scientific research. [p.9]
12. Be able to define and apply the following terms: *control group, variable,* and *sampling error.* [p.10]
13. Describe the butterfly experiments of Kapan as examples of the nature of biological inquiry. [pp.10–11]
14. Explain how the methods of science differ from answering questions by using subjective thinking and systems of belief. [p.12]

## Integrating and Applying Key Concepts

1. For parents to produce generally identical offspring for generation after generation, something must be copied. From information given in the chapter, speculate on what might be copied within organisms to allow generations of the same organism to continue.
2. What might happen if all decomposers were removed from the world of life?
3. What sorts of topics do scientists usually regard as untestable by applications of the scientific method?

# 2

# MOLECULES OF LIFE

## INTRODUCTION

This chapter looks at basic chemistry including atoms, molecules, and chemical bonding. It also looks at water and its life-sustaining properties and at important biological molecules.

## FOCAL POINTS

- Figure 2.2 [p.17] diagrams the shell model of atoms.
- Figure 2.3 [p.18] describes chemical bonding.
- Figure 2.7 [p.22] shows the pH values of many common materials.
- Figure 2.12 [p.25] diagrams the major functional groups found in organic compounds.
- Figures 2.14–2.17 [pp.26, 27] illustrate the diversity of carbohydrates.
- Figures 2.18–2.21 [pp.28, 29] show the structures of common lipids.
- Figures 2.22–2.25 [pp.30–31, 32] look at protein formation and protein structure.
- Figures 2.28 and 2.29 [pp.34, 35] diagram nucleotide and nucleic acid structure.
- Table 2.1 [p.36] lists important biochemicals and their functions.

## Interactive Exercises

*Science or Supernatural?* [p.15]

### 2.1. ATOMS AND THEIR INTERACTIONS [pp.16–17]
### 2.2. BONDS IN BIOLOGICAL MOLECULES [pp.18–19]

**Selected Words:** neutrons [p.16], elements [p.16], *trace* elements [p.16], *radio*isotopes [p.16], *tracer*, or *probe* [p.16], PET [p.16], orbital [p.16], energy levels [p.17], *chemical bond* [p.17], *covalent bond* [p.18], *double* covalent bond [p.19], *triple* covalent bond [p.19], *nonpolar* covalent bond [p.18], *polar* covalent bond [p.19]

**Boldfaced Terms**

[p.16] proton _____

_____

[p.16] electrons _____

[p.16] isotopes _____

[p.17] shell model _____

[p.17] chemical bonding _____

[p.17] compounds _____

[p.17] mixture _____

[p.18] ion _____

[p.18] ionic bond _____

[p.18] covalent bond _____

[p.19] hydrogen bond _____

## Identification

1. Identify each element in the following figure by the number of protons and neutrons given. Then fill in the correct number of electrons for each shell (example: helium, 2e⁻). [p.17]

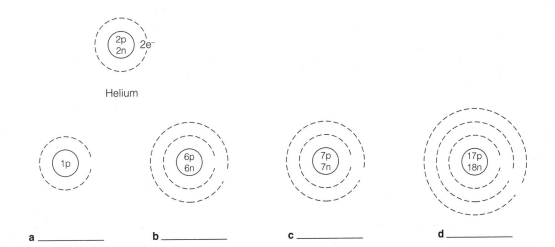

a _____   b _____   c _____   d _____

Molecules of Life  **11**

2. Complete the diagram of this chemical reaction by showing with arrows what happens to the electrons in the outer shell of magnesium. In the diagram of the final molecule, enter the number of electrons in the outer shell of the two chloride ions. Then enter the charge of each ion in the parentheses. [p.18]

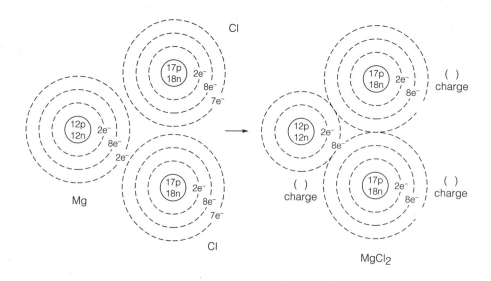

*Choice*

Match each drawing with the appropriate description(s). (Some descriptions may require more than one letter.) [p.19]

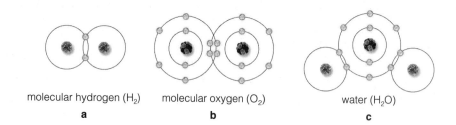

3. _____ Molecules that contain single covalent bonds
4. _____ Molecules exhibiting one or more double covalent bonds
5. _____ Molecules containing polar covalent bonds
6. _____ Molecules that have nonpolar covalent bonds

*Short Answer*

7. Describe the role hydrogen bonds play in the structure of such large molecules as DNA. [p.19] _____

_____

_____

## 2.3. WATER'S LIFE-GIVING PROPERTIES [pp.20–21]
## 2.4. ACIDS AND BASES [pp.22–23]

*Selected Words:* dissolved [p.21], "spheres of hydration" [p.21], surface tension [p.21], *acidic* solutions [p.22], *basic* solutions [p.22], "alkaline" solutions [p.22], *acid stomach* [p.22], *chemical burns* [p.22], *acid rain* [p.22], *acute respiratory acidosis* [p. 23], *coma* [p.23], *alkalosis* [p.23], *tetany* [p.23]

### Boldfaced Terms

[p.20] hydrophilic _____

[p.20] hydrophobic _____

[p.20] temperature _____

[p.21] evaporation _____

[p.21] solute _____

[p.21] cohesion _____

[p.22] pH scale _____

[p.22] acids _____

[p.22] bases _____

[p.22] salt _____

[p.23] buffer system _____

### Fill in the Blanks

Because they dissolve so well in water, polar compounds are often called (1) _____ [p.20] while nonpolar compounds, like oils, are classified as (2) _____ [p.21]. This occurs because water can form (3) _____ _____ [p.20] with polar, but not nonpolar, compounds. Because of its (3), water can absorb a great deal of energy before its temperature rises. Water is a (4) _____ _____ [p.21]. When liquid water molecules escape as a gas, a process called (5) _____ [p.21], the water left behind (6) _____ [p.21], which helps to keep some animals from overheating. Another property of water, called (7) _____ [p.21], is that water molecules resist separating, they stick to one another. This property helps the movement of water in plants.

In any container of water, some molecules dissociate into hydrogen and (8) _____ [p.22] ions. The concentration of hydrogen ions is the basis of the (9) _____ [p.22] scale. Solutions with a (9) less than 7 are considered to be (10) _____ [p.22] while readings above 7 are considered to be (11) _____ [p.22]. A change of one unit in this scale represents a (12) _____ [p.22] change in the hydrogen ion concentration. It is important that organisms maintain a constant pH and most maintain it with (13) _____ _____ [p.23].

## 2.5. MOLECULES OF LIFE—FROM STRUCTURE TO FUNCTION [pp.23–25]

***Selected Words:*** hydrocarbons [p.23], *hydroxyl* groups [p.24], *monomer* [p.25], *polymer* [p.25], *functional-group transfer* [p.25], *electron transfer* [p.25], *rearrangement* [p.25], *condensation* [p.25], *cleavage* [p.25]

### Boldfaced Terms

[p.23] organic compounds _____

_____

[p.23] functional group _____

_____

[p.25] reactions _____

_____

[p.25] enzymes _____

_____

[p.25] condensation reaction _____

_____

[p.25] hydrolysis _____

_____

### Fill in the Blanks

The molecules of life are (1) _____ [p.23] compounds, which contain carbon and hydrogen atoms. The (2) _____ [p.23] are hydrogen atoms covalently bonded to carbon. Methane, ethane, and ethylene are examples. Hydroxyl, methyl, carbonyl, carboxyl, amino, and phosphate are all examples of (3) _____ [pp.24–25] groups found on organic compounds. Such groups dictate features of carbohydrates, lipids, proteins, and nucleic acids. Living organisms consist mainly of oxygen, hydrogen, and (4) _____ [p.23]. (4) has a versatile bonding behavior as each of its atoms can covalently bond with as many as (5) _____ [p.23] other atoms. Carbon atoms often join to produce "backbones" to which hydrogen, oxygen, and other elements are attached. The three-dimensional shapes of organic compounds start with these bonds.

Cells build big molecules mainly from four families of small organic compounds: simple (6) _____ [p.25], (7) _____ acids [p.25], amino (8) _____ [p.25], and (9) _____ [p.25]. They continually use these as energy sources and subunits, or (10) _____ [p.25] of larger molecules, or polymers. The reactions by which a cell builds, rearranges, and splits apart all organic molecules require more than energy inputs; they also require proteins known as (11) _____ [p.25] that make substances react faster than they would on their own.

## Identification

12. Identify the metabolic reactions below by writing *condensation* or *hydrolysis* in the blank. [p.25]

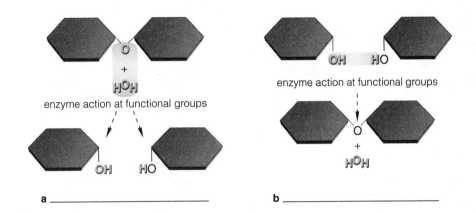

a _____    b _____

## 2.6. THE TRULY ABUNDANT CARBOHYDRATES [pp.26–27]

**Selected Words:** *mono*saccharides [p.26], *oligo*saccharide [p.26], *di*saccharides [p.26], "complex" carbohydrates [p.26], *poly*saccharides [p.26], starch [p.26], cellulose [p.27], glycogen [p.27], chitin [p.27]

## Matching

Match each carbohydrate with its description. [pp.26–27]

1. _____ cellulose
2. _____ chitin
3. _____ deoxyribose
4. _____ glucose
5. _____ glycogen
6. _____ lactose
7. _____ ribose
8. _____ starch
9. _____ sucrose

a. Most plentiful sugar in nature; transport form of carbohydrates in plants; table sugar; formed from glucose and fructose
b. Five-carbon sugar occurring in DNA
c. Instant energy source and building block of cells; a parent molecule for many other compounds
d. Tough structural material of plant cell walls; formed from glucose chains
e. Five-carbon sugar occurring in RNA
f. Sugar in milk; formed from one glucose and one galactose unit
g. Muscle and liver cells tap stores of this compound for a rapid burst of energy by degrading it to release glucose units
h. Strengthens external skeletons and other hard body parts of some animals and cell walls of fungi
i. Storage form for photosynthetically produced sugars

Molecules of Life  **15**

## 2.7. GREASY, FATTY—MUST BE LIPIDS [pp.28–29]

**Selected Words:** *unsaturated* fatty acids [p.28], *saturated* fatty acids [p.28], "neutral" fats [p.28], cholesterol [p.29], steroids [p.29]

### Boldfaced Terms

[p.28] lipids _____

_____

[p.28] fats _____

_____

[p.28] fatty acid _____

_____

[p.28] triglycerides _____

_____

[p.29] phospholipid _____

_____

[p.29] sterols _____

_____

[p.29] waxes _____

_____

### Labeling

1. Label the molecules shown as *saturated* or *unsaturated*. For the unsaturated molecules, circle the region(s) that make them unsaturated. [p.28]

oleic acid

a _____

stearic acid

b _____

linolenic acid

c _____

2. Label each sketch as a *fatty acid, triglyceride, phospholipid,* or *sterol.* [pp.28–29]

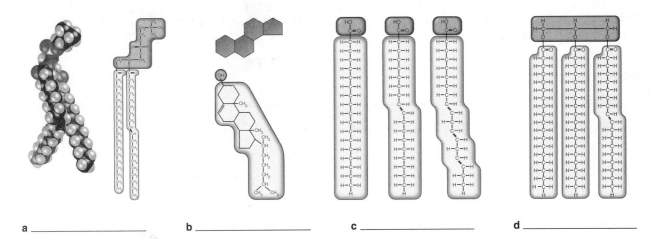

a _____   b _____   c _____   d _____

*Choice*

Choose from the following:

      a. fatty acids    b. triglycerides    c. phospholipids    d. waxes    e. sterols

3. _____ Richest source of body energy [p.28]
4. _____ Main structural component of honeycombs [p.29]
5. _____ Cholesterol [p.29]
6. _____ Animal fats such as butter and lard [p.28]
7. _____ Lack fatty acid tails [p.29]
8. _____ The main cell membrane component [p.29]
9. _____ All possess a rigid backbone of four fused carbon rings [p.29]
10. _____ Plant cuticles [p.29]
11. _____ Cholesterol remodeled into vitamin D, steroids, and bile salts [p.29]
12. _____ Unsaturated tails [p.28]
13. _____ Vegetable oil [p.28]
14. _____ Cushions and insulates vertebrate bodies [pp.28–29]
15. _____ Furnishes lubrication for skin and hair [p.29]
16. _____ Neutral fats [p.28]

## 2.8. PROTEINS—DIVERSITY IN STRUCTURE AND FUNCTION [pp.30–31]
## 2.9. WHY IS PROTEIN STRUCTURE SO IMPORTANT? [pp.32–33]

***Selected Words:*** R group [p.30], *peptide* bond [p.30], *primary* structure [p.30], *secondary* structure [p.30], *tertiary* structure [p.30], *quaternary* structure [p.30], *glyco*proteins [p.31], *lipo*proteins [p. 31], hemoglobin [p. 32], sickle-cell anemia [p.32], albumin [p.32]

## Boldfaced Terms

[p.30] amino acid _____

_____

[p.30] polypeptide chain _____

_____

[p.32] denaturation _____

_____

## Labeling

1. Label the sketches below by entering the correct term in the blank corresponding to the letter of the sketch. Choose from *linear primary structure* [p.30], *coiled secondary structure* [p.30], *sheetlike secondary structure* [p.30], *tertiary structure* (used twice) [p.30], and *quaternary structure* [p.30].

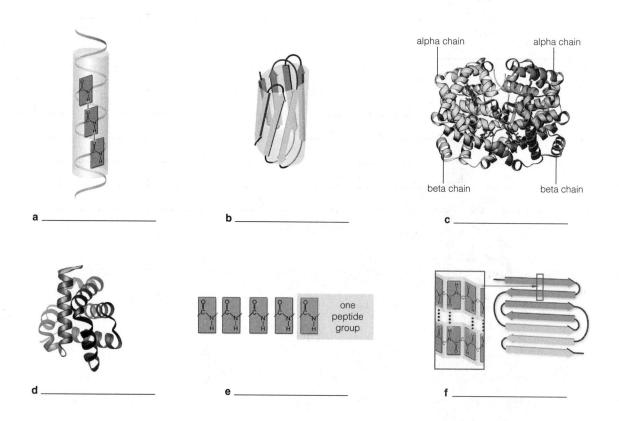

a _____  b _____  c _____

d _____  e _____  f _____

## Short Answer

2. Describe the effect of just one amino acid substitution in hemoglobin. [pp.32–33] _____

_____

_____

_____

_____

3. What are some of the functions of proteins in organisms? [pp.30–33] _____
_____
_____
_____

## 2.10. NUCLEOTIDES AND THE NUCLEIC ACIDS [pp.34–35]

*Selected Words:* nucleotides [p.34], NAD$^+$ and FAD [p.34], cAMP [p.34], adenine [p.35], guanine [p.35], thymine [p.35], cytosine [p.35]

*Boldfaced Terms*

[p.34] ATP _____
_____

[p.34] coenzymes _____
_____

[p.34] DNA _____
_____

[p.34] nucleic acids _____
_____

[p.35] RNAs _____
_____

*Identification*

1. In the diagram of a single-stranded nucleic acid molecule below, encircle as many complete nucleotides as possible. How many complete nucleotides are present? _____ [p.34]

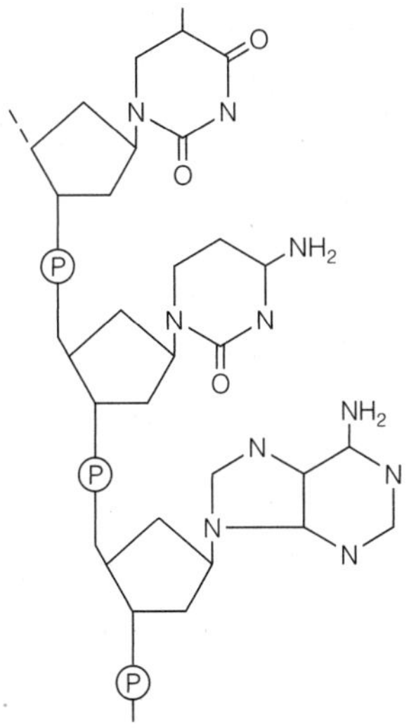

Molecules of Life   19

*Matching*

Choose the most appropriate answer for each term.

2. _____ adenosine triphosphate or ATP [p.34]
3. _____ RNAs [p.35]
4. _____ coenzymes [p.34]
5. _____ nucleic acids [p.34]
6. _____ DNA [p.34]
7. _____ base pairing [p.35]

a. Usually a single strand of nucleotides, one base is uracil, not thymine; some deliver DNA's instructions to cytoplasm, others are involved in translating
b. Transfers a phosphate group to energize acceptor molecules
c. Synthesized from four kinds of nucleotides; maintains cells with instructions in these double-stranded molecules
d. Enzyme helpers that have different metabolic roles; NAD+, FAD, and cAMP are examples
e. Every one is thought of as one rung on the DNA ladder
f. Nucleotides that retrieve heritable information stored in all cells

## Self-Quiz

_____ 1. A molecule is _____ . [p.17]
   a. a combination of two or more atoms
   b. less stable than its constituent atoms separated
   c. electrically charged
   d. a carrier of one or more extra neutrons
   e. identical to an isotope

_____ 2. Substances that are nonpolar and repelled by water are known as _____ . [p.20]
   a. hydrolyzed
   b. nonpolar
   c. hydrophilic
   d. hydrophobic
   e. condensed

_____ 3. An ionic bond is one in which _____ . [p.18]
   a. electrons are shared equally
   b. electrically neutral atoms have a mutual attraction
   c. two charged atoms have a mutual attraction due to electron transfer
   d. electrons are shared unequally

_____ 4. A covalent bond is one in which _____ . [p.18]
   a. electrons are shared
   b. electrically neutral atoms have a mutual attraction
   c. two charged atoms have a mutual attraction due to electron transfer
   d. electrons are lost

_____ 5. The shapes of large molecules are controlled by _____ bonds. [p.19]
   a. hydrogen
   b. ionic
   c. covalent
   d. inert
   e. single

_____ 6. A solution with a pH of 10 is _____ times as basic as one with a pH of 8. [p.22]
   a. 2
   b. 3
   c. 10
   d. 100
   e. 1,000

_____ 7. Proteins _____. [p.30]
   a. act as weapons against pathogens
   b. are composed of nucleotide subunits
   c. translate protein-building instructions into actual protein structures
   d. lack diversity of structure and function

_____ 8. Lipids _____. [p.28]
   a. include fats that are broken down into one fatty acid molecule and three glycerol molecules
   b. are composed of monosaccharides
   c. include triglycerides that serve as rich energy sources
   d. include cartilage and chitin

_____ 9. DNA _____. [p.30]
   a. is one of the adenosine phosphates
   b. is one of the nucleotide coenzymes
   c. contains protein-building instructions
   d. is composed of monosaccharides

_____ 10. Most of the chemical reactions in cells must have _____ present before they can occur. [p.25]
   a. RNA
   b. salt
   c. enzymes
   d. fats

## Chapter Objectives/Review Questions

1. Explain the source of the "powers" claimed by the oracle of Delphi that allowed the god Apollo to speak to ancient Greeks through her. [p.15]
2. Define *element, trace element, proton, electron,* and *neutron.* [p.16]
3. Explain the relationship of orbitals to shells. [pp.16–17]
4. How does a mixture differ from a compound? [p.17]
5. An association of two oppositely charged ions is formed by a(n) _____ bond. [p.18]
6. In a(n) _____ bond, two atoms share electrons. [p.18]
7. In a(n) _____ bond, the slight positive charge of a covalently bonded hydrogen atom weakly attracts an atom that is taking part in a different covalent bond and that bears a slight negative charge. [p.19]
8. Polar molecules attracted to water are _____; all nonpolar molecules are _____ and are repelled by water. [p.20]
9. Distinguish a solvent from a solute; explain the meaning of "spheres of hydration." [p.21]
10. Define *ionization* and *pH scale;* distinguish acids, from bases. [p.22]
11. Define *buffer system;* cite an example and describe how buffers operate. [pp.22–23]
12. Each carbon atom can share pairs of electrons with as many as _____ other atoms. [p.23]
13. Carboxyl, amino, phosphate, hydroxyl, methyl, and carbonyl are all examples of _____ groups. [pp.24–25]
14. List the four families of organic compounds from which cells build big molecules. [p.25]
15. _____ make specific metabolic reactions proceed faster than they would on their own, and different _____ mediate different reactions. [p.25]
16. Know the five types of reactions cells use to construct molecules. [p.25]
17. Define *carbohydrates;* be able to list their two general functions. [p.26]
18. Define and be able to cite important examples of *monosaccharides, oligosaccharides* (disaccharides), and *polysaccharides.* [pp.26–27]
19. Describe the structure of a fatty acid; distinguish a saturated fatty acid from an unsaturated fatty acid. [p.28]
20. Define phospholipid; state where they are abundant in a cell. [p.29]
21. Describe the structure of proteins and list their general functions. [p.30]
22. Using sickle-cell anemia as an example, explain why protein structure is so important. [pp.32–33]
23. Be able to describe the three parts of every nucleotide. [p.34]
24. Describe the general cellular function of ATP. [p.34]
25. The nucleic acids _____ and _____, built of nucleotides, are the basis of inheritance, reproduction, and protein construction. [pp.34–35]

## Integrating and Applying Key Concepts

1. Explain what would happen if water were a nonpolar molecule instead of a polar molecule. Would water be a good solvent for the same kinds of substances? Would the nonpolar molecule's specific heat likely be higher or lower than that of water? Would surface tension be affected? Cohesive nature? Ability to form hydrogen bonds? Is it likely that the nonpolar molecules could form unbroken columns of liquid? What implications would that hold for trees?

2. Humans can obtain energy from many different food sources. Do you think this ability is an advantage or a disadvantage in terms of long-term survival? Why?

# 3
# HOW CELLS ARE PUT TOGETHER

## INTRODUCTION
Chapter 3 looks at the structure and function of cells. Prokaryotic and eukaryotic cells are compared and contrasted and the origin of eukaryotes is discussed.

## FOCAL POINTS
- Figures 3.6 and 3.7 [p.42] describe the structure and function of the plasma membrane.
- Figure 3.8 [p.44] depicts a prokaryotic cell.
- Table 3.1 [p.45] describes the functions of various cell organelles.
- Figure 3.9 [pp.46–47] looks at the endomembrane system.
- Figures 3.10 and 3.11 [p.48] depict mitochondria and chloroplasts.
- Figures 3.12 and 3.13 [p.49] diagram typical plant and animal cells.
- Figure 3.20 [p.54] looks at plant cell walls.
- Figure 3.22 [p.55] describes common cell junctions.

## Interactive Exercises

*Animalcules and Cells Fill'd With Juices* [p.38]

### 3.1. WHAT IS "A CELL"? [p.39]
### 3.2. MOST CELLS ARE *REALLY SMALL* [pp.40–41]

*Selected Words:* eukaryotic [p.39], prokaryotic [p.39], archaea [p.39], bacteria [p.39], *stained* [p.40], *transmission* electron microscope [p.40], *scanning* electron microscope [p.40]

### Boldfaced Terms

[p.39] cell theory _____

[p.39] cell _____

[p.39] plasma membrane _____

[p.39] nucleus _____

_____

[p.39] nucleoid _____

_____

[p.39] cytoplasm _____

_____

[p.39] ribosomes _____

_____

[p.41] surface-to-volume ratio _____

_____

## Fill in the Blanks [p.38]

In the seventeenth century, (1) _____ _____ manufactured a simple microscope that he used to observe the details of insect eyes. Later, the term "cell" was coined after (2) _____ _____ saw small compartments, which he called *cellulae,* in thin slices of cork. The microscope was greatly improved by (3) _____ _____ _____, a Dutch shopkeeper, who had great skill at producing lenses. In the early nineteenth century, (4) _____ _____ was the first to identify the plant cell nucleus. (5) _____ _____ was first to report that cells had a life of their own and (6) _____ _____ determined that every cell comes from one that already exists.

## Short Answer

7. What are the three points of the cell theory? [p.39]. _____

_____

_____

8. What is a surface-to-volume ratio? How does it affect the size and shape of cells? [p.41]

_____

_____

## Choice

9. _____ Prokaryotic cells lack _____. [p.38]

    a. a nucleus    b. a plasma membrane    c. both a. and b.    d. neither a. nor b.

10. _____ Which of the following are prokaryotic? [p.38]

    a. archaea    b. bacteria    c. both a. and b.    d. neither a. nor b.

11. _____ The surface of the nuclear envelope is best visualized using a _____ microscope. [p.40]

    a. light    b. scanning electron    c. transmission electron

12. _____ How many micrometers are in one meter? [p.41]

    a. 0.001   b. 100   c. 10,000   d. 1,000,000

13. _____ Egg cells are _____ normal metabolically active cells. [p.41]

    a. larger than   b. smaller than   c. the same size as

*Fill in the Blanks*

14. All cells have a _____ that helps maintain the cell as a distinct entity separate from the environment. [p.39]

15. All cells also have _____ that are needed for protein synthesis. [p.39]

16. Instead of glass lenses, electron microscopes use _____ lenses to focus the beam of electrons. [p.40]

17. When you double the diameter of a cell, its surface area increases by _____ while its volume increases by _____. [p.41]

## 3.3. THE STRUCTURE OF CELL MEMBRANES [pp.42–43]
## 3.4. A CLOSER LOOK AT PROKARYOTIC CELLS [pp.44–45]
## 3.5. A CLOSER LOOK AT EUKARYOTIC CELLS [pp.45–49]
## 3.6. WHERE DID ORGANELLES COME FROM? [pp.50–51]

***Selected Words:*** phospholipids [p.42], hydrophilic [p.42], hydrophobic [p.42], *mosaic* [p.42], *fluid* [p.42], adhesion proteins [p.42], communication proteins [p.42], receptor proteins [p.43], recognition proteins [p.43], passive transporters [p.43], active transporters [p.43], pili [p.44], microenvironment [p.45], *secretory* pathway [p.45], *endocytic* pathway [p.45], *rough* ER [p.46], *smooth* ER [p.47], ATP [p.48], NADPH [p.48], stroma [p.48], centrioles [p.49], cytoskeleton [p.49], cell wall [p.49]

***Boldfaced Terms***

[p.42] lipid bilayer _____

_____

[p.42] fluid mosaic model _____

_____

[p.45] cytoskeleton _____

_____

[p.45] organelles _____

_____

[p.46] chromosome _____

_____

[p.46] chromatin _____

_____

[p.46] nuclear envelope _____

_____

[p.46] endomembrane system _____

_____

[p.46] endoplasmic reticulum (ER) _____

_____

[p.47] Golgi bodies _____

_____

[p.47] lysosomes _____

_____

[p.47] peroxisomes _____

_____

[p.47] central vacuole _____

_____

[p.48] mitochondria _____

_____

[p.48] chloroplasts _____

_____

[p.48] thylakoid _____

_____

[p.50] endosymbiosis _____

_____

## Fill in the Blanks

(1) _____ [p.42] are the most abundant components of cell membranes. Their (2) _____ _____ [p.42] are hydrophobic while their (3) _____ [p.42] is hydrophilic. In water, these molecules spontaneously form a (4) _____ _____ [p.42] sheet. Other common components of cell membranes are (5) _____, _____, and _____ [p.42]. The various components are often held together by (6) _____ _____ [p.42]. This model of the membrane is often called the (7) _____ _____ [p.42] model. The phospholipids serve as a barrier for most (8) _____-_____ [p.43] materials, but these materials can enter with the help of transport (9) _____ [p.43]. (10) _____ [p.43] also help in recognition of self vs. nonself and serve as signal receptors.

26  Chapter Three

*Choice*

Choose one of the letters for each statement. [pp.44–46]

        a. prokaryotic cells    b. eukaryotic cells    c. both prokaryotic and eukaryotic cells

11. \_\_\_\_ Cells are smaller
12. \_\_\_\_ Cells have ribosomes
13. \_\_\_\_ Cells have a nucleoid region
14. \_\_\_\_ Cells have a nucleus
15. \_\_\_\_ Cells have a plasma membrane
16. \_\_\_\_ DNA is circular
17. \_\_\_\_ Cells have membrane bound organelles (e.g., chloroplasts, mitochondria)

*Choice*

18. \_\_\_\_ The capsule of prokaryotic cells is usually composed of _____. [p.44]

        a. lipid    b. polysaccharide    c. protein    d. simple sugar

19. \_\_\_\_ These often project out of prokaryotic cells and are used for locomotion. [p.44]

        a. cilia    b. flagella    c. pili    d. pseudopods

20. \_\_\_\_ These also project out of prokaryotic cells and are used to attach the cells to various surfaces including other cells. [p.44]

        a. cilia    b. flagella    c. pili    d. pseudopods

21. \_\_\_\_ The nucleus of a eukaryotic cell _____. [p.46]

        a. isolates the DNA for protection    c. both a. and b.
        b. restricts/controls access to information    d. neither a. nor b.

22. \_\_\_\_ Chromatin in eukaryotic cells contains _____. [p.46]

        a. DNA    b. protein    c. both a. and b.    d. neither a. nor b.

23. \_\_\_\_ How many phospholipid bilayers are found in the nuclear envelope? [p.46]

        a. 0    b. 1    c. 2    d. 3

24. \_\_\_\_ Ribosomes are found on _____ endoplasmic reticulum (ER). [p.46]

        a. rough    b. smooth    c. both a. and b.    d. neither a. nor b.

## Matching

For questions 25–30, match an organelle with each function.

25. \_\_\_\_ chloroplast [p.48]
26. \_\_\_\_ Golgi body [p.47]
27. \_\_\_\_ lysosomes [p.47]
28. \_\_\_\_ mitochondrion [p.48]
29. \_\_\_\_ peroxisome [p.47]
30. \_\_\_\_ smooth ER [p.47]

a. Converts food energy to ATP energy
b. Converts light energy to ATP energy
c. Packages proteins for secretion
d. Produces membrane lipids
e. Vesicles with digestive enzymes
f. Vesicles that detoxify alcohol

For questions 31–44, match an organelle with one or more letters from the diagrams. (Organelles found in both plant and animal cells will have two letters.) [p.49]

31. \_\_\_\_ cell wall
32. \_\_\_\_ central vacuole
33. \_\_\_\_ centrioles
34. \_\_\_\_ chloroplasts
35. \_\_\_\_ cytoskeleton
36. \_\_\_\_ Golgi body
37. \_\_\_\_ lysosome
38. \_\_\_\_ mitochondrion
39. \_\_\_\_ nuclear envelope
40. \_\_\_\_ nucleolus
41. \_\_\_\_ nucleus
42. \_\_\_\_ plasma membrane
43. \_\_\_\_ rough ER
44. \_\_\_\_ smooth ER

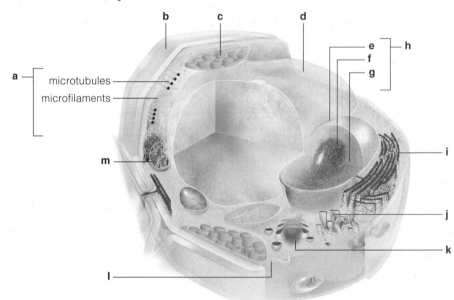

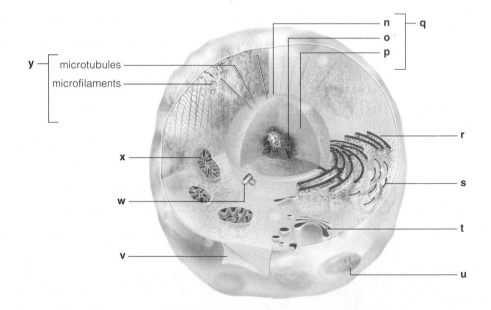

28  Chapter Three

*Short Answer*

45. What is the endosymbiotic theory [p.50]? How are mitochondria [p.48] and chloroplasts [p.48] like cells?

_____

_____

_____

## 3.7. THE DYNAMIC CYTOSKELETON [pp.52–53]
## 3.8. CELL SURFACE SPECIALIZATIONS [pp.54–55]

***Selected Words:*** colchicine [p.52], taxol [p.52], *cell typing* [p.52], dynein [p.53], cellulose [p.54], pectin [p.54], lignin [p.54], cuticle [p.55], *tight* junctions [p.55], *adhering* junctions [p.55], *gap* junctions [p.55]

***Boldfaced Terms***

[p.52] cytoskeleton _____

_____

[p.52] microtubules _____

_____

[p.52] microfilaments _____

_____

[p.52] cell cortex _____

_____

[p.52] intermediate filaments _____

_____

[p.52] motor proteins _____

_____

[p.53] flagella _____

_____

[p.53] cilia _____

_____

[p.53] pseudopods _____

_____

[p.54] cell wall _____

_____

[p.54] primary wall _____

_____

[p.54] secondary wall _____

[p.55] cell junctions _____

## Fill in the Blanks

The (1) _____ [p.52] is an interconnected system of (2) _____ [p.52] filaments needed to reinforce, organize, and move the cell. (3) _____ [p.52] are long, hollow cylinders. Inside the cell, they move cell components such as (4) _____ [p.52] to specific locations. (5) _____ [p.52] and (6) _____ [p.52] are chemicals that block microtubule assembly and are used by physicians. (7) _____ [p.52] reinforce the (8) _____ _____ [p.52] and are also needed for (9) _____ [p.52] contraction. (10) _____ _____ [p.52] strengthen and help maintain cell structure. Researchers use these filaments to identify the origin of cells in various forms of (11) _____ [p.52]. (12) _____ [p.53] and (13) _____ [p.53] are motile structures that extend from the surface of many eukaryotic cells. They are needed to move sperm and to move bacteria and airborne particles away from the lungs. Both share an internal structure of (14) _____ [p.53] and (15) _____ _____ [p.53]. Another mechanism for locomotion is a (16) _____ [p.53], an irregular lobe extending from the body of the cell.

## True–False

If the statement is true, write a T in the blank. If the question is false, correct it by changing the underlined word(s) and writing the correct word(s) in the answer blank.

_____ 17. Cell walls are formed <u>inside</u> the plasma membrane. [p.54]

_____ 18. <u>Plants, fungi, and some protists</u> have cell walls. [p.54]

_____ 19. <u>Cellulose</u> is a glue-like substance that binds the fibers of cell walls and binds adjacent cell walls together. [p.54]

_____ 20. The secondary cell wall is formed on the <u>outside</u> of the primary cell wall. [p.54]

_____ 21. In woody plants, the <u>primary</u> cell wall is up to 25 percent lignin. [p.54]

## Matching

Match each of the following junctions with its function. [p.55]

22. _____ adhering
23. _____ gap
24. _____ tight

a. link cells of tissue, especially epithelium, to prevent leakage
b. link cells of organs subject to stretching
c. allow easy communication between adjacent cells

## Self-Quiz

_____ 1. If a spherical cell's diameter increases 4 times, its volume increases _____ times while its surface area increases _____ times. [p.41]
  a. 4, 16
  b. 16, 4
  c. 16, 64
  d. 64, 16

_____ 2. The primary component of a cell membrane is a _____ bilayer. [p.42]
  a. carbohydrate
  b. glycolipid
  c. phospholipid
  d. protein

_____ 3. Membrane proteins are needed for _____. [p.43]
  a. cell/cell recognition
  b. transport
  c. both a. and b.
  d. neither a. nor b.

_____ 4. Prokaryotic cells lack _____. [p.44]
  a. mitochondria
  b. nuclei
  c. both a. and b.
  d. neither a. nor b.

_____ 5. _____ prepare proteins and lipids for secretion. [pp.45, 47]
  a. Chloroplasts
  b. Golgi bodies
  c. Mitochondria
  d. Rough ER

_____ 6. Chromatin contains _____. [p.46]
  a. DNA
  b. protein
  c. both a. and b.
  d. neither a. nor b.

_____ 7. _____ are vesicles that contain digestive enzymes. [p.47]
  a. Lysosomes
  b. Peroxisomes
  c. Both a. and b.
  d. Neither a. nor b.

_____ 8. Which of the following have their own DNA and replicate independently of the cell? [p.51]
  a. chloroplasts
  b. Golgi bodies
  c. both a. and b.
  d. neither a. nor b.

_____ 9. The cytoskeleton is made up primarily of _____. [p.52]
  a. carbohydrates
  b. lipids
  c. nucleic acids
  d. proteins

_____ 10. Which of the following is found in primary cell walls? [p.54]
  a. cellulose
  b. pectin
  c. both a. and b.
  d. neither a. nor b.

## Chapter Objectives/Review Questions

1. What are the three points of the cell theory? [p.39]
2. Compare and contrast light and electron microscopes. [p.40]
3. Why are cells always small? [p.41]
4. Describe the basic structure of a cell membrane. [pp.42–43]
5. How are prokaryotic and eukaryotic cells the same? How do they differ? [pp.44–48]
6. What are the major organelles of eukaryotic cells? What are the structure and function of each? [pp.46–48]
7. What is the endosymbiotic theory? What is the evidence that supports this theory? [pp.50–51]
8. What are the three main components of the cytoskeleton? What are the structure and function of each? [pp.52–53]

9. How do plant cell walls form? [p.54]
10. Describe the junctions between animal cells. Why are these different junctions needed? [p.55]

## Integrating and Applying Key Concepts

Organisms with eukaryotic cells are very common. What evolutionary advantages do eukaryotic cells have over prokaryotic cells?

# 4

# HOW CELLS WORK

## INTRODUCTION

Chapter 4 discusses cellular metabolism. It covers general energy flow, enzymes and their cofactors, and movement of material into and out of cells.

## FOCAL POINTS

- Figure 4.2 [p.59] illustrates the ATP/ADP cycle.
- Figure 4.3 [p.60] contrasts exergonic and endergonic reactions.
- Figure 4.4 [p.60] looks at electron transfer reactions.
- Figure 4.6 [p.62] diagrams the energy hill for reactions with and without enzymes.
- Figures 4.8 and 4.9 [p.64] describe controls of enzyme activity.
- Figure 4.13 [p.66] illustrates passive transport (facilitated diffusion) across the plasma membrane.
- Figure 4.14 [p.67] diagrams active transport across the plasma membrane.
- Figures 4.16 and 4.17 [p.69] look at osmosis.
- Figures 4.18 and 4.19 [p.70] compare endocytosis and exocytosis.

## Interactive Exercises

*Beer, Enzymes, and Your Liver* [p.58]

### 4.1. INPUTS AND OUTPUTS OF ENERGY [pp.59–60]

**Selected Words:** ethanol [p.58], catalase [p.58], alcohol dehydrogenase [p.58], acetaldehyde [p.58], alcoholic hepatitis [p.58], alcoholic cirrhosis [p.58], binge drinking [p.58], *endergonic* [p.60], *exergonic* [p.60]

*Boldfaced Terms*

[p.58] metabolism _____

[p.59] energy _____

[p.59] first law of thermodynamics _____

[p.59] second law of thermodynamics _____

[p.59] ATP _____

[p.59] phosphorylations _____

[p.59] ATP/ADP cycle _____

[p.60] oxidation–reduction reactions _____

[p.60] electron transfer chains _____

## Fill in the Blanks [p.58]

The (1) _____ is the organ most involved in protecting the body from alcohol. Most of the alcohol is broken down into harmless (2) _____ . There is approximately the same amount of alcohol in (3) _____ oz. of beer, (4) _____ oz. of wine, and (5) _____ oz. of eighty-proof liquor. If more alcohol is consumed than can be broken down, (6) _____ _____, which results in permanent scarring, or (7) _____ _____, which results in inflammation and tissue destruction, can occur. (8) _____ _____, consuming large amounts of alcohol in a short time period, is now the most serious drug problem on U.S. campuses. Excessive alcohol consumption can actually stop your (9) _____ .

## Choice

10. _____ Approximately how many of the 17,600 students studied admitted to drinking four or more alcoholic drinks per day? [Calculate using information on p.58]

    a. 17,600   b. 9850   c. 7550   d. 1760   e. 44

11. _____ The cooling of a cup of coffee is an example of the _____ law of thermodynamics. [p.59]

    a. first   b. second

12. _____ A glowing lightbulb is an example of the _____ law of thermodynamics. [p.59]

    a. first   b. second

13. _____ The most common way that energy escapes into the environment during energy conversions is as _____ . [p.59]

    a. electricity   b. heat   c. light   d. mechanical energy   e. nuclear energy

14. _____ The conversion of glucose to $H_2O$ and $CO_2$ is classified as an _____ reaction. [p.60]

    a. endergonic    b. exergonic

15. _____ A molecule that gains electrons during a chemical reaction becomes _____ . [p.60]

    a. oxidized    b. reduced

*Short Answer*

16. Faced with the second law of thermodynamics, how can life maintain its organization? [p.59]
_____
_____

17. What are some of the problems associated with binge drinking? [p.58]
_____
_____

## 4.2. INPUTS AND OUTPUTS OF SUBSTANCES [p.61]
## 4.3. HOW ENZYMES MAKE SUBSTANCES REACT [pp.62–64]

***Selected Words:*** *biosynthetic* pathways [p.61], *degradative* pathways [p.61], *reactants* [p.61], *intermediates* [p.61], *products* [p.61], *energy carriers* [p.61], *enzymes* [p.61], *cofactors* [p.61], coenzymes [p.61], *transport proteins* [p.61], reversible reactions [p.61], activation energy [p.62], "binding energy" [p.63], *free radicals* [p.63], activator [p.64], inhibitor [p.64], *allosteric* site [p.64]

### Boldfaced Terms

[p.61] metabolic pathways _____
_____

[p.61] chemical equilibrium _____
_____

[p.62] enzymes _____
_____

[p.62] substrates _____
_____

[p.62] active sites _____
_____

[p.62] transition state _____
_____

[p.63] cofactors _____
_____

How Cells Work    35

[p.63] antioxidant _____

[p.64] feedback inhibition _____

## Fill in the Blanks

Substances that enter a reaction are called (1) _____ [p.61] and are converted to (2) _____ [p.61], the material left at the end. (3) _____ [p.61], often helped by (4) _____ [p.61], speed up these chemical reactions in cells. Many chemical reactions are (5) _____ [p.61], which means they can run in either the forward or reverse direction. When the rates of the forward and reverse reaction are about the same, the reaction is said to be in (6) _____ [p.61]. The reaction sequences used to build and break down substances in cells are called (7) _____ _____ [p.61]. When smaller molecules are assembled into more complex molecules, the process is called a (8) _____ [p.61] pathway. Pathways that break down large molecules, often to supply energy to the cell, are called (9) _____ [p.61] pathways.

An (10) _____ [p.63] is able to neutralize damaging free radicals in cells thus helping to protect cells from harm. Regulating the rate of enzyme reactions is important to cells. (11) _____ [p.64] are able to decrease an enzyme's activity, often shutting off the enzyme completely. When it is the end product of a metabolic pathway that does this, the process is called (12) _____ _____ [p.64].

## Matching

Match each of the words/phrases below with a letter from the diagram. [p.62]

13. _____ Activation energy with enzyme
14. _____ Activation energy without enzyme
15. _____ Energy released by the reaction
16. _____ Reactants
17. _____ Products

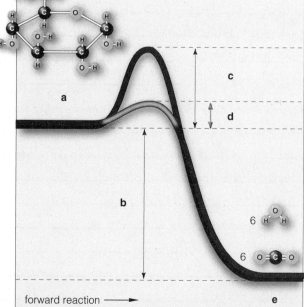

36  Chapter Four

*Short Answer*

18. What are the four most common types of enzyme reactions? [p.62]

_____

_____

19. What are the four most common ways that enzymes speed up reactions? [p.63]

_____

_____

## 4.4. DIFFUSION AND METABOLISM [pp.65–66]
## 4.5. WORKING WITH AND AGAINST DIFFUSION [pp.66–67]
## 4.6. WHICH WAY WILL WATER MOVE? [pp.68–69]
## 4.7. CELL BURPS AND GULPS [p.70]

*Selected Words:* solutes [p.65], facilitated diffusion [p.66], *net* movement [p.66], *tonicity* [p.68], *turgor* [p.68], *receptor-mediated* endocytosis [p.70], *phagocytosis* [p.70], *bulk-phase* endocytosis [p.70]

*Boldfaced Terms*

[p.65] selective permeability _____

[p.65] concentration gradient _____

[p.65] diffusion _____

[p.66] electric gradient _____

[p.66] pressure gradient _____

[p.66] passive transport _____

[p.67] active transport _____

[p.67] calcium pump _____

[p.67] sodium–potassium pump _____

[p.68] bulk flow _____

[p.68] osmosis _____

_____

[p.68] hypotonic _____

_____

[p.68] hypertonic _____

_____

[p.68] isotonic _____

_____

[p.68] hydrostatic pressure _____

_____

[p.69] osmotic pressure _____

_____

[p.70] exocytosis _____

_____

[p.70] endocytosis _____

_____

## Choice

1. Indicate whether each of the following can or cannot diffuse without help through the lipid bilayer of the cell membrane. [p.65]

   Calcium ion, carbon dioxide, glucose, oxygen, protein, water

   Can diffuse _____

   Cannot diffuse _____

2. _____ Which of the following is necessary to help ions diffuse across a membrane? [p.65]

   a. carbohydrate    b. lipid    c. nucleic acid    d. protein

3. _____ Which of the following best describes the movement of molecules during diffusion? [p.65]

   a. Molecules move down the concentration gradient.
   b. Molecules move up the concentration gradient.
   c. Molecules move in all directions but net movement is down the concentration gradient.
   d. Molecules move in all directions but net movement is up the concentration gradient.

## Short Answer

4. What is selective permeability? Why is it important to cell membranes? [p.65]

_____

_____

*38*  Chapter Four

5. What are five common factors that affect the speed of diffusion? [p.66]

_____

_____

*Fill in the Blanks*

When a protein acts as a channel to help solutes diffuse across the membrane, the process is called (6) _____ _____ [p.66]. (7) _____ _____ [p.67] can move solutes across the membrane against the concentration gradient. This requires the cell to expend (8) _____ [p.67]. One form of this is the sodium–potassium pump which moves (9) _____ [p.67] out of the cell while moving (10) _____ [p.67] into the cell.

*Choice*

In the diagram to the right, sides a and b are separated by a membrane that is permeable to water but not permeable to sucrose. Predict the outcomes of the following experiments. [pp.68–69]

*Experiment I.* 50 ml of 5% sucrose in water is placed in a, 50 ml of pure water is placed in b.

11. _____ Which of the following best describes the movement of the water?

   a. Net movement is from a to b.
   b. Net movement is from b to a.
   c. There is no net movement.

12. _____ Which of the following best describes the movement of the sucrose?

   a. Net movement is from a to b.
   b. Net movement is from b to a.
   c. There is no net movement.

13. _____ What would you expect to happen to the fluid levels in the apparatus?

   a. Side a would rise while side b goes down.
   b. Side b would rise while side a goes down.
   c. Levels would not change.

14. _____ Side a is considered to be _____ to side b.

   a. hypertonic   b. hypotonic   c. isotonic

15. _____ Side b is considered to be _____ to side a.

   a. hypertonic   b. hypotonic   c. isotonic

How Cells Work   **39**

*Experiment II.* 50 ml 1% sucrose is placed in both sides a and b.

16. _____ Which of the following best describes the movement of the water?

    a. Net movement is from a to b.
    b. Net movement is from b to a.
    c. There is no net movement.

17. _____ Which of the following best describes the movement of the sucrose?

    a. Net movement is from a to b.
    b. Net movement is from b to a.
    c. There is no net movement.

18. _____ What would you expect to happen to the fluid levels in the apparatus?

    a. Side a would rise while side b goes down.
    b. Side b would rise while side a goes down.
    c. Levels would not change.

19. _____ Side a is considered to be _____ to side b.

    a. hypertonic    b. hypotonic    c. isotonic

20. _____ Side b is considered to be _____ to side a.

    a. hypertonic    b. hypotonic    c. isotonic

## Fill in the Blanks

The pressure that a fluid, like water, exerts against a wall is called (21) _____ [p.68] pressure. In plants, this pressure against cell walls is called (22) _____ [p.68] pressure and helps keep a plant erect. In the process called (23) _____ [p.70], vesicles inside the cell fuse with the plasma membrane and release their contents into the surroundings. By the reverse process, (24) _____ [p.70], the cell engulfs materials near the surface of the membrane. White blood cells use (25) _____ [p.70] to remove viruses, bacteria, and other threats.

---

## Self-Quiz

_____ 1. What organ is most involved in protecting the body against alcohol? [p.58]
    a. liver
    b. pancreas
    c. spleen
    d. stomach

_____ 2. The chemical that acts as energy currency in cells is _____ . [p.59]
    a. ATP
    b. fat
    c. glucose
    d. starch

___ 3. Transport _____ help move solutes across the plasma membrane. [p.61]
   a. carbohydrates
   b. lipids
   c. nucleic acids
   d. proteins

___ 4. The most common biosynthetic pathway is _____ . [p.61]
   a. cell respiration
   b. fermentation
   c. glycolysis
   d. photosynthesis

___ 5. Enzymes lower the _____ energy of chemical reactions. [p.62]
   a. activation
   b. electrical
   c. mechanical
   d. potential

___ 6. Feedback _____ stop metabolic pathways. [p.64]
   a. activators
   b. cofactors
   c. enzymes
   d. inhibitors

___ 7. Movement of substances down a concentration gradient without the input of cellular energy is called _____ . [p.65]
   a. active transport
   b. diffusion
   c. electron transfer
   d. endocytosis

___ 8. Molecules that cross the plasma membrane most easily are usually _____ . [p.65]
   a. large nonpolar
   b. large polar
   c. small nonpolar
   d. small polar

___ 9. Active transport includes which of the following? [p.67]
   a. calcium pump
   b. facilitated diffusion
   c. both a. and b.
   d. neither a. nor b.

___ 10. Phagocytosis is a form of _____ . [p.70]
   a. active transport
   b. diffusion
   c. endocytosis
   d. exocytosis

## Chapter Objectives/Review Questions

1. How much beer and wine are equivalent to 1.5 oz. of eighty-proof vodka? [p.58]
2. What is the first law of thermodynamics? [p.59]
3. Are energy conversions ever 100% efficient? [p.59]
4. Describe the ATP/ADP cycle. Why is it important? [p.59]
5. Why are oxidation and reduction reactions always coupled? [p.60]
6. What are electron transfer chains? Why are they important? [p.60]
7. What are enzymes? What do they do? [pp.61–62]
8. Why are antioxidants important to living organisms? [p.63]
9. What is feedback inhibition? Why is it important to cells? [p.64]
10. What causes the distinct color pattern seen in Siamese cats? [p.64]
11. Why is selective permeability of membranes important to cells? [p.65]
12. What happens to a human blood cell when it is placed in water? In 5% sucrose? [p.69]
13. Relate exocytosis to secretion. [p.70]

## Integrating and Applying Key Concepts

1. Feedback inhibition is very common in organisms. Why do you think it is important?
2. Cystic fibrosis is caused by a defective ion channel protein in cells. Some symptoms are thickened mucus, increased respiratory infection, and digestive disturbances. Relate the symptoms to the defective protein.

# 5

# WHERE IT STARTS—PHOTOSYNTHESIS

## INTRODUCTION
This chapter outlines the steps needed to convert light energy to usable energy in the form of ATP and food.

## FOCAL POINTS
- Figure 5.4 [p.75] describes the sites of photosynthesis.
- Figure 5.7 [p.77] diagrams the steps in the conversion of light energy to ATP energy—the light-dependent reactions.
- Figure 5.8 [p.78] illustrates the Calvin–Benson cycle of fixing carbon dioxide into carbohydrate—the light-independent reactions.
- Figures 5.9 and 5.10 [p.79] contrast carbon fixation in C3, C4, and CAM plants.
- Figure 5.11 [p.80] summarizes the light-dependent and light-independent reactions.

## Interactive Exercises

*Sunlight and Survival* [p.73]

### 5.1. THE RAINBOW CATCHERS [pp.74–75]

**Selected Words:** *photo*autotroph [p.73], *chemo*autotrophs [p.73], *accessory* pigments [p.74], bacteriorhodopsin [p.74], ATP [p.74], NADP$^+$ [p.75], NADPH [p.75]

### *Boldfaced Terms*

[p.73] autotrophs _____

_____

[p.73] heterotrophs _____

_____

[p.73] photosynthesis _____

_____

[p.74] wavelength _____

_____

[p.74] pigments _____

[p.74] chlorophyll *a* _____

[p.74] chlorophyll *b* _____

[p.74] carotenoids _____

[p.74] xanthophylls _____

[p.74] anthocyanins _____

[p.74] phycobilins _____

[p.75] light-dependent reactions _____

[p.75] light-independent reactions _____

[p.75] chloroplasts _____

[p.75] stroma _____

[p.75] thylakoid membrane _____

## Fill in the Blanks

Organisms that make their own food are called (1) _____ [p.73] while you, and all other organisms that must feed to obtain energy, are called (2) _____ [p.73]. Plants and many other (1) make food via (3) _____ [p.73] with no more than sunlight, water, and carbon dioxide. (4) _____ _____ [p.74] is the primary photosynthetic pigment in plants, green algae, and some bacterial species. It absorbs red and blue-violet wavelengths and reflects yellow and (5) _____ [p.74] ones. This is the reason that plant parts loaded with this pigment look (6) _____ [p.74] to us. Accessory pigments include chlorophyll *b* and (7) _____ [p.74], including beta-carotene, which reflect (8) _____ [p.74], red, and orange light. Other plant pigments, the (9) _____ [p.74], color many fruits including cherries and blueberries. (10) _____ [p.74] are accessory pigments and are major pigments of red algae and cyanobacteria.

Some bacterial lineages use unusual photopigments like (11) _____ [p.74]. Photopigments usually have an array of alternating single and double bonds that act as a(n) (12) _____ [p.74] for receiving light energy. When the pigments absorb energy from the sun, the array becomes unstable and some electrons move to a (13) _____ [p.74] energy level. They emit extra energy as they quickly return to a lower level, and the array stabilizes. Photosynthetic organisms snag some of that emitted energy and convert it to energy in (14) _____ [p.74].

15. In the spaces below, supply the missing information to complete the summary equation for photosynthesis:

$$12 \underline{\phantom{xx}} + \underline{\phantom{xx}} CO_2 \rightarrow \underline{\phantom{xx}} O_2 + C_6H_{12}O_6 + 6 \underline{\phantom{xx}} \text{ [p.75]}$$

16. Supply the appropriate information to state the equation (above) for photosynthesis in words:
(a) _____ molecules of water plus six molecules of (b) _____ _____ (in the presence of pigments, enzymes, and visible light) yields six molecules of (c) _____ plus one molecule of (d) _____ plus (e) _____ molecules of water. [p.75]

## Labeling

The diagram below shows the location of the light-dependent reactions and the light-independent reactions within a chloroplast. Replace the numbers on the diagram with the missing words. Refer to text Figure 5.4. [p. 75]

17. _____ _____
18. _____
19. _____
20. _____
21. _____
22. _____ _____
23. _____ - _____
24. _____
25. _____
26. _____ - _____
27. _____
28. _____
29. _____

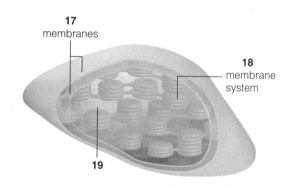

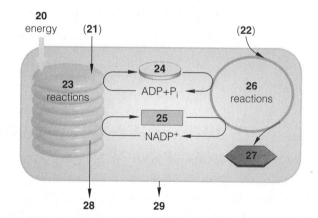

44  Chapter Five

*Short Answer*

30. Explain how Earth's atmosphere, once devoid of all oxygen, accumulated oxygen; explain the evolutionary significance of the presence of oxygen. [p.75]

_____
_____
_____
_____
_____

## 5.2. LIGHT-DEPENDENT REACTIONS [pp.76–77]

*Selected Words:* photon [p.76], cyclic pathway [p.77], noncyclic pathway [p.77]

*Boldfaced Terms*

[p.76] photosystems _____
_____

[p.76] ATP synthases _____
_____

*Fill in the Blanks*

When light is absorbed by a (1) _____-_____ _____ [p.76], the (1) passes the energy onto a (2) _____ [p.76]. The (2) then releases a(n) (3) _____ [p.76] which immediately enters a(n) (4) _____ _____ _____ [p.76]. The (4) also transfers (5) _____ [p.76] across the (6) _____ [p.76] membrane. This forms a concentration gradient but (5) cannot diffuse across the (7) _____ [p.76] bilayer. Instead, it diffuses through channels in the protein (8) _____ _____ [p.76]. The flow of (5) powers the conversion of (9) _____ [p.76] to (10) _____ [p.76] by (8). In the non-cyclic pathway, spent electrons are used to reduce the cofactor (11) _____ [p.77] to (12) _____ [p.77]. Replacement electrons are stripped from (13) _____ [p.77] producing hydrogen ions and (14) _____ [p.77] gas. When excess (11) is produced, the cells can continue to make (15) _____ [p.77] via the (16) _____ [p.77] pathway.

## 5.3. LIGHT-INDEPENDENT REACTIONS [pp.78–79]
## 5.4. PASTURES OF THE SEAS [p.80]

*Selected Words:* ribulose biphosphate (RuBP) [p.78], phosphoglycerate (PGA) [p.78], phosphoglyceraldehydes (PGAL) [p.78], sucrose [p.78], starch [p.78], oxaloacetate [p.79]

## Boldfaced Terms

[p.78] Calvin–Benson cycle _____

[p.78] rubisco _____

[p.78] carbon fixation _____

[p.79] stomata (singular, stoma) _____

[p.79] C3 plants _____

[p.79] C4 plants _____

[p.79] CAM plants _____

## Label-Match

Identify each part of the illustration below by using abbreviations where available. Complete the exercise by matching and entering the letter of the proper function description in the parentheses following each label. [p.78]

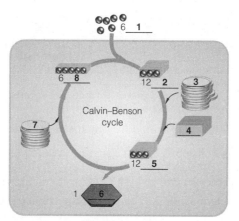

1. _____ ( )
2. _____ ( )
3. _____ ( )
4. _____ ( )
5. _____ ( )
6. _____ ( )
7. _____ ( )
8. _____ ( )

a. Donates phosphates to PGA
b. Formed by the combination of two PGAL molecules
c. Donates phosphate groups to PGAL
d. Results from ATP donating phosphates to PGA plus hydrogens and electrons from NADPH
e. A compound regenerated by synthesis reactions involving PGAL and ATP
f. In air spaces inside a leaf; diffuses into photosynthetic cells
g. Donates hydrogen and electrons to PGA
h. Formed by the enzyme rubisco attaching a carbon atom of carbon dioxide to RuBP

46 Chapter Five

*Choice*

Choose from the following [p.79]:

a. C3 plants     b. C4 plants     c. CAM plants

9. _____ Succulents such as cacti, which have juicy, water-storing tissues and thick surface layers
10. _____ Refers to the three-carbon PGA, the first stable intermediate of the Calvin–Benson cycle
11. _____ Four-carbon oxaloacetate is the first to form
12. _____ They open their stomata and fix carbon at night.
13. _____ Two types of photosynthetic cells occur in these plants.
14. _____ Do not grow well without irrigation in hot, dry climates
15. _____ These plants survive prolonged droughts by closing stomata even at night.
16. _____ $CO_2$ is fixed by repeated turns of a type of C4 cycle, then it enters the Calvin–Benson cycle the next day in the same cell.
17. _____ These plants lose less water and make more sugar than the C3 plants can when days are hot, bright, and dry.
18. _____ Their rubisco puts oxygen, not $CO_2$, in the Calvin–Benson cycle when the oxygen level is high and garbles the reactions.

*Fill in the Blanks* [p.80]

Marine photoautotrophs are mainly (19) _____ and (20) _____ . Collectively, they fix about (21) _____ _____ tons of (22) _____ annually. Without these photoautotrophs, (22) would accumulate more quickly in the environment and accelerate (23) _____ _____ . They also form the bases of vast marine (24) _____ _____ .

# Self-Quiz

_____ 1. Of the following, which one is *not* "self-nourishing?" [p.73]
   a. chemoautotrophs
   b. photoautotrophs
   c. autotrophs
   d. heterotrophs

_____ 2. Plants need _____ and _____ as the raw materials to carry on photosynthesis. [p.75]
   a. oxygen; water
   b. oxygen; $CO_2$
   c. $CO_2$; $H_2O$
   d. sugar; water
   e. $CO_2$; NADPH

_____ 3. Chlorophyll is found _____ . [p.75]
   a. on the outer chloroplast membrane
   b. inside the mitochondria
   c. inside the stroma
   d. in the thylakoid membranes
   e. on the membranes of the stroma

_____ 4. The cyclic pathway functions mainly to _____ . [p.76]
   a. make NADPH
   b. make PGAL
   c. set up conditions for making ATP
   d. regenerate RuBP
   e. break down $CO_2$

_____ 5. During the noncyclic pathway, the origin of the electrons passed to $NADP^+$ is _____. [pp.76–77]
   a. $CO_2$
   b. glucose
   c. sunlight
   d. water
   e. ATP

_____ 6. Two products of the light-dependent reactions are required to drive the chemistry of the light-independent reactions. They are _____ and _____. [p.78]
   a. $O_2$; NADPH
   b. $CO_2$; $H_2O$
   c. $O_2$; inorganic phosphate
   d. ATP; NADPH
   e. RuBP; PGA

_____ 7. ATP synthases are involved in _____. [p.76]
   a. creating electron transport chains
   b. photosystems
   c. attaching inorganic phosphates to ADP during the cyclic route
   d. the noncyclic pathway
   e. allowing $H^+$ flow to attach inorganic phosphate to ADP

_____ 8. The Calvin–Benson cycle uses _____. [p.78]
   a. $CO_2$
   b. hydrogen and electrons from NADPH
   c. phosphate group transfers from ATP
   d. an enzyme known as rubisco
   e. all of the above

_____ 9. In the Calvin–Benson cycle, RuBP is regenerated by _____. [p.78]
   a. rearrangement of PGAL molecules
   b. the combination of two PGAL molecules
   c. PGA receiving phosphate groups from ATPs
   d. glucose entering other reactions that form carbohydrates
   e. rubisco attaching carbon atoms to RuBP

_____ 10. C4 plants have an advantage in hot, dry conditions because _____. [p.79]
   a. their leaves are covered with thicker wax layers than those of C3 plants
   b. their stomates open wider than those of C3 plants, thus cooling their surfaces
   c. special leaf cells possess a means of capturing $CO_2$ even in stress conditions
   d. they carry on normal photosynthesis at very high oxygen levels
   e. they can carry on carbon fixation at night

## Chapter Objectives/Review Questions

1. Define the following terms and cite examples of each: *autotrophs, photoautotrophs, heterotrophs,* and *chemoautotrophs*. [p.73]
2. List the major stages of photosynthesis and briefly tell what happens in each. [pp.74–75]
3. Be able to reproduce the general equation for photosynthesis from memory. [p.75]
4. What is the reason photosynthesis has changed the biosphere? [p.75]
5. Be able to locate and name the sites of the major stages of photosynthesis beginning with a green leaf and continuing to chloroplast structure as in Figure 5.4 of the main text. [p.75]
6. Describe the structure of a photosystem; relate this to photon energy. [p.76]
7. List the components of an electron transport chain and describe how it functions. [p.76]
8. Be able to "tell a story" about how the cyclic and noncyclic reactions proceed; use the correct terms as they occur in the story. [pp.76–77]
9. What are the two energy-carrying molecules produced in the light-dependent reactions? Why are these molecules necessary for the light-independent reactions? [p.78]
10. Be able to describe the Calvin–Benson cycle and include the key molecular components. [p.78]

11. Describe the mechanism by which C4 plants thrive in stressful, hot, dry conditions. Contrast this with the $CO_2$-capturing chemistry of C3 plants. [p.79]
12. Describe the mechanisms that allow CAM plants such as cacti to survive desert conditions; tell how this chemical mechanism differs from that of C4 plants. [p.79]
13. List the types of organisms that are included in "pastures of the seas." What benefits do these organisms provide? How do industrial wastes, fertilizers in runoff, and raw sewage affect these photoautotrophs? [p.80]

## Integrating and Applying Key Concepts

A scientist once proposed that human cells be injected with chloroplasts extracted from living plant cells. Speculate about that possibility in terms of changes in human anatomy, physiology, and behavior. Could this be successful? Why or why not?

# 6

# HOW CELLS RELEASE CHEMICAL ENERGY

## INTRODUCTION

This chapter describes the ways cells extract energy from food. Glycolysis, which begins both aerobic respiration and fermentation, is outlined. This is followed by aerobic respiration and the related electron transfer phosphorylation. The chapter concludes with an anaerobic process, fermentation.

## FOCAL POINTS

- The introductory section [p.82] looks at the importance of mitochondria.
- Figure 6.2 [pp.84–85] describes glycolysis.
- Figure 6.3 [p.86] diagrams the Krebs cycle.
- Figures 6.4 and 6.5 [pp.86, 87] look at mitochondria and electron transfer phosphorylation.
- Figures 6.6 and 6.7 [pp.88, 89] describe fermentation pathways.
- Figure 6.8 [p.90] relates food intake to aerobic respiration.

## Interactive Exercises

*When Mitochondria Spin Their Wheels* [p.82]

### 6.1. OVERVIEW OF ENERGY-RELEASING PATHWAYS [pp.83–84]

*Selected Words:* Luft's syndrome [p.82], Friedreich's ataxia [p.82], *anaerobic* [p.83]

*Boldfaced Terms*

[p.82] free radicals _____

[p.83] fermentation _____

[p.83] aerobic respiration _____

[p.83] glycolysis _____

[p.83] pyruvate _____

[p.84] Krebs cycle _____

[p.84] NAD⁺ _____

[p.84] FAD _____

[p.84] electron transfer phosphorylation _____

*Short Answer*

1. What do Luft's syndrome, Friedreich's ataxia, type 1 diabetes, atherosclerosis, amyotrophic lateral sclerosis, and Parkinson's, Alzheimer's, and Huntington's diseases have in common? [p.82]

*Fill in the Blanks*

The numbers in this illustration represent missing information in the following narrative, which gives an overview of the three stages of aerobic respiration. [pp.83–84]

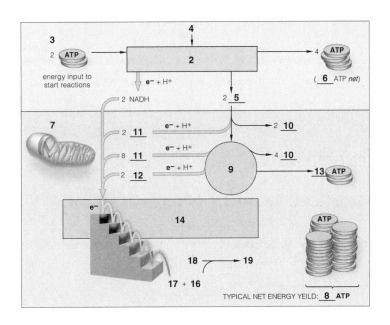

The initial stage of the reactions in aerobic respiration is called (2) _____ [p.83] and it occurs in the (3) _____ [p.83]. Here, with no need for oxygen, enzymes cleave and rearrange a (4) _____ [p.83] molecule into two molecules of (5) _____ [p.83]. During glycolysis, a net yield of (6) _____ [p.83] ATP is produced. Once glycolysis ends, though, the energy-releasing pathways differ. Only the aerobic pathway continues in a (7) _____ [p.83], where oxygen

How Cells Release Chemical Energy   51

removes the electrons that drove the reactions. By contrast, anaerobic pathways that produce a low-yield of ATP start and end in the cytoplasm.

Aerobic respiration makes the most out of a glucose molecule, typically (8) _____ [p.83] or more ATP are produced. The second stage is a cyclic pathway called the (9) _____ _____ [p.84] and a few steps preceding it. Here enzymes break down pyruvate to (10) _____ _____ [p.84] and water. Two coenzymes, abbreviated NAD$^+$ and FAD, pick up electrons and hydrogen released in the reactions. When reduced, we abbreviate these coenzymes as (11) _____ [p.84] and (12) _____ [p.84]. Krebs cycle reactions and those just preceding it produce (13) _____ (number) [p.83] ATP molecules. The big energy harvest comes in the third stage, after coenzymes give up electrons and hydrogen to electron transfer chains. These chains are the machinery for (14) _____ _____ _____ [p.84]. They create H$^+$ concentration gradients and electric gradients that drive the formation of (15) _____ [p.84] ATP molecules. It is in this final stage of aerobic respiration that so many ATP molecules are produced. As it ends, (16) _____ [p.84] inside the mitochondrion accepts the "spent" (17) _____ [p.84] from the last component of each transport system. At the same time, oxygen (taken into an organism from the atmosphere) also picks up (18) _____ [p.84] ions and thereby forms (19) _____ [p.84].

## Fill in the Blanks

Using a glucose molecule as the reactant, complete this equation, which summarizes the degradative pathway known as aerobic respiration:

(20) _____ + _____ O$_2$ → 6 _____ + 6 _____ [p.83]

(21) Now state the equation for aerobic respiration (question 20) in words: One molecule of glucose plus six molecules of _____ (in the presence of appropriate enzymes) yield _____ molecules of carbon dioxide plus _____ molecules of metabolic water [p.83]. (Review Figure 6.1, p.83, in the main text and locate the components of this basic equation.)

## 6.2. GLYCOLYSIS—GLUCOSE BREAKDOWN STARTS [pp.84–85]

*Selected Words:* *energy-requiring* [p.84], glucose–6–phosphate [p.84], *energy-releasing* step [p.84], *net* energy yield [p.85]

## Boldfaced Term

[p.84] **substrate-level phosphorylation** _____

_____

## Sequence

Arrange the following steps of glycolysis, the first stage of aerobic respiration, in correct chronological sequence. Write the letter of step 1 next to 1, the letter of step 2 next to 2, and so on. [pp.84–85]

52    Chapter Six

1. _____
2. _____
3. _____
4. _____
5. _____
6. _____
7. _____
8. _____

a. Each PGAL gives up two electrons and a hydrogen atom to NAD⁺, forming two NADH.
b. Intermediates form; each releases a hydrogen atom and an –OH group. These combine as water. Two molecules of PEP form by the reactions.
c. First one ATP molecule transfers a phosphate group to glucose, then another; atoms are rearranged as the cell has now invested two ATP molecules already present.
d. Each PEP transfers a phosphate group to ADP; once again, two ATP have formed by substrate-level phosphorylation.
e. Each PGAL also picks up an inorganic phosphate and transfers a phosphate group to ADP, forming ATP (substrate-level phosphorylation); the original cell investment of two ATP is now paid off.
f. In sum, the net energy yield is two ATP for each molecule of glucose entering glycolysis; two molecules of pyruvate are the end product.
g. Glucose molecules are present in the cytoplasm.
h. The rearranged and phosphorylated glucose splits into two PGAL molecules, each with a three-carbon backbone.

## 6.3. SECOND AND THIRD STAGES OF AEROBIC RESPIRATION [pp.86–88]

***Selected Words:*** acetyl–CoA [p.87], oxaloacetate [p.87], citrate [p.87], FADH$_2$ [p.87], *electron transfer phosphorylation* [p.87], ATP synthases [p.88], *final acceptor of the electrons* [p.88]

### Boldfaced Term

[p.86] mitochondrion (plural, mitochondria) _____

_____

### Labeling

Identify each numbered structure or location in the sketch. [p.86]

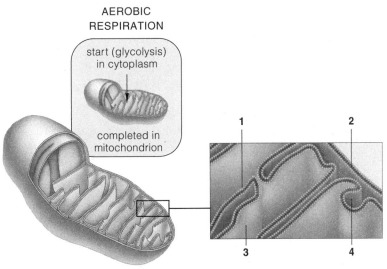

1. _____ _____ membrane
2. _____ _____ membrane
3. _____ compartment
4. _____ compartment

*Fill in the Blanks*

The numbers in this illustration represent missing information in the following narrative about aerobic respiration.

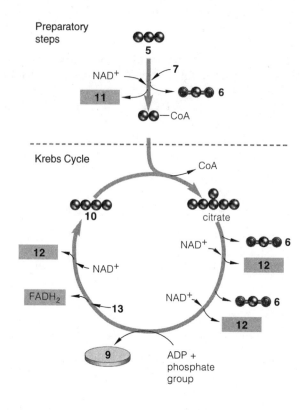

Two (5) _____ [p.86] molecules formed by glycolysis leave the cytoplasm and enter a mitochondrion. In this organelle, the second (preparatory steps) and third (electron transfer phosphorylation) stages of the aerobic pathway are completed. It is during the second stage that glucose is broken down to carbon dioxide and water. Six carbon atoms, three from each pyruvate, enter these reactions. Six carbon atoms also depart, as six (6) _____ _____ [pp.86–87] molecules (recall that the Krebs cycle turns twice for each glucose molecule entering glycolysis—all respiration chemistry begins with glycolysis).

In a few preparatory reactions, an enzyme removes a carbon from each pyruvate. (7) _____ [pp.86–87] becomes (8) _____-_____ [pp.86–87] by joining with the two-carbon fragment which enters the Krebs cycle (also called the citric acid cycle). Two (9) _____ [pp.86–87] form by substrate-level phosphorylations in the cycle and the intermediate compounds get rearranged into (10) _____ [pp.86–87], the cycle's entry point. Cells have only so much (10), so it must be regenerated to keep the reactions going. The two ATP formed in the Krebs cycle don't add much to the small yield from glycolysis. The big payoff is ten coenzymes loaded with electrons and hydrogen. To summarize these for two turns of the Krebs cycle and its preparatory stage: two $NAD^+$ are reduced to (11) _____ [pp.86–87] in the preparatory stage; six $NAD^+$ are reduced to (12) _____ [pp.86–87]. Two (13) _____ [pp.86–87] are reduced to $FADH_2$ during the Krebs cycle.

The numbers in this illustration represent missing information in the following narrative about electron transport phosphorylation.

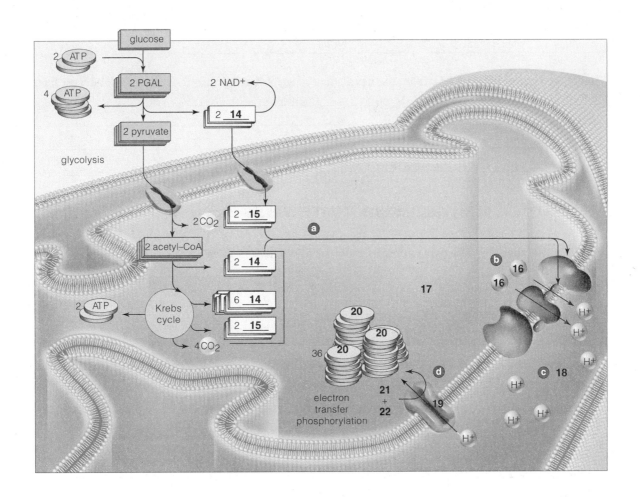

In the aerobic pathway's third stage, electron transfer phosphorylation, ATP formation gets into high gear. (14) _____ [p.87] and (15) _____ [p.87] that formed during the first and second stages deliver electrons and (16) _____ [p.87] to electron transfer chains, which are at the membrane dividing a mitochondrion's interior into two compartments. As the electrons are transferred through the chains, they attract $H^+$ from the (17) _____ [p.87] compartment. These ions get shuttled across the membrane, into the (18) _____ [p.87] compartment. Very soon, the $H^+$ concentration is greater in the outer compartment than the inner one. $H^+$ and electric gradients have become established across the membrane. $H^+$ follows the gradients, through the interior of (19) _____ _____ [p.87]. $H^+$ follows the gradients through the interior of (18). Energy released by the flow drives formation of (20) _____ [p.87] from (21) _____ [p.87] and unbound (22) _____ [p.87]. Hence the name, electron transfer phosphorylation.

How Cells Release Chemical Energy   55

*Short Answer*

23. What role does free oxygen play in electron transport phosphorylation? [p.88]

_____

_____

_____

24. What is the typical yield of ATP from the third stage of aerobic respiration? The net ATP harvest from one molecule of glucose? [p.88]

_____

_____

_____

## 6.4. ANAEROBIC ENERGY-RELEASING PATHWAYS [pp.88–89]

*Selected Words:* ethanol [p.89], *Saccharomyces cerevisiae* [p.89], *Lactobacillus* [p.89]

**Boldfaced Terms**

[p.89] alcoholic fermentation _____

_____

[p.89] lactate fermentation _____

_____

*Choice*

Choose from the following:

    a. alcoholic fermentation     b. lactate fermentation     c. applies to both types of fermentation

1. _____ Glycolysis is the first stage [p.88]
2. _____ Yeasts are famous for their use of this pathway [p.89]
3. _____ Humans, rabbits, and many other animals use this for a quick fix of energy [p.89]
4. _____ Pyruvate and NADH form, and the net energy yield is 2 ATP [p.88]
5. _____ Muscle cells use this pathway but not for long; diverting glucose into this pathway would waste too much of its energy for too little ATP [p.89]
6. _____ The final steps simply regenerate $NAD^+$, the coenzyme that assists the breakdown reactions [p.88]
7. _____ Each pyruvate molecule that formed in glycolysis is converted to the intermediate acetaldehyde [p.89]
8. _____ These reactions do not completely degrade glucose to $CO_2$ and $H_2O$ [p.88]
9. _____ *Saccharomyces cerevisiae* makes bread dough rise [p.89]
10. _____ Produce no more ATP beyond the rather tiny yield from glycolysis [p.88]
11. _____ Drunken birds such as tipsy wild turkeys and robins [p.89]

12. _____ *Lactobacillus* and some other bacteria use only this anaerobic pathway [p.89]
13. _____ NADH transfers electrons and hydrogen to this form and thereby converts it to ethanol [p.89]
14. _____ Yields enough energy to sustain many single-celled anaerobic organisms [p.88]
15. _____ The transfer by NADH converts each pyruvate to a three-carbon compound found in animals and other organisms who carry on this process [p.89]

## 6.5. ALTERNATIVE ENERGY SOURCES IN THE BODY [pp.89–91]
## 6.6. CONNECTIONS WITH PHOTOSYNTHESIS [pp.91–92]

*Choice*

Choose from the following:

a. fatty acids   b. triglycerides   c. amino acids   d. PGAL   e. glucose–6–phosphate
f. glucose   g. glycogen   h. acetyl–CoA

1. _____ At mealtime, a rise in blood concentration of this molecule prompts the pancreas to secrete insulin [p.89]
2. _____ The ammonia that forms undergoes conversions and becomes urea [p.91]
3. _____ When you aren't eating, the level in blood declines [p.90]
4. _____ A conversion to this compound allows cells to trap incoming glucose [p.89]
5. _____ A storage polysaccharide created by a glucose–6–phosphate pathway [p.90]
6. _____ Muscle and liver cells maintain the largest stores of this compound [p.90]
7. _____ Makes up 1 percent or so of the body's total energy reserves [p.91]
8. _____ Enzymes split dietary protein into these units [p.91]
9. _____ An excess of these molecules ends up as excess fat [p.91]
10. _____ Following removal of $-NH_3^+$ groups, the carbon backbones may be converted to fats or carbohydrates or they may enter the Krebs cycle [p.91]
11. _____ Yields more ATP than glucose [p.91]
12. _____ Enzymes in the liver convert glycerol to these molecules [p.91]
13. _____ Muscle cells will not give it up [p.91]
14. _____ Accumulate inside the fat cells of adipose tissue at strategic body regions [p.91]
15. _____ Supply about half of the ATP that muscle, liver, and kidney cells require between meals [p.91]
16. _____ Molecules resulting from enzymes cleaving fatty acid backbones; can enter the Krebs cycle [p.91]

*Short Answer*

17. Through the evolution of life over a very long period of time, photosynthesis and aerobic respiration became "linked" on a global scale. In terms of your knowledge of the two chemical processes, explain what forms this linkage. [pp.91–92]

_____
_____
_____

# Self-Quiz

_____ 1. Anaerobic respiration pathways begin and end in the _____ . [p.83]
   a. thylakoid of a chloroplast
   b. plasma membrane of the cell
   c. inner membrane of the mitochondrion
   d. cytoplasm
   e. outer compartment of the mitochondrion

_____ 2. The first energy-releasing step in glycolysis splits activated glucose into two molecules of _____ . [p.84]
   a. $NAD^+$
   b. PGAL
   c. ATP
   d. pyruvate
   e. PEP

_____ 3. The *net* energy yield of glycolysis is _____ ATP molecules [p.85]
   a. three
   b. four
   c. thirty-two
   d. two
   e. eight

_____ 4. Pyruvic acid is regarded as the end-product of _____ . [pp.84–85]
   a. glycolysis
   b. acetyl–CoA formation
   c. fermentation
   d. electron transfer phosphorylation
   e. the Krebs cycle

_____ 5. During which of the following phases of aerobic respiration is ATP produced directly by substrate-level phosphorylation? [pp.86–87]
   a. glucose formation
   b. the Calvin–Benson Cycle
   c. the Krebs cycle
   d. formation of acetyl–CoA
   e. electron transfer phosphorylation

_____ 6. Select the process by which NADH and $FADH_2$ transfers electrons along a chain of acceptors to oxygen so as to form water and set up conditions for producing a large number of ATP molecules. [p.87]
   a. glycolysis
   b. the Krebs cycle
   c. acetyl–CoA formation
   d. fermentation pathways
   e. electron transfer phosphorylation

_____ 7. The most efficient means of ATP production during aerobic respiration involves _____ . [pp.87–88]
   a. concentration of $H^+$ and electric gradients across a membrane
   b. ATP synthases
   c. formation of ATP in the inner mitochondrial compartment
   d. NADH and $FADH_2$
   e. all of the above

_____ 8. The total number of ATP molecules produced by the complete degradation of one glucose molecule is often thirty-six. Which of the following is not a contribution to that number? [pp.83–88]
   a. glycolysis produces 2 ATP
   b. steps preparatory to the Krebs cycle produce 2 ATP
   c. Krebs cycle produces 2 ATP
   d. electron transfer phosphorylation produces 32 ATP
   e. none of the above are correct

_____ 9. What factor allows fermentation reactions to continue and produce a low yield of ATP when a cell has no oxygen? [pp.88–89]
   a. the presence of ethanol molecules
   b. the presence of lactic acid
   c. ATP itself
   d. regeneration of the coenzyme $NAD^+$ during glycolysis
   e. the Krebs cycle

____ 10. Of the following, which one is not available as an alternative energy source for the human body? [pp.89–91]
   a. fats
   b. glycogen
   c. proteins
   d. carbohydrates
   e. all of the above are available

____ 11. The basic reason that we have an accumulation of oxygen in our atmosphere is _____. [pp.91–92]
   a. that oxygen produces more oxygen
   b. cyclic photophosphorylation
   c. that oxygen is released in electron transfer phosphorylation
   d. the noncyclic pathway
   e. the production of ATP

## Chapter Objectives/Review Questions

1. Define *free radicals* and be able to list at least three age-related mitochondrial disorders. [p.82]
2. Be able to summarize the events occurring during the three stages of aerobic respiration as illustrated in Figure 6.1 of the main text; be familiar with necessary terms. [pp.82–83]
3. Be able write the general equation for aerobic respiration, and be able state the equation in words. [p.83]
4. Describe the energy-requiring steps of glycolysis. [p.84]
5. Describe the energy-releasing steps of glycolysis. [p.84–85]
6. What is the role of $NAD^+$ in glycolysis? [p.85]
7. How many ATP molecules does the cell invest to jump-start glycolysis for each glucose molecule? [p.85]
8. How many NADH and ATP molecules form during glycolysis? What is the net energy yield from the process in terms of ATP molecules? [p.85]
9. For every glucose molecule entering glycolyis, two _____ molecules are produced as an end product. [p.85]
10. Be able to describe exactly where in the mitochondria the second and third stages of aerobic respiration occur. [pp.86–87]
11. The total net harvest of ATP, involving all stages of aerobic respiration, is _____ (number) molecules of ATP. [p.88]
12. Be able to summarize the biochemistry of alcoholic fermentation and lactate fermentation and tell in what types of organisms these processes occur. [pp.88–89]
13. The purpose of fermentation is the regeneration of _____ needed for further glycolysis. [pp.89–91]
14. Be able to summarize the metabolism processes of glucose at and between meals. [pp.89–91]
15. Know how the body metabolizes fats and proteins if glucose is not available. [pp.89–91]
16. Briefly relate how the pathways of photosynthesis and respiration are biochemically linked on a global scale. [pp.91–92]

## Integrating and Applying Key Concepts

Evaluate the following statement: Every atom in your body has first passed through the cell(s) of a photosynthetic organism.

# 7
# HOW CELLS REPRODUCE

## INTRODUCTION

Chapter 7 describes mitotic and meiotic cell division. It also looks at the relationship of cell division to variation and to cancer.

## FOCAL POINTS

- Figure 7.2 [p.96] diagrams the cell cycle.
- Figure 7.5 [pp.98–99] illustrates mitotic cell division.
- Figure 7.6 [p.100] contrasts cytoplasmic division in animals and plants.
- Figure 7.9 [pp.102–103] diagrams meiotic cell division.
- Figure 7.10 [p.104] demonstrates crossing over.
- Figure 7.11 [p.105] illustrates the results of random alignments during metaphase I of meiosis.
- Figure 7.14 [p.107] contrasts egg and sperm production.
- Figure 7.17 [p.109] compares benign and malignant tumors.
- Figure 7.18 [p.110] contrasts meiosis and mitosis.

## Interactive Exercises

*Henrietta's Immortal Cells* [p.94]

### 7.1. OVERVIEW OF CELL DIVISION MECHANISMS [p.95]
### 7.2. INTRODUCING THE CELL CYCLE [p.96]

**Selected Words:** *HeLa* cells [p.94], *somatic* cells [p.95], *germ* cells [p.95], prokaryotic fission [p.95], histones [p.95], G1, S, G2 [p.96]

*Boldfaced Terms*

[p.95] mitosis _____

_____

[p.95] meiosis _____

_____

[p.95] chromosome _____

[p.95] sister chromatids _____

[p.95] nucleosome _____

[p.95] centromere _____

[p.96] cell cycle _____

[p.96] interphase _____

## Short Answer

1. Describe the origin and the significance of HeLa cells. [p.94]

## Matching

Match each term with its description. [p.95]

2. _____ centromere
3. _____ chromosome
4. _____ somatic cells
5. _____ sister chromatids
6. _____ germ cells
7. _____ nucleosome
8. _____ gametes
9. _____ mitosis
10. _____ meiosis
11. _____ histones
12. _____ prokaryotic fission

a. Sex cells, such as sperm and eggs
b. Composed of a chromosome and its copy attached to each other
c. Cell divisions that give rise to gametes or spores; functions in sexual reproduction
d. An organizational unit composed of a histone–DNA spool
e. Chromosomal proteins; appear rather like beads on a string
f. Asexual reproduction as observed in the prokaryotic cells of archaebacteria and eubacteria
g. Body cells
h. Constricted region on a chromosome; docking site for microtubules during nuclear division
i. Cell divisions by which an organism grows, replaces cells, and repairs tissues; used by single-celled organisms for asexual reproduction
j. Composed of DNA and its proteins
k. Immature reproductive cells

## Identification

Identify each stage in the cell cycle indicated by the numbers in the diagram. [p.96]

13. _____
14. _____
15. _____
16. _____
17. _____
18. _____
19. _____
20. _____ _____
21. _____
22. _____
23. _____ _____

## Matching

Match each time span below with the most appropriate number in the preceding diagram. [p.96]

24. _____ Interval following DNA replication; cell prepares to divide
25. _____ The complete period of nuclear division, followed by cytoplasmic division (a separate event)
26. _____ Interval of cell growth, when DNA replication is completed (chromosomes duplicated)
27. _____ Interval of cell growth, before DNA duplication (chromosomes unduplicated)
28. _____ Usually the longest part of a cell cycle
29. _____ Interval of cytoplasmic division
30. _____ Period that includes G1, S, G2

## 7.3. MITOSIS MAINTAINS THE CHROMOSOME NUMBER [pp.97–99]
## 7.4. DIVISION OF THE CYTOPLASM [pp.100–101]

**Selected Words:** sex chromosomes [p.97], *prophase* [p.97], *metaphase* [p.97], *anaphase* [p.97], *telophase* [p.97], *centrioles* [p.98], *centrosomes* [p.98], spindle poles [p.98], cell plate [p.100]

## Boldfaced Terms

[p.97] chromosome number _____
_____

[p.97] diploid _____
_____

[p.97] mitotic spindle _____

[p.100] cleavage _____

[p.100] cleavage furrow _____

[p.100] cell plate formation _____

## Identification

Identify each mitotic stage shown. (Select from *late prophase, transition to metaphase, cell at interphase, metaphase, early prophase, telophase, interphase–daughter cells,* and *anaphase.*) Complete the exercise by entering the letter of the correct phase description in the parentheses following each label. [pp.98–99]

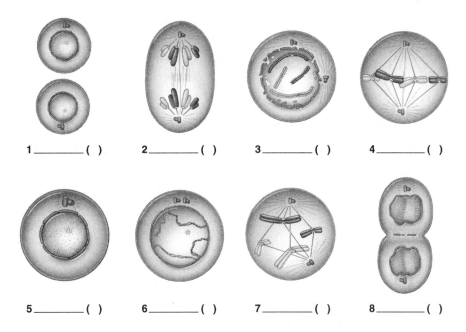

1._____( )  2._____( )  3._____( )  4._____( )

5._____( )  6._____( )  7._____( )  8._____( )

a. Attachments between two sister chromatids of each chromosome break; the two are now separate chromosomes that microtubules move to opposite spindle poles.
b. Microtubules penetrate the nuclear region and collectively form the bipolar spindle apparatus; microtubules become attached to the two sister chromatids of each chromosome.
c. The DNA and its associated proteins have started to condense.
d. All the chromosomes are now fully condensed and lined up at the equator of the fully formed microtubular spindle; chromosomes are now in their most tightly condensed form.
e. The cell duplicates its DNA and prepares for nuclear division.
f. Two daughter cells have formed; each is diploid with two of each type of chromosome, just like the parent cell's nucleus.
g. Chromosomes continue to condense. New microtubules become assembled. They move one of the two centriole pairs to the opposite end of the cell. The nuclear envelope begins to break up.
h. There are two clusters of chromosomes, which decondense. Patches of new membrane fuse to form a new nuclear envelope. Mitosis is completed.

*Choice*

For questions 9–17 on cytoplasmic division, choose from the following:

a. plant cells   b. animal cells

9. _____ Formation of a cell plate [pp.100–101]
10. _____ A cleavage furrow [p.100]
11. _____ Vesicles packed with wall-building materials fuse with one another and with remnants of the microtubular spindle [p.100]
12. _____ Cellulose deposits build up at the plate [p.100]
13. _____ Their cytoplasm cannot be pinched in two [p.100]
14. _____ A patch of plasma membrane sinks inward [p.100]
15. _____ Cytoskeletal elements just under the plasma membrane pull inward until there are two daughter cells [p.100]
16. _____ Just under the plasma membrane, a band of microfilaments generates contractile force for the cut [p.100]
17. _____ A wall bridges the cytoplasm and divides the parent cell in two [p.101]

## 7.5. MEIOSIS AND SEXUAL REPRODUCTION [pp.101–103]
## 7.6. HOW MEIOSIS PUTS VARIATION IN TRAITS [pp.104–105]

**Selected Words:** *clones* [p.101], *fertilization* [p.101], *diploid (2n)* [p.101], *hom–* [p.101], *homologue to homologue* [p.102], *chromosome shufflings* [p.104], *"nonsister" chromatids* [pp.104–105], *maternal chromosomes* [p.105], *paternal chromosomes* [p.105]

## Boldfaced Terms

[p.101] asexual reproduction _____

_____

[p.101] sexual reproduction _____

_____

[p.101] haploid _____

_____

[p.101] homologous chromosomes _____

_____

[p.101] gene _____

_____

[p.101] alleles _____

_____

[p.104] crossing over _____

_____

*Choice*

Choose from the following [p.101]:

a. asexual reproduction    b. sexual reproduction

1. \_\_\_\_\_ One parent alone produces offspring by mitotic divisions
2. \_\_\_\_\_ Each offspring inherits the same number and kinds of genes as its parent
3. \_\_\_\_\_ Commonly involves two parents
4. \_\_\_\_\_ The production of "clones"
5. \_\_\_\_\_ Involves meiosis, formation of gametes, and fertilization
6. \_\_\_\_\_ Offspring are genetically identical copies of the parent
7. \_\_\_\_\_ Agents of selection act on offspring and this invites evolutionary change
8. \_\_\_\_\_ Change can only occur by rare mutations
9. \_\_\_\_\_ The first cell of a new individual has pairs of genes on pairs of homologous chromosomes that are of maternal and paternal origin
10. \_\_\_\_\_ New gene combinations introduce variations in traits among offspring

*Dichotomous Choice*

Circle one of two possible answers given in parentheses in each statement. To answer these questions refer to the cited text pages and Figure 7.9. [pp.101–103]

11. If a cell has a diploid number ($2n$), it has (one/two) of the same type of chromosome. [p.101]
12. A pair of each type of chromosome having the same length, shape, and genes (with the exception of the nonidentical sex chromosomes) would be (homologous/nonhomologous) chromosomes. [p.101]
13. Gametes possess a (diploid/haploid) number of chromosomes. [p.101]
14. As long as two DNA molecules and their proteins remain attached at the centromere, they are (sister chromosomes/sister chromatids). [p.102]
15. With meiosis, chromosomes proceed through (one/two) consecutive divisions to yield four haploid nuclei. [p.102]
16. During meiosis I, each duplicated (chromosome/chromatid) lines up with its partner, homologue to homologue, and then the partners are moved apart from one another. [p.102]
17. Two attached sister chromatids represent (one/two) chromosome(s). [p.103]
18. Two different forms of the same gene on a pair of homologous chromosomes would be called (homologues/alleles). [p.101]
19. One pair of duplicated chromosomes would be composed of (two/four) chromatids. [p.102]
20. Cytoplasmic division following meiosis I results in two (diploid/haploid) daughter cells. [p.102]
21. DNA is replicated during (prophase I of meiosis I/interphase preceding meiosis I). [pp.101–102]
22. During (meiosis I/meiosis II), the two sister chromatids of each chromosome are separated from each other, and each sister chromatid is now a separate chromosome. [p.103]
23. If human body cell nuclei contain twenty-three pairs of homologous chromosomes, each resulting gamete will contain (twenty-three/forty-six) chromosomes. [p.101]

*Matching*

Following careful study of the major stages of meiosis shown in Figure 7.9 of the main text [pp.102–103], apply what you have learned by matching the following written descriptions with the appropriate sketch. Assume that the cell in this model initially has only one pair of homologous chromosomes (one from a paternal source and one from a maternal source) and crossing over does not occur. Complete the exercise by indicating the diploid (2n = 2) or haploid (n = 1) chromosome number of the cell in the parentheses.

24. _____ (   ) A pair of homologous chromosomes prior to S of interphase in a diploid germ cell [pp.102–103]
25. _____ (   ) While the germ cell is in S of interphase, chromosomes are duplicated through DNA replication; the two sister chromatids are attached at the centromere [p.102–103]
26. _____ (   ) During meiosis I, each duplicated chromosome lines up with its partner, homologue to homologue [p.102]
27. _____ (   ) Also during meiosis I, the chromosome partners separate from each other in anaphase I; cytokinesis occurs, and each chromosome goes to a different cell [p.102]
28. _____ (   ) During meiosis II (in two cells), the sister chromatids of each chromosome are separated from each other; four haploid nuclei form; cytokinesis results in four cells (potential gametes) [p.103]

*Short Answer*

29. List the two major meiotic mechanisms responsible for genetic variation and the resulting huge numbers of new gene combinations. [pp.104–105]

_____

_____

## 7.7. FROM GAMETES TO OFFSPRING [pp.106–107]
## 7.8. THE CELL CYCLE AND CANCER [pp.108–109]

**Selected Words:** *sporophyte* [p.106], *gametophytes* [p.106], primary spermatocyte [p.106], spermatids [p.106], secondary oocyte [p.106], first polar body [p.106], second polar body [p.106], ovum [p.106], checkpoint proteins [p.108], *growth factors* [p.108], *benign* [p.109], *malignant* [p.109], *metastasis* [p.109]

*Boldfaced Terms*

[p.106] spore _____

_____

[p.106] sperm _____

_____

[p.106] oocyte _____

_____

[p.106] egg _____

_____

[p.106] fertilization _____

_____

[p.108] neoplasms _____

_____

[p.109] cancers _____

_____

## Choice

Choose from the following: [p.106]

a. animal life cycle   b. plant life cycle   c. both animal and plant life cycles

1. _____ Meiosis results in the production of haploid resting spores
2. _____ A zygote divides by mitosis
3. _____ Meiosis results in the production of haploid gametes
4. _____ Haploid gametes fuse in fertilization to form a diploid zygote
5. _____ A zygote divides by mitosis to form a diploid sporophyte
6. _____ A spore divides by mitosis to produce a haploid gametophyte
7. _____ A haploid gametophyte divides by mitosis to produce haploid gametes
8. _____ A haploid spore divides by mitosis to produce a gametophyte
9. _____ A diploid body forms from mitosis of a zygote
10. _____ A gamete-producing body and a spore-producing body develop during the life cycle

## Sequence

Arrange the following entities in correct order of development, entering a 1 by the stage that appears first and a 5 by the stage that completes the process of spermatogenesis.

Complete the exercise by indicating in the parentheses whether each cell is *n* or *2n*. [pp.107–108]

11. _____ (   ) primary spermatocyte
12. _____ (   ) sperm
13. _____ (   ) spermatid
14. _____ (   ) spermatogonium
15. _____ (   ) secondary spermatocyte

## Matching

Match each cell type with its description. [pp.107–108]

16. _____ primary oocyte
17. _____ oogonium
18. _____ secondary oocyte
19. _____ egg (ovum) and three polar bodies
20. _____ first polar body

a. The cell in which synapsis, crossing over, and recombination occur
b. A cell that is equivalent to a diploid germ cell
c. A haploid cell formed after division of the primary oocyte that does not form an ovum at second division
d. Four haploid cells, but only one of which functions as an egg (ovum)
e. A haploid cell formed after division of the primary oocyte, the division of which forms a functional haploid egg (ovum)

## Short Answer

21. List the various mechanisms that contribute to the huge number of new gene combinations that may result from fertilization. [pp.106–107]

22. List the four characteristics displayed by *all* cancer cells. [p.109]

## Self-Quiz

_____ 1. The replication of DNA occurs _____. [p.96]
   a. between the growth phases of interphase
   b. immediately before prophase of mitosis
   c. during prophase of mitosis
   d. during anaphase of mitosis
   e. in daughter cells during telophase

_____ 2. Each histone-DNA spool is a single structural unit called a _____. [p.95]
   a. kinetochore
   b. motor protein
   c. centromere
   d. nucleosome
   e. gene

_____ 3. In the cell life cycle of a particular cell, _____. [p.96]
   a. mitosis occurs immediately after G1
   b. G2 precedes S
   c. G1 precedes S
   d. mitosis and S precede G1
   e. S occurs immediately prior to mitosis

_____ 4. The correct order of the stages of mitosis is _____. [pp.96–97]
   a. prophase, metaphase, telophase, anaphase
   b. telophase, anaphase, metaphase, prophase
   c. telophase, prophase, metaphase, anaphase
   d. anaphase, prophase, telophase, metaphase
   e. prophase, metaphase, anaphase, telophase

_____ 5. During _____, sister chromatids of each chromosome are separated from each other, and those former partners, now chromosomes, are moved toward opposite spindle poles. [p.99]
   a. prophase
   b. metaphase
   c. anaphase
   d. telophase
   e. transition to metaphase

_____ 6. In the process of cytokinesis, cleavage furrows are associated with _____ cell division, and cell plate formation is associated with _____ cell division. [pp.100–101]
   a. animal; animal
   b. plant; animal
   c. plant; plant
   d. animal; plant

_____ 7. Which of the following does not occur in prophase I of meiosis? [p.102]
   a. a cytoplasmic division
   b. a cluster of four chromatids
   c. homologues pairing tightly
   d. crossing over
   e. condensation of chromosomes

_____ 8. Crossing over is one of the most important events in meiosis because _____. [pp.104–105]
   a. it leads to genetic recombination
   b. homologous chromosomes must be separated into different daughter cells
   c. the number of chromosomes allotted to each daughter cell must be halved
   d. homologous chromatids must be separated into different daughter cells
   e. nonsister chromatids do not break during the process

_____ 9. Which of the following does not increase genetic variation? [pp.104–107]
   a. crossing over
   b. random fertilization
   c. prophase of mitosis
   d. random homologue alignments at metaphase I
   e. activities occurring during prophase I

_____ 10. Which of the following is the most correct sequence of events in animal life cycles? [p.106]
   a. meiosis → fertilization → gametes → diploid organism
   b. diploid organism → meiosis → gametes → fertilization
   c. fertilization → gametes → diploid organism → meiosis
   d. diploid organism → fertilization → meiosis → gametes

_____ 11. The cell in the diagram is a diploid that has three pairs of chromosomes. From the number and pattern of chromosomes, the cell _____. [p.103]

   a. could be in the first division of meiosis
   b. could be in the second division of meiosis
   c. could be in prophase of mitosis
   d. could be in neither mitosis nor meiosis, because this stage is not possible in a cell with three pairs of chromosomes
   e. could be in anaphase of mitosis

## Chapter Objectives/Review Questions

1. Define *HeLa cells*; explain their origin. [p.94]
2. Mitosis and meiosis refer to the division of the cell's _____. [p.95]
3. Be able to define *somatic cells, gametes, prokaryotic fission, chromosome, nucleosome, sister chromatids,* and *centromere*. [p.95]
4. Be able to list and describe, in order, the various activities occurring in the eukaryotic cell cycle. [p.96]
5. Be able to give a detailed description of the cellular events occurring in the prophase, metaphase, anaphase, and telophase of mitosis. [pp.97–99]
6. Compare and contrast cytokinesis as it occurs in plant and animal cell division; use the following concepts: cleavage furrow, microfilaments at the cell's midsection, and cell plate formation. [pp.100–101]
7. Be able to list and describe the characteristics of sexual and asexual reproduction; define *zygote*. [p.101]
8. What is the difference between diploid and haploid? [pp.97, 101]
9. Be able to tell the story of the stages of meiosis; include proper terminology and chromosome movement descriptions. [pp.102–103]

10. Describe the ways that meiosis adds variation in traits. [pp.104–105]
11. Be able to relate the terms used to describe gamete formation in plants and animals. [p.106]
12. Be able to define *growth factors, neoplasms, benign, malignant,* and *metastasis*. [pp.108–109]
13. List the four characteristics displayed by *all* cancer cells. [p.109]

## Integrating and Applying Key Concepts

Runaway cell division is characteristic of cancer. Imagine the various points of the mitotic process that might be sabotaged in cancerous cells in order to halt their multiplication. Then try to imagine how one might discriminate between cancerous and normal cells to guide those methods of sabotage most effective in combating cancer.

# 8

# OBSERVING PATTERNS IN INHERITED TRAITS

## INTRODUCTION

Chapter 8 explores the various patterns of inheritance that exist. Dominance, codominance, incomplete dominance, gene interactions, continuous variation, linked genes, and sex linkage are discussed.

## FOCAL POINTS

- Figures 8.5 and 8.6 [p.115] look at simple monohybrid inheritance.
- Figures 8.7 and 8.8 [pp.116, 117] describe dihybrid inheritance and independent assortment.
- Figures 8.14 and 8.15 [pp.120, 121] look at continuous variation.
- Figure 8.18 [p.123] illustrates a human karyotype.
- Figure 8.20 [p.125] describes linkage.
- Figures 8.22 and 8.23 [pp.126, 127] show human pedigrees.
- Figures 8.26 and 8.28 [pp.128, 129] look at sex-linked traits.
- Figure 8.30 [p.131] demonstrates nondisjunction.

# Interactive Exercises

*Menacing Mucus* [p.112]

## 8.1. TRACKING TRAITS WITH HYBRID CROSSES [pp.113–117]

***Selected Words:*** *cystic fibrosis* (CF) [p.112], *homozygous* [p.113], *heterozygous* [p.113], *dominant* [p.113], *recessive* [p.113], *gene expression* [p.113], *genotype* [p.113], *phenotype* [p.113], $F_1$, $F_2$ [p.113], *Pisum sativum* [p.113]

***Boldfaced Terms***

[p.113] genes _____

_____

[p.113] alleles _____

_____

[p.113] hybrids _____

_____

[p.113] homozygous dominant _____

_____

[p.113] homozygous recessive _____

_____

[p.113] heterozygous _____

_____

[p.114] monohybrid experiments _____

_____

[p.114] probability _____

_____

[p.115] Punnett-square method _____

_____

[p.115] segregation _____

_____

[p.116] dihybrids _____

_____

[p.116] dihybrid experiment _____

_____

[p.117] theory of independent assortment _____

_____

As you study each section of the text, become very familiar with all the "selected words" and "boldfaced terms." Try the genetics problems where they are provided. The best way to understand genetics is through the problem-solving process.

## Problems

1. In garden pea plants, tall (*T*) is dominant over dwarf (*t*). Use the Punnett-square method (refer to the main text, Figures 8.4, 8.5, and 8.6, pp.114–115) to determine the genotype and phenotype probabilities of offspring from the above cross, *Tt* × *tt*:

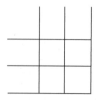

2. If *W* = purple and *w* = white, predict the genotype and phenotype probabilities of the cross *Ww* × *Ww*: [pp.114–115]

3. Two white dogs mated. They produced seven puppies, 5 white and 2 black. How did this happen? [pp.114–115] _____
_____
_____

4. When working genetics problems that deal with two gene pairs, use a fork-line device to visualize the independent assortment of gene pairs located on nonhomologous chromosomes into gametes.

   Assume that in humans, pigmented eyes (*B*) are dominant (an eye color other than blue) over blue (*b*), and right-handedness (*R*) is dominant over left-handedness (*r*). Cross the parents *BbRr* × *BbRr*. A sixteen-block Punnett square is required with gametes from each parent arrayed on two sides of the Punnett square (refer to Figures 8.7 and 8.8 in the main text, pp.116–117).

   Place the gametes shown on two sides of the Punnett square; combine these haploid gametes to form diploid zygotes within the squares. In the blank spaces below, enter the probability ratios derived within the Punnett square for the phenotypes listed: [pp.116–117]

   a. _____ pigmented eyes, right-handed

   b. _____ pigmented eyes, left-handed

   c. _____ blue-eyed, right-handed

   d. _____ blue-eyed, left-handed

## 8.2. NOT-SO-STRAIGHTFORWARD PHENOTYPES [pp.117–119]

**Selected Words:** *codominance* [p.117], *ABO blood typing* [p.118], *transfusion reaction* [p.118], *incomplete dominance* [p.118], *interaction among products of two or more gene pairs* [p.118], *albinism* [p.119], *Marfan syndrome* [p.119]

### Boldfaced Terms

[p.118] multiple allele system _____
_____

[p.119] pleiotropy _____
_____

## Genetics Problems

Study the example of ABO blood typing as an example of a multiple allele system with codominance and then solve the following problems. [pp.117–118]

1. $I^A i \times I^A I^B =$ _____

2. $I^B i \times I^A i =$ _____

3. $I^A I^B \times I^A I^B =$ _____

4. Study the examples of snapdragons as an example of incomplete dominance and then determine the phenotypes and genotypes of the offspring of the following crosses. R = red flower color, R' = white flower color, and RR' is pink. [p.118]

| Cross | Phenotypes | Genotypes |
|---|---|---|
| a. RR × R'R' = | | |
| b. RR' × RR' = | | |

5. In the inheritance of the coat (fur) color of Labrador retrievers, allele B specifies black that is dominant to brown (chocolate), b. Allele E permits full deposition of color pigment but the presence of two recessive alleles, ee, reduces deposition, and a yellow coat results.

   Predict the phenotypes of the coat color and their proportions resulting from the following cross:

   BbEe × Bbee = [p.118] _____

   _____

6. In poultry, an interaction occurs in which two genes produce a phenotype that neither gene can produce alone. The two interacting genes (R and P) produce comb shape in chickens. The possible genotypes and phenotypes are: [pp.118–119]

   | Genotypes | Phenotypes |
   |---|---|
   | R_P_ | walnut comb |
   | R_pp | rose comb |
   | rrP_ | pea comb |
   | rrpp | single comb |

[Hint: Where a blank appears in the genotypes above, either the dominant or the recessive symbol in that blank yields the same phenotype.]

What are the genotype and phenotype ratios of the offspring of a heterozygous walnut-combed male and a single-combed female? [p.119]

_____

_____

_____

74 Chapter Eight

## 8.3. COMPLEX VARIATIONS IN TRAITS [pp.120–122]

*Selected Words:* camptodactyly [p.120], environmental effects [p.122]

### Boldfaced Terms

[p.120] continuous variation _____
_____

[p.121] bell curve _____
_____

### Choice

Choose from the following primary contributing factors:

        a. environment     b. a number of genes affect a trait

1. _____ Height of human beings [p.121]
2. _____ Continuous variation in a trait [p.120]
3. _____ Flower color in *Hydrangea macrophylla* [p.122]
4. _____ The range of eye colors in the human population [p.120]
5. _____ Heat-sensitive version of one of the enzymes required for melanin production in Himalayan rabbits [pp.121–122]
6. _____ Yarrow plants grown at different elevations [p.122]

## 8.4. THE CHROMOSOMAL BASIS OF INHERITANCE [pp.123–124]
## 8.5. IMPACT OF CROSSING OVER ON INHERITANCE [p.125]

*Selected Words:* gene [p.123], locus [p.123], alleles [p.123], *wild-type* allele [p.123], *mutant* allele [p.123], *crossing over* [p.123], *independent assortment* [p.123], colchicine [p.123], *centrifuge* [p.123], SRY [p.124], linkage groups [p.125], recombinant genotypes [p.125]

### Boldfaced Terms

[p.123] karyotyping _____
_____

[p.124] sex chromosomes _____
_____

[p.124] autosomes _____
_____

### Dichotomous Choice

Choose the word in parentheses that more correctly completes each sentence.

1. The most common form of a gene is called the (wild-type/mutant) allele. [p.123]
2. Crossing over occurs between nonsister chromatids of (homologous/nonhomologous) chromosomes. [p.123]

3. Colchicine prevents (crossing over/cell division). [p.123]
4. Male humans transmit their Y chromosome only to their (sons/daughters). [p.124]
5. Male humans receive their X chromosome only from their (mothers/fathers). [p.124]
6. Human mothers and fathers each provide an X chromosome for their (sons/daughters). [p.124]

*Short Answer*

7. State the significance of the *SRY* gene. [p.124] _____

_____

_____

*Problem*

8. Circle the option below that represents a chromosome that has undergone crossover and recombination. It is assumed that the organism involved is heterozygous with the genotype. [p.125]   $\begin{vmatrix}A\\B\end{vmatrix}$  $\begin{vmatrix}a\\b\end{vmatrix}$

   a. $\begin{vmatrix}A\\B\end{vmatrix}$   b. $\begin{vmatrix}a\\b\end{vmatrix}$   c. $\begin{vmatrix}B\\A\end{vmatrix}$   d. $\begin{vmatrix}a\\B\end{vmatrix}$

## 8.6. HUMAN GENETIC ANALYSIS [pp.126–127]
## 8.7. EXAMPLES OF HUMAN INHERITANCE PATTERNS [pp.127–129]

**Selected Words:** *genetic abnormality* [p.126], *genetic disease* [p.126], *Huntington's disease* [p.127], *expansion mutations* [p.127], *achondroplasia* [p.127], *galactosemia* [p.128], *color blindness* [p.129], *Duchenne muscular dystrophy* [p.129], *hemophilia A* [p.129]

*Boldfaced Terms*

[p.126] pedigrees _____

_____

[p.126] genetic disorder _____

_____

[p.126] syndrome _____

_____

[p.126] disease _____

_____

*Problems*

1. The autosomal allele that causes galactosemia, (*g*), is recessive to the allele for normal galactose metabolism, (*G*). A normal woman whose father is galactosemic marries a man with galactosemia whose parents are normal. They have three children, two normal and one affected with galactosemia. List the genotypes for each person involved. [p.128]

_____

_____

2. Huntington's disease is a rare form of autosomal dominant inheritance, *H*; the normal gene is *h*. The disease causes progressive degeneration of the nervous system with onset exhibited near middle age. An apparently normal man in his early twenties learns that his father has recently been diagnosed as having Huntington's disease. What are the chances that the son will develop this disorder? [p.127]

3. A color-blind man and a woman with normal vision whose father was color blind have a son. Color blindness, in this case, is caused by an X-linked recessive gene. If only the male offspring are considered, what is the probability that their son is color blind? [p.129]

4. Hemophilia A is caused by an X-linked recessive gene. A woman who is seemingly normal but whose father was a hemophiliac marries a normal man. What proportion of their sons will have hemophilia? What proportion of their daughters will have hemophilia? What proportion of their daughters will be carriers? [p.129]

5. The accompanying pedigree shows the pattern of inheritance of color blindness in a family (persons with the trait are indicated by black circles). What is the chance that the third-generation female indicated by the arrow (below) will have a color-blind son if she marries a normal male? A color-blind male? [p.129]

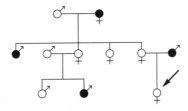

## 8.8. STRUCTURAL CHANGES IN CHROMOSOMES [p.130]
## 8.9. CHANGE IN THE NUMBER OF CHROMOSOMES [pp.131–133]

*Selected Words:* cri-du-chat [p.130], trisomy [p.131], monosomy [p.131], *Down syndrome* [p.131], *Turner syndrome* [p.132], *XXX condition* [p.132], *Klinefelter syndrome* [p.132], *XYY condition* [p.132]

### Boldfaced Terms

[p.130] duplications _____

[p.130] deletion _____

[p.130] inversion _____

[p.130] translocation _____

Observing Patterns in Inherited Traits 77

[p.131] aneuploidy _____

[p.131] polyploidy _____

[p.131] nondisjunction _____

## Labeling

On rare occasions, chromosome structure becomes abnormally rearranged. Such changes may have profound effects on the phenotype of an organism. Label the following diagrams of abnormal chromosome structure as a deletion, a duplication, an inversion, or a translocation.

1 _____    2 _____    3 _____    4 _____

## Short Answer

5. Which of the abnormal chromosomal rearrangements in the preceding diagrams results in the cri-du-chat syndrome? [p.130] _____

6. If a nondisjunction occurs at anaphase I of the first meiotic division, what will be the proportion of abnormal gametes (for the chromosomes involved in the nondisjunction)? [p.131] _____

7. Contrast the effects of polyploidy in plants and humans. [p.131] _____

## Choice

Choose from the following:

      a. Down syndrome   b. Turner syndrome   c. Klinefelter syndrome
      d. XYY condition   e. XXX condition

8. ____ Although female, most don't have functional ovaries; lack of hormones affects secondary sexual traits [p.132]

9. ____ Males whose testes and prostate are smaller than average, have sparse body hair, and whose breasts are a bit enlarged [p.132]

10. ____ Could only be caused by a nondisjunction in males; once thought to be connected with criminal behavior [pp.132–133]

11. ____ Most affected individuals show moderate to severe mental impairment and heart defects; they tend to be cheerful and sociable people [p.131]

12. \_\_\_\_ Phenotypically female; adults are often an inch or so taller and more slender than average; fertile; most fall within the normal range of social behavior [p.132]

## 8.10. SOME PROSPECTS IN HUMAN GENETICS [pp.133–134]

*Selected Words:* genetic counseling [p.133], prenatal diagnosis [p.133], embryo [p.133], fetus [p.133], amniocentesis [p.133], chorionic villi sampling (CVS) [p.134], fetoscopy [p.134], preimplantation diagnosis [p.134], "test-tube" babies [p.134], phenylketonuria, or PKU [p.134], "pro-life" [p.134], "pro-choice" [p. 134]

*Boldfaced Terms*

[p.133] abortion _____

[p.134] in-vitro fertilization _____

*Matching*

1. \_\_\_\_ phenotypic treatments [p.134]
2. \_\_\_\_ abortion [p.133]
3. \_\_\_\_ preimplantation diagnosis [p.134]
4. \_\_\_\_ genetic counseling [p.133]
5. \_\_\_\_ prenatal diagnosis [p.133]

a. Detects genetic disorders before birth; may use biochemical tests, amniocentesis, CVS, and fetoscopy; an example is a pregnancy at risk in a mother forty-five years old
b. A controversial option if prenatal diagnosis reveals a serious genetic disorder
c. If a severe heritable problem exists, it includes diagnosis of parental genotypes, pedigrees, and genetic testing for hundreds of known metabolic disorders; geneticists may be contacted for predictions
d. Relies on in-vitro fertilization; a fertilized egg mitotically divides into a ball of eight cells that provides a cell whose genes can be analyzed for genetic defects
e. Suppressing or minimizing symptoms of genetic disorders by surgical intervention, dietary control, or environmental adjustment; PKU is an example

## Self-Quiz

\_\_\_\_ 1. The best statement of Mendel's principle of independent assortment is that _____ . [pp.116–117]
   a. one allele is always independently dominant to another
   b. independent hereditary units from the male and female parents are blended in the offspring
   c. the two hereditary units that influence certain independent traits separate during gamete formation
   d. genes on pairs of homologous chromosomes are distributed into one gamete or another independently of genes on pairs of other chromosomes

\_\_\_\_ 2. In the $F_2$ generation of a cross between a red-flowered snapdragon (homozygous) and a white-flowered snapdragon, the expected phenotypic ratio of the offspring is _____ . [p.118]
   a. 3/4 red, 1/4 white
   b. 100 percent red
   c. 1/4 red, 1/2 pink, 1/4 white
   d. 100 percent pink

Observing Patterns in Inherited Traits

_____ 3. The tendency for dogs to bark while trailing is determined by a dominant gene, S, while silent trailing is due to the recessive gene, s. In addition, erect ears, D, is dominant over drooping ears, d. What combination of offspring would be expected from a cross between two erect-eared barkers who are heterozygous for both genes? [pp.116–117]
   a. 1/4 erect barkers, 1/4 drooping barkers, 1/4 erect silent, 1/4 drooping silent
   b. 9/16 erect barkers, 3/16 drooping barkers, 3/16 erect silent, 1/16 drooping silent
   c. 1/2 erect barkers, 1/2 drooping barkers
   d. 9/16 drooping barkers, 3/16 erect barkers, 3/16 drooping silent, 1/16 erect silent

_____ 4. A man with type A blood could be the father of _____ . [p.117–118]
   a. a child with type A blood
   b. a child with type B blood
   c. a child with type O blood
   d. a child with type AB blood
   e. all of the above

_____ 5. Suppose two individuals, each heterozygous for the same characteristic, are crossed. The characteristic involves complete dominance. The expected genotypic ratio of their progeny is _____ . [pp.114–115]
   a. 1:2:1
   b. 1:1
   c. 100 percent of one genotype
   d. 3:1

_____ 6. All the genes located on a given chromosome compose a _____ . [p.125]
   a. karyotype
   b. bridging cross
   c. wild-type allele
   d. linkage group

_____ 7. The farther apart two genes are on a chromosome, _____ . [pp.125–126]
   a. the less likely that crossing over and recombination will occur between them
   b. the greater will be the frequency of crossing over and recombination between them
   c. the more likely they are to be in two different linkage groups
   d. the more likely they are to be segregated into different gametes when meiosis occurs

_____ 8. Red–green color blindness is a sex-linked recessive trait in humans. A color-blind woman and a man with normal vision have a son. What are the chances that the son is color blind? If the parents ever have a daughter, what is the chance for each birth that the daughter will be color blind? (Consider only the female offspring.) [p.129]
   a. 100 percent, 0 percent
   b. 50 percent, 0 percent
   c. 100 percent, 100 percent
   d. 50 percent, 100 percent
   e. none of the above

_____ 9. Suppose that a hemophilic male (X-linked recessive allele) and a female carrier for the hemophilic trait have a nonhemophilic daughter with Turner syndrome. Nondisjunction could have occurred in _____ . [pp.129, 132]
   a. both parents
   b. neither parent
   c. the father only
   d. the mother only
   e. the nonhemophilic daughter

_____ 10. Of all phenotypically normal males in prisons, the type once thought to be genetically predisposed to becoming criminals was the group with _____ . [pp.132–133]
   a. XXY disorder
   b. XYY disorder
   c. Turner syndrome
   d. Down syndrome
   e. Klinefelter syndrome

# Chapter Objectives/Review Questions

1. Name the genetic disease caused by an abnormal copy of the CFTR gene; list the symptoms. [p.112]
2. Explain Mendel's laws of segregation and independent assortment; include Mendel's experimental methods with *Pisum sativum*. [pp.115–117]
3. Be able to solve simple Mendelian monohybrid and dihybrid crosses by use of a Punnett square and understand the role of probability in genetics. [pp.114–117]
4. Distinguish between complete dominance, incomplete dominance, and codominance; cite examples of each type of inheritance and be able to solve simple problems involving each. [pp.117–118]
5. Define *multiple allele system,* cite an example, and be able to solve problems involving multiple alleles. [pp.117–118]
6. Give several possible explanations for less predictable trait variations. [pp.120–122]
7. List two human traits that are explained by continuous variation. [pp.120–121]
8. Himalayan rabbits and garden hydrangeas are good examples of _____ effects on gene expression. [pp.121–122]
9. State the circumstances required for crossing over and describe the results. [p.123]
10. Demonstrate the pattern of sex determination in humans. [p.124]
11. Define *karyotyping;* briefly describe its preparation and value. [p.123]
12. Define *linkage group* and how it is applied. [pp.125–126]
13. Describe the relationship between crossover frequency and the location of genes on a chromosome; be able to solve a simple crossover problem. [pp.125–126]
14. Describe a pedigree chart and its purpose. [pp.126–127]
15. Describe the patterns of human inheritance. [pp.127–129]
16. Define duplication, inversion, deletion, and translocation. [p.130]
17. Cite evidence supporting the idea that chromosome structure does evolve. [p.130]
18. Know the mechanisms that change chromosome numbers. [p.131]
19. Describe the characteristics and the genetic cause of Down syndrome, Turner syndrome, XXX condition, XXY males, and XYY males. [pp.132–133]
20. Discuss some of the ethical considerations that might be associated with a decision to induce abortion. [pp.133–134]
21. Describe purpose, tools, and techniques involved in phenotypic treatments, prenatal diagnosis, genetic counseling, and preimplantation diagnosis. [pp.133–134]

# Integrating and Applying Key Concepts

Solve these following genetics problems:

1. A husband sues his wife for divorce, arguing that she has been unfaithful. His wife gave birth to a girl with a fissure in the iris of her eye, an X-linked recessive trait. Both parents have normal eye structure. Can the genetic facts be used to argue for the husband's suit? Explain your answer.

2. In a hospital, three women gave birth to sons on the same day and the babies got mixed up in the nursery. Match each couple with the correct child.

    Couple 1: Mother type O, Father type AB     Child X: type O

    Couple 2: Mother type A, Father type B      Child Y: type B

    Couple 3: Mother type A, Father type A      Child Z: type AB

# 9

# DNA STRUCTURE AND FUNCTION

## INTRODUCTION
This chapter describes the experiments leading up to the understanding of DNA's structure and function. The chapter also covers cloning.

## FOCAL POINTS
- Figure 9.1 [p.138] illustrates Griffith's experiments with *Streptococcus pneumoniae*.
- Figure 9.2 [p.139] diagrams the Hershey-Chase experiments.
- Figure 9.4 [p.140] shows the structure of the four DNA nucleotides.
- Figure 9.6 [p.141] describes DNA structure.
- Figure 9.7 [p.143] summarizes DNA replication and repair.

## Interactive Exercises

*Here Kitty, Kitty, Kitty, Kitty, Kitty* [p.137]

### 9.1. THE HUNT FOR FAME, FORTUNE, AND DNA [pp.138–139]

**Selected Words:** Ian Wilmut [p.137], Dolly [p.137], Johann Miescher [p.138], Frederick Griffith [p.138], *Streptococcus pneumoniae* [p.138], Oswald Avery [p.138], Linus Pauling [p.138], Hershey and Chase [p.139], James Watson [p.139], Francis Crick [p.139]

### Boldfaced Terms

[p.137] clone _____

[p.138] deoxyribonucleic acid, or DNA _____

[p.138] bacteriophages _____

*Short Answer*

1. The cloned sheep, Dolly, developed health problems and died after five years of life. List some of the health problems typically exhibited by surviving cloned mammals. [p.137]

_____
_____
_____

*Matching*

Match each scientist or group of scientists with a discovery leading up to the structure of DNA. [pp.138–139]

2. _____ Meischer
3. _____ Griffith
4. _____ Avery
5. _____ Pauling
6. _____ Hershey and Chase
7. _____ Watson and Crick

a. In 1928, discovered the transforming principle in *Streptococcus pneumoniae*; live harmless cells were mixed with dead S cells, R cells became S cells
b. Worked with radioactive sulfur (found in protein) and phosphorus (found in DNA) labels; T4 bacteriophage and *E. coli* cells demonstrated that labeled phosphorus was in bacteriophage DNA and contained hereditary instructions for new bacteriophages
c. Discovered the double helix structure of DNA
d. The first to isolate DNA
e. Described the alpha helix structure of proteins
f. Reported that the transforming substance in Griffith's bacteria experiments was probably DNA

## 9.2. DNA STRUCTURE AND FUNCTION [pp.140–142]

*Selected Words:* pyrimidines [p.140], purines [p.140], Erwin Chargaff [p.140], Rosalind Franklin [p.140], Maurice Wilkins [p.141], double helix [p.142], complementary [p.142]

*Boldfaced Terms*

[p.140] nucleotide _____
_____

[p.140] adenine _____
_____

[p.140] guanine _____
_____

[p.140] thymine _____
_____

[p.140] cytosine _____
_____

[p.140] x-ray diffraction image _____
_____

## Short Answer

1. List the three parts of every nucleotide. [p.140] _____

## Labeling

For the four nucleotides shown in questions 2–5, label each nitrogen-containing base correctly as guanine, thymine, cytosine, or adenine. In the parentheses following each blank, indicate whether that nucleotide base is a purine (pu) or a pyrimidine (py). [p.140]

2. _____ (   )   3. _____ (   )   4. _____ (   )   5. _____ (   )

For questions 6–12, label each numbered part of the DNA model as *phosphate, purine, pyrimidine, nucleotide,* or *deoxyribose.* [p.141]

6. _____
7. _____
8. _____
9. _____
10. _____
11. _____
12. _____

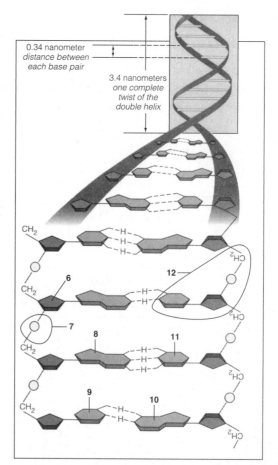

## 9.3. DNA REPLICATION AND REPAIR [pp.142–143]

*Selected Words:* *semiconservative* replication [p.142], mutation [p.143]

### Boldfaced Terms

[p.142] DNA replication _____

_____

[p.142] DNA polymerases _____

_____

[p.142] DNA ligases _____

_____

### Fill in the Blanks

Until Watson and Crick presented their model, no one could explain DNA (1) _____ [p.142], or how the molecule of inheritance is duplicated before a cell divides. (2) _____ [p.142] break the hydrogen bonds between the two-nucleotide strands of a DNA molecule. After the hydrogen bonds are broken, one (3) _____ [p.142] unwinds from the other and exposes stretches of its nucleotide bases. Cells contain stockpiles of free (4) _____ [p.142] that can pair with the exposed bases.

Each parent strand stays intact, and a companion strand is assembled on each one according to this base-pairing rule: A to (5) _____ [p.142], and G to (6) _____ [p.142]. As soon as a stretch of a new, partner strand forms on a stretch of the parent strand, the two twist up together into a (7) _____ _____ [p.142]. Because the parent DNA strand is (8) _____ [p.142] during the replication process, half of every double-stranded DNA molecule is "old" and half is "new." The process is called (9) _____ [p.142] replication.

DNA replication uses a team of molecular workers. Enzymes called DNA (10) _____ [p.142] catalyze the formation of a new strand using the parent strand as a template. Free nucleotides themselves provide energy for strand assembly. Each has three (11) _____ [p.142] groups. DNA polymerase splits off two, and it is this release of (12) _____ [p.142] that drives the attachments.

DNA (13) _____ [p.142] fill in the tiny gaps between the new short stretches to form one continuous strand. Then enzymes wind the template strand and complementary strand together to form a DNA (14) _____ _____ [p.142].

Errors are occasionally introduced during replication. Very often (15) _____ _____ [p.143] themselves correct them. If the mistakes are not repaired, the DNA has acquired a (16) _____ . [p.143]

*Labeling*

17. The term *semiconservative replication* refers to the fact that each new DNA molecule resulting from the replication process is "half-old, half-new." In the adjacent illustration, complete the replication required in the middle of the molecule by adding the required letters representing the missing nucleotide bases. Recall that ATP energy and the appropriate enzymes are actually required to complete this process. [pp.142–143]

T– ___         ___ –A
G– ___         ___ –C
A– ___         ___ –T
C– ___         ___ –G
C– ___         ___ –G
C– ___         ___ –G
old  new       new  old

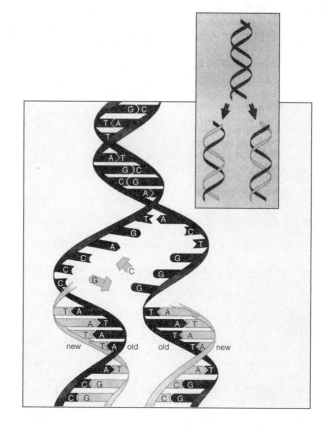

## 9.4. USING DNA TO CLONE MAMMALS [pp.143–144]

**Selected Words:** embryo cloning [p.143], "artificial twinning" [p.143], nuclear transfer [p.144]

*Choice*

Choose from these cloning methods:

    a. embryo cloning    b. nuclear transfers    c. both embryo cloning and nuclear transfers

1. _____ The clones are identical to one another but not to their sexually reproducing parents. [p.143]
2. _____ Cloning an adult [p.144]
3. _____ Involves tricking a differentiated cell into rewinding the developmental clock [p.144]
4. _____ Tiny cell clusters are implanted into surrogate mothers, where they grow and develop. [pp.143–144]
5. _____ The first cloned cat, CC, did not have the exact fur color and patterning of her genetic donor; some environmental factors must be at work. [p.144]
6. _____ The procedures most likely invite genetic defects. [p.144]

86 Chapter Nine

# Self-Quiz

1. _____ was the first scientist to isolate DNA. [p.138]
   a. Oswald Avery
   b. Linus Pauling
   c. James Watson
   d. Johann Miescher
   e. Frederick Griffith

2. The scientist who demonstrated that harmless pneumonia-causing bacteria cells had become permanently transformed into pathogens through a change in the bacterial hereditary material was _____ . [p.138]
   a. Oswald Avery
   b. Linus Pauling
   c. James Watson
   d. Johann Miescher
   e. Frederick Griffith

3. _____ demonstrated that radioactively labeled bacteriophages transfer their DNA but not their protein coats to their host bacteria. [p.139]
   a. Watson and Crick
   b. Hershey and Chase
   c. Hershey and Griffith
   d. Watson and Pauling
   e. Wilkins and Chargaff

4. In 1953, _____ built a model of DNA that fit all the pertinent biochemical rules and insights they had gleaned from other sources. [p.139]
   a. Watson and Crick
   b. Hershey and Chase
   c. Wilkins and Franklin
   d. Watson and Pauling
   e. Wilkins and Chargaff

5. Adenine and guanine are _____ . [p.140]
   a. double-ringed purines
   b. single-ringed purines
   c. double-ringed pyrimidines
   d. single-ringed pyrimidines
   e. neither purines nor pyrimidines

6. _____ discovered that in DNA, the amounts of A = T and the amounts of G = C. [p.140]
   a. Maurice Wilkins
   b. Rosalind Franklin
   c. Erwin Chargaff
   d. Hershey and Chase
   e. Watson and Crick

7. Rosalind Franklin's x-ray diffraction research on DNA established _____ . [p.141]
   a. phosphate groups project outward from the molecule
   b. one pair of chains
   c. one pair of chains oriented in opposite directions
   d. a helical structure
   e. all of the above

8. A single strand of DNA with the base-pairing sequence C–G–A–T–T–G is compatible only with the sequence _____ . [p.142]
   a. C–G–A–T–T–G
   b. G–C–T–A–A–G
   c. T–A–G–C–C–T
   d. G–C–T–A–A–C
   e. G–C–T–A–T–C

9. Enzymes known as _____ catalyze the formation of new DNA strands using the parent strands as templates. [p.142]
   a. semiconservative enzymes
   b. DNA ligases
   c. conservative enzymes
   d. DNA polymerases
   e. replication enzymes

10. Nuclear transfer was used to produce _____ . [pp.143–144]
    a. Dolly, a sheep
    b. CC, a cat
    c. both Dolly and CC
    d. neither Dolly nor CC

## Chapter Objectives/Review Questions

1. Describe the cloning technique used to clone Dolly the sheep and list the problems that have arisen as a result of cloning mammals with this technique. [p.137]
2. Summarize the research carried out by Miescher; Griffith; Avery; and colleagues Pauling, Hershey, and Chase; state the specific advances made by each in the understanding of DNA. [pp.138–139]
3. The two scientists who assembled the clues to DNA structure and produced the first model were _____ and _____ . [p.139]
4. Be able to name the four nucleotides in DNA and indicate if each is a purine or a pyrimidine. [p.140]
5. Be able to sketch a DNA molecule, using letters to represent the nitrogenous bases, phosphates, sugars, and hydrogen bonds. [p.141]
6. List the pieces of information about DNA structure that Rosalind Franklin discovered through her x-ray diffraction research. [pp.140–142]
7. Assume that the two parent strands of DNA have been separated and that the base sequence on one parent strand is A–T–T–C–G–C; the base sequence that will complement that parent strand is _____ . [p.142]
8. Describe how double-stranded DNA replicates from stockpiles of nucleotides. [pp.142–143]
9. Define and give examples of *embryo-splitting* and *nuclear transfers* as cloning methods. [pp.143–144]

## Integrating and Applying Key Concepts

Write a review of the stages of mitosis and meiosis, as well as the process of fertilization. Include what has now been learned about DNA replication and the relationship of DNA to a chromosome. As you cover the stages, be sure each cell receives the proper number of DNA threads.

# 10

# GENE EXPRESSION AND CONTROL

## INTRODUCTION

Chapter 10 looks at how information in DNA results in the production of protein. The chapter also discusses how this is controlled.

## FOCAL POINTS

- Figure 10.1 [p.147] compares DNA and RNA.
- Figure 10.2 [p.148] diagrams transcription of RNA.
- Figure 10.3 [p.149] looks at the modification of pre-mRNA to form mature mRNA.
- Figures 10.4 and 10.5 [p.150] look at the genetic code and how it is translated into the amino acids of protein.
- Figure 10.8 [p.152] diagrams protein synthesis.
- Figure 10.9 [p.154] describes the results of mutation.
- Figure 10.13 [p.157] illustrates the lactose operon.
- Figure 10.14 [p.158] diagrams the places where controls of protein formation can be found.
- Figure 10.18 [p.160] summarizes information flow from DNA to protein.

## Interactive Exercises

*Ricin and Your Ribosomes* [p.146]

### 10.1. MAKING AND CONTROLLING THE CELL'S PROTEINS [p.147]
### 10.2. HOW IS RNA TRANSCRIBED FROM DNA? [pp.147–149]

**Selected Words:** ricin [p.146], RNA (ribonucleic acid) [p.147], alternative splicing [p.148], pre-mRNA [p.149], guanine "cap" [p.149], poly-A tail [p.149]

**Boldfaced Terms**

[p.147] gene _____

[p.147] transcription _____

[p.147] translation _____

[p.147] messenger RNA (mRNA) _____

[p.147] ribosomal RNA (rRNA) _____

[p.147] transfer RNA (tRNA) _____

[p.147] uracil _____

[p.148] RNA polymerase _____

[p.148] promoter _____

[p.148] introns _____

[p.148] exons _____

## Fill in the Blanks

The only two toxins known to be more potent than ricin are (1) _____ [p.146] and (2) _____ _____ [p.146]. Ricin is extracted from seeds of the (3) _____ _____ [p.146] plant and is toxic because it damages (4) _____ [p.146], cellular organelles needed to produce (5) _____ [p.146].

Transcription, the production of (6) _____ [p.147], occurs in the (7) _____ [p.147] of a eukaryotic cell. Three kinds of RNA are produced: (8) _____ [p.147], (9) _____ [p.147], and (10) _____ [p.147]. The enzyme that is needed for transcription is called (11) _____ _____ [p.148] and begins copying DNA at a sequence called the (12) _____ [p.148].

## Short Answer

13. What is the purpose of each of the three types of RNA in protein synthesis? [p.147]

_____

_____

14. How do DNA and RNA differ in structure and function? [p.147]

15. The "pre-mRNA" of eukaryotes must be modified before it can be used as mRNA. What are these three modifications? [pp.148–149]

16. How can a cell use the same "pre-mRNA" to produce different proteins? [pp.148–149]

## 10.3. DECIPHERING mRNA [pp.150–151]
## 10.4. FROM mRNA TO PROTEIN [pp.152–153]

*Selected Words:* triplets [p.150], START signal [p.150], STOP signals [p.150], "wobble" [p.151], translation [p.152], *initiation* [p.152], *elongation* [p.152], *termination* [p.152], release factors [p.153], polysome [p.153]

### Boldfaced Terms

[p.150] codons _____

[p.150] genetic code _____

[p.151] anticodon _____

### Fill in the Blanks

If a DNA code is ATCCAGTA, the mRNA transcribed would be (1) _____ [p.150]. Because three bases are needed to code for each amino acid, the genetic code is often referred to as a (2) _____ [p.150] code. There are (3) _____ [p.150] codons in this code and (4) _____ [p.150] of the codons actually code for amino acids. The start codon is (5) _____ [p.150] and the stop codons are (6) _____ [p.150], (7) _____ [p.150], and (8) _____ [p.150].

Use the chart to answer questions 9–14 [p.150]

9. How many codons code for tryptophan (trp)?
_____

10. What are they?
_____
_____

11. How many codons code for isoleucine (ile)?
_____

12. What are they?
_____
_____

13. How many codons code for serine (ser)?
_____

14. What are they?
_____
_____

| first base | \- second base \- | | | | third base |
|---|---|---|---|---|---|
| | U | C | A | G | |
| U | phenylalanine | serine | tyrosine | cysteine | U |
| | phenylalanine | serine | tyrosine | cysteine | C |
| | leucine | serine | STOP | STOP | A |
| | leucine | serine | STOP | tryptophan | G |
| C | leucine | proline | histidine | arginine | U |
| | leucine | proline | histidine | arginine | C |
| | leucine | proline | glutamine | arginine | A |
| | leucine | proline | glutamine | arginine | G |
| A | isoleucine | threonine | asparagine | serine | U |
| | isoleucine | threonine | asparagine | serine | C |
| | isoleucine | threonine | lysine | arginine | A |
| | methionine (or START) | threonine | lysine | arginine | G |
| G | valine | alanine | aspartate | glycine | U |
| | valine | alanine | aspartate | glycine | C |
| | valine | alanine | glutamate | glycine | A |
| | valine | alanine | glutamate | glycine | G |

*True–False*

If the statement is true, write a T in the blank. If the question is false, correct it by changing the underlined word(s) and writing the correct word(s) in the answer blank. [pp.150–153]

_____ 15. Humans and bacteria use <u>the same</u> genetic code.

_____ 16. In <u>Golgi bodies</u> a slightly different code is used.

_____ 17. According to the wobble theory, as long as the anticodon for the <u>second and third</u> bases of the codon are complementary, the remaining base makes less difference.

_____ 18. A polysome is <u>one ribosome connected to several mRNAs.</u>

92   Chapter Ten

*Matching*

Match each of the following components of protein synthesis with the proper descriptive phrase. [pp.152–153]

19. _____ AUG
20. _____ anticodon
21. _____ A site
22. _____ codon
23. _____ large ribosomal subunit
24. _____ mRNA
25. _____ release factor
26. _____ small ribosomal subunit
27. _____ UAA

a. the code for the protein to be made
b. any three-base code found on mRNA
c. any three-base code found on tRNA
d. combines with initiator tRNA and mRNA to begin initiation
e. is recognized by initiator tRNA
f. no tRNA recognizes this codon
g. tRNA binds to ribosomes here
h. its binding completes the initiation complex
i. binds to ribosomes in place of a tRNA when certain codons are present

## 10.5. MUTATED GENES AND THEIR PROTEIN PRODUCTS [pp.154–155]
## 10.6. CONTROLS OVER GENE EXPRESSION [pp.156–159]

**Selected Words:** *frameshift* mutations [p.154], Barbara McClintock [p.154], thymine dimer [p.155], signaling molecules [p.156], hormones [p.156], methylation [p.156], *repressor* [p.156], *lactose intolerance* [p.156], *antennapedia* [p.158], "mosaic" [p.159], *anhydrotic ectodermal dysplasia* [p.159]

*Boldfaced Terms*

[p.154] base-pair substitutions _____

[p.154] insertions _____

[p.154] deletions _____

[p.154] transposons _____

[p.154] ionizing radiation _____

[p.155] nonionizing radiation _____

[p.155] alkylating agents _____

[p.156] negative control _____

Gene Expression and Control 93

[p.156] positive control _____

[p.156] promoter _____

[p.156] operators _____

[p.156] lac operon _____

[p.157] differentiation _____

[p.158] homeotic genes _____

[p.159] X chromosome inactivation _____

[p.159] Barr body _____

[p.159] dosage compensation _____

## Choice

1. _____ Which of the following changes in an mRNA codon would not lead to a change in the protein being synthesized? [pp.150, 154, chart on p.105 of the study guide]

    a. UAC → UAA    b. UAC → UAU    c. UAC → UAG    d. UAC → UGC

2. _____ Sickle-cell anemia is caused by _____ . [p.154]

    a. a base-pair substitution    b. a one base addition    c. a one base deletion    d. a transposon

3. _____ Which of the following mutations shift the reading frame of the mRNA? [p.154]

    a. adding one base    b. deleting one base    c. both a. and b.    d. neither a. nor b.

4. _____ Specialized DNA polymerases are able to repair _____ of the mutations made during DNA replication. [p.154]

    a. none    b. a few    c. most    d. all

5. _____ Cigarette smoke causes mutations because it contains _____ . [p.155]

    a. alkylating agents    b. ionizing radiation    c. nonionizing radiation    d. transposons

*Short Answer*

6. How do transposable elements cause mutations? [pp.154–155]

_____

_____

*Fill in the Blanks*

Regulatory proteins include (7) _____ [p.156] and other (8) _____ _____ [p.156]. In a (9) _____ [p.156] control system, regulatory proteins slow or stop transcription. A noncoding DNA sequence called a (10) _____ [p.156] signals the start of transcription and serves as a binding site for the enzyme, (11) _____ _____ [p.156]. (12) _____ [p.156] are binding sites for (13) _____ [p.156] and flank the promoter. When a (13) is bound, transcription is stopped.

The control system for the enzymes that metabolize lactose was the first regulatory system to be worked out. When lactose is present, some is converted to (14) _____ [p.156]. This sugar binds to the repressor, altering its (15) _____ [p.156] and preventing it from binding to the operator. Transcription of the genes can then proceed. The operator, promoter, and genes coding for the proteins are collectively called the lactose (16) _____ [p.156]. In some humans, the intestine stops producing (17) _____ [p.156], the enzyme needed to degrade lactose, in early childhood. This leads to (18) _____ _____ [p.156].

In eukaryotes, different genes are active in cells in different parts of the body. This causes cells to become (19) _____ [p.157] in composition, structure, and function. This gives rise to different cell lines, a process called (20) _____ _____ [p.157]. By some estimates, only (21) _____ to _____ [p.157] percent of genes are active in any cell line.

*Choice*

Indicate whether each of the following controls is transcriptional, transcript processing, translational, or posttranslational. [pp.156–157]

22. _____ chemical modification of the chromosome
23. _____ DNA rearrangement
24. _____ gene duplication
25. _____ length of poly-A tail
26. _____ protein modification
27. _____ snipping of introns
28. _____ splicing of exons

*Short Answer*

29. What is a homeotic gene? [p.159] _____

_____

_____

*Fill in the Blanks*

In mammalian females, (30) _____ [p.159] X chromosome(s) is/are inactivated in each cell. These permanently condensed X chromosomes are called (31) _____ _____ [p.159]. Because this is a random event, females have different patches of cells with different active X chromosomes. In effect, she is a (32) _____ [p.159]. An example is (33) _____ _____ _____ [p.159] where some patches of the affected female's skin have no sweat glands while others have the normal number. (32) also gives rise to the (34) _____ [p.159] color pattern in cats. Only (35) _____ [p.159] cats can show this pattern.

# Self-Quiz

_____ 1. The production of an RNA copy of DNA is called _____ . [p.147]
  a. replication
  b. transcription
  c. translation
  d. transduction

_____ 2. The class of RNA that contains the actual information to make proteins is _____ . [p.147]
  a. mRNA
  b. nRNA
  c. rRNA
  d. tRNA

_____ 3. RNA has the sugar _____ and the base _____ . [p.147]
  a. deoxyribose; thymine
  b. deoxyribose; uracil
  c. ribose; thymine
  d. ribose; uracil

_____ 4. The excision of introns and the splicing together of exons takes place in _____ . [pp.148–149]
  a. eukaryotes
  b. prokaryotes
  c. both a. and b.
  d. neither a. nor b.

_____ 5. Which of the following is not a stop signal in mRNA? [p.150]
  a. UAA
  b. UAG
  c. UGA
  d. UGG

_____ 6. During termination, a _____ binds to ribosome. [pp.152–153]
  a. terminator tRNA
  b. terminator mRNA
  c. protein release factor
  d. lipid release factor

_____ 7. _____ mutations can shift the reading frame of the mRNA. [p.154]
  a. Addition
  b. Deletion
  c. both a. and b.
  d. neither a. nor b.

_____ 8. Free radicals are formed by _____ . [p.155]
  a. ionizing radiation
  b. nonionizing radiation
  c. both a. and b.
  d. neither a. nor b.

____ 9. In negative control, a/an _____ protein binds to the _____, stopping transcription. [p.156]
   a. enhancer; operator
   b. enhancer; promoter
   c. repressor; operator
   d. repressor; promoter

____ 10. Posttranscriptional controls are seen in _____. [p.158]
   a. eukaryotes
   b. prokaryotes
   c. both a. and b.
   d. neither a. nor b.

## Chapter Objectives/Review Questions

1. How does ricin cause death? [p.146]
2. What are the three classes of RNA? What is the role of each in protein synthesis? [p.147]
3. Describe how pre-mRNA must be modified to become active (mature) RNA. [p.148]
4. Describe the genetic code. [p.150]
5. What is the "wobble effect" and why is it important? [p.151]
6. Describe the translation of mRNA into protein. [pp.152–153]
7. What are some of the ways that mutations arise? [pp.154–155]
8. How does *E. coli* control transcription of the genes for enzymes needed to metabolize lactose? [p.156]
9. What is lactose intolerance? How does it arise? [p.156]
10. Relate selective gene expression to differentiation. [p.157]
11. What is X chromosome inactivation? How is it related to the mosaicism and dosage compensation? [pp.158–159]

## Integrating and Applying Key Concepts

1. Do you think bioterrorism is going to be a major threat? Why or why not?
2. Describe the production of a protein starting with the information on the DNA.
3. Why do eukaryotes need more levels of control than prokaryotes?

# 11

# STUDYING AND MANIPULATING GENOMES

## INTRODUCTION

Chapter 11 looks at recombinant DNA and its uses in modern genetics. It also discusses the uses of restriction fragments.

## FOCAL POINTS

- Figure 11.1 [p.163] illustrates how the restriction sites and sticky ends produced are specific for each restriction enzyme.
- Figure 11.3 [p.164] diagrams formation of recombinant DNA cloning vectors.
- Figure 11.5 [p.165] shows how probes can be used to find specific recombinant DNA molecules.
- Figure 11.6 [p.166] illustrates PCR.
- Figure 11.8 [p.168] diagrams the steps of DNA fingerprinting.
- Figure 11.12 [p.171] shows gene transfer in plants.

## Interactive Exercises

*Golden Rice or Frankenfood?* [p.162]

**11.1. A MOLECULAR TOOLKIT** [pp.163–164]

**11.2. HAYSTACKS TO NEEDLES** [pp.165–166]

**11.3. DNA SEQUENCING** [p.167]

**11.4. FIRST JUST FINGERPRINTS, NOW DNA FINGERPRINTS** [p.168]

*Selected Words:* frankenfood [p.162], vitamin A [p.162], beta-carotene [p.162], "sticky ends" [p.163], DNA ligase [p.163], recombinant DNA [p.163], *genomic library* [p.165], *cDNA library* [p.165], *Thermus aquaticus* [p.166], *Taq* polymerase [p.166], *DNA fingerprinting* [p.168]

*Boldfaced Terms*

[p.163] restriction enzymes _____

_____

[p.163] plasmids _____

_____

[p.163] cloning vector _____

[p.164] reverse transcriptase _____

[p.164] cDNA _____

[p.165] library _____

[p.165] genome _____

[p.165] probe _____

[p.165] nucleic acid hybridization _____

[p.165] PCR _____

[p.166] primers _____

[p.167] DNA sequencing _____

[p.167] gel electrophoresis _____

[p.168] DNA fingerprint _____

[p.168] tandem repeats _____

## Fill in the Blanks

WHO estimates that (1) _____ _____ [p.162] children are vitamin A deficient which can lead to (2) _____ [p.162] and other disorders. Scientists have transferred genes to rice that allow the plant to produce (3) _____ _____ [p.162], a vitamin A precursor. Since these plants are the product of (4) _____ _____ [p.162], opponents of this process have labeled this rice and other genetically modified organisms (5) _____ [p.162].

## Choice

6. _____ Restriction enzymes are isolated from _____. [p.163]

    a. algae    b. bacteria    c. fungi    d. plants

7. _____ Restriction enzymes are able to cleave _____. [p.163]

   a. DNA   b. RNA   c. both a. and b.   d. neither a. nor b.

8. _____ Complimentary single-stranded tails formed by restriction enzymes are called _____. [p.163]

   a. compends   b. comptails   c. sticky ends   d. sticky tails

9. _____ The enzyme that joins DNA fragments back together is _____. [p.163]

   a. DNA polymerase   b. ligase   c. primase   d. RNA polymerase

10. _____ A small circular piece of DNA that contains only a few genes is called a _____. [p.163]

    a. clone   b. genome   c. plasmid   d. ribosome

11. _____ Reverse transcriptase uses _____ as a template to make a complimentary copy of _____. [p.164]

    a. DNA; DNA   b. DNA; RNA   c. RNA; DNA   d. RNA; RNA

12. _____ cDNA is good for insertion into bacteria because it lacks _____ which bacteria would be unable to remove from transcribed RNA. [p.164]

    a. exons   b. introns   c. neutrons   d. protons

## Matching

Match each phrase or statement with the correct diagram. [p.164]

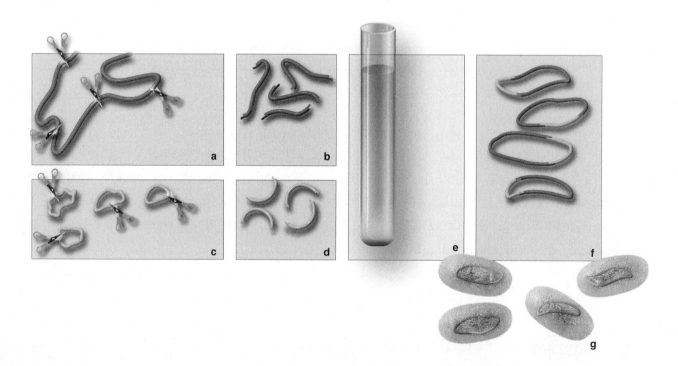

13. _____ Bacterial cells able to divide quickly take up the recombinant plasmids
14. _____ A collection of recombinant plasmids
15. _____ DNA or cDNA fragments with sticky ends
16. _____ Plasmid DNA and DNA or cDNA fragments mixed together with DNA ligase
17. _____ Plasmid DNA with sticky ends
18. _____ Restriction enzymes cut the chromosomal or cDNA at sites where a specific base sequence is found.
19. _____ Restriction enzymes cut the plasmid DNA at sites where a specific base sequence is found.

Match the letter from the appropriate statement with a number on the figure below [p.165]

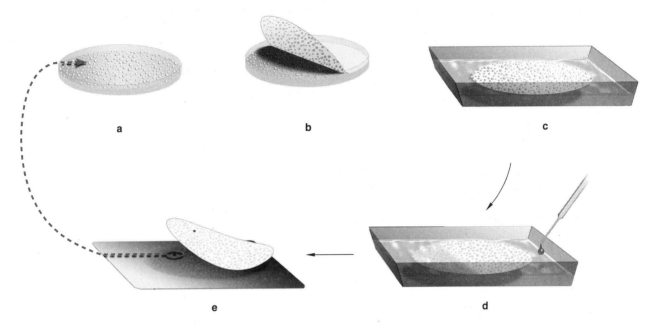

20. _____ Some cells from each colony adhere to a nitrocellulose filter placed on the plate.
21. _____ DNA is denatured to single strands and a radioactive probe binds to fragments with a complimentary sequence.
22. _____ Bacterial colonies are grown on a culture plate.
23. _____ The probe's location is identified by exposing the filter to X-ray film.
24. _____ The filter is put into a solution that causes the cells to rupture; DNA sticks to the filter.

*Short Answer*

25. Outline the steps needed to amplify DNA using PCR. [pp.165–166]

_____

_____

_____

*Fill in the Blanks* [pp.165–166]

PCR requires (26) _____ and (27) _____-_____ enzymes. These enzymes have been isolated from bacteria like (28) _____ _____ which lives in hot springs. (29) _____ _____ isolated from this bacteria is an enzyme that can function at the high temperatures needed to denature DNA into single strands. In addition to (26) and enzymes, (30) _____ _____ and the (31) _____ to be replicated are needed to begin the process. Each round of PCR (32) _____ the number of DNA molecules.

*Short Answer*

33. Outline the steps needed to sequence DNA. [p.167]

_____
_____
_____

*True–False*

If the statement is true, write a T in the blank. If the question is false, correct it by changing the underlined word(s) and writing the correct word(s) in the answer blank. [p.168]

_____ 34. Although there are few differences, DNA fingerprinting <u>can</u> differentiate between full siblings.

_____ 35. <u>Ninety</u> percent of the DNA in all humans is identical.

_____ 36. The variable regions of the human chromosome are made up of <u>triplet codons</u>.

_____ 37. Gel electrophoresis is used to separate DNA fragments by their <u>base sequence</u>.

## 11.5. TINKERING WITH THE MOLECULES OF LIFE [pp.169–170]
## 11.6. PRACTICAL GENETICS [pp.171–173]
## 11.7. WEIGHING THE BENEFITS AND RISKS [pp.173–174]

**Selected Words:** Human Genome Project [p.170], *structural* genomics [p.170], *comparative* genomics [p.170], *Agrobacterium tumefaciens* [p.171], Ti plasmid [p.171], "knockout mice" [p.172], SCID-X1 [p.173], *eugenics* [p.173]

*Boldfaced Terms*

[p.170] genomics _____
_____

[p.170] human gene therapy _____
_____

[p.171] transgenic _____
_____

[p.174] xenotransplantation _____

## Fill in the Blanks

In 1972, Paul Berg and his colleagues were the first to make (1) _____ _____ [p.169]. Over the next several years, there were many advances so that in 1986, scientists began working to determine the DNA sequence of all human genes, an initiative called the (2) _____ _____ _____ [p.170]. In 2003, it was announced that (3) _____ [p.170] percent of the coding regions had been sequenced. This project is part of an emerging research field called (4) _____ _____ [p.170]. A related field, (5) _____ _____ [p.170] looks at evolutionary relationships by comparing genomes. The hope is that the research will lead to effective human (6) _____ _____ [p.170], the insertion of genes to correct genetic defects.

## Matching

Match each plant with one or more engineered traits. [pp.171–172]

7. _____ aspen
8. _____ cotton
9. _____ papaya
10. _____ melons
11. _____ tobacco
12. _____ tomato
13. _____ wheat

a. Resistance to herbicides
b. Resistance to pathogens
c. Delayed ripening
d. Reduced nicotine
e. Tolerance of salty soil
f. Increased per-acre yield
g. Reduced lignin production

## Choice

14. _____ The Ti plasmid used in producing transgenic plants is derived from _____. [p.171]

   a. *Agrobacterium tumefaciens*   b. *E. coli*   c. *Haemophilus influenzae*   d. *Thermus aquaticus*

15. _____ Which of the following have bacteria been engineered to produce? [p.172]

   a. insulin   b. HGH (somatotropin)   c. both a. and b.   d. neither a. nor b.

16. _____ Bacteria have been engineered to do which of the following? [p.172]

   a. break down crude oil   b. absorb heavy metals   c. both a. and b.   d. neither a. nor b.

## Fill in the Blanks

The first mammals used in gene transfer studies were (17) _____ [p.172]. The first gene transferred was the gene for (18) _____ [p.172] which caused the recipients to grow much larger than normal. Gene transfers have also been used to engineer animals that produce human proteins. Goats now

Studying and Manipulating Genomes    103

produce proteins used to treat (19) _____ _____ [p.172]. In addition, cattle have been engineered to produce (20) _____ [p.172] which is used to repair cartilage, bone, and skin and rabbits produce IL-2 which is used to stimulate division of (21) _____ _____ [p.172].

Although this research is in its infancy, children suffering from (22) _____ [p.173], in which the immune system is severely compromised, have been treated and of the eleven treated initially, (23) _____ [p.173] developed at least partially functional immune systems. However, three later developed (24) _____ [p.173] which may have been triggered by the gene transfer.

(25) _____ [p.174] is the transplantation of organs from one species to another. This can lead to immediate rejection by the immune system. Scientists are developing pigs which lack (26) _____-_____ [p.174] chemicals on their cell surfaces.

## Short Answer

27. What is eugenics? [pp.173–174]

_____

_____

_____

28. How is genetic engineering being used to improve xenotransplantation? [p.174]

_____

_____

_____

_____

## Self-Quiz

_____ 1. The enzymes used to create DNA fragments are called _____. [p.163]
a. DNA polymerases
b. ligases
c. restriction enzymes
d. RNA polymerases

_____ 2. A small circle of DNA independent from the chromosomes is called a _____. [p.163]
a. genome
b. plasmid
c. ribosome
d. zygote

_____ 3. DNA without introns produced by reverse transcriptase is called _____. [p.164]
a. cDNA
b. hDNA
c. rDNA
d. zDNA

_____ 4. To introduce genes into an organism, _____ can be used as vectors. [pp.164, 170]
a. plasmids
b. viruses
c. both a. and b.
d. neither a. nor b.

104 Chapter Eleven

_____ 5. In PCR, double-stranded DNA is separated into single strands by _____ treatment. [p.166]
   a. acid
   b. enzyme
   c. heat
   d. primer

_____ 6. A person's DNA fingerprint is usually identical to his/her _____ . [p.168]
   a. parents'
   b. siblings'
   c. both a. and b.
   d. neither a. nor b.

_____ 7. An organism with genes added from another species is called _____ . [p.171]
   a. degenerate
   b. genomic
   c. hybrid
   d. transgenic

_____ 8. Most insulin currently in use by diabetics is produced by _____ . [p.172]
   a. cows
   b. *E. coli*
   c. humans
   d. pigs

_____ 9. What human genetic problem has been ameliorated using gene therapy? [p.173]
   a. cystic fibrosis
   b. diabetes
   c. hemophilia
   d. SCID-X1

_____ 10. Approximately what percentage of the processed food in the United States contains material from genetically engineered plants? [p.174]
   a. 0%
   b. 30%
   c. 70%
   d. 100%

## Chapter Objectives/Review Questions

1. Golden rice could provide what benefit? What problems could be associated with it? [p.162]
2. What is the relationship between restriction enzymes and sticky ends? Why are sticky ends helpful in genetic engineering? [p.163]
3. What is a cloning vector? [p.163]
4. Why is cDNA better for transfer into bacterial cells than native human DNA? How is cDNA produced? [p.164]
5. Describe the steps of PCR. Why is PCR important? [pp.165–166]
6. What is an automated DNA sequencer? How does it work? [p.167]
7. How is DNA fingerprinting beneficial to humanity? What are its drawbacks? [p.168]
8. What are some of the human proteins and other chemicals produced by engineered bacteria? [p.172]
9. What is meant by barnyard biotech? How could this be beneficial to humans? [p.172]

## Integrating and Applying Key Concepts

1. Frankenfood is an emotionally charged term. What are some of the benefits and drawbacks of using genetically modified organisms (GMOs) for food? Do you think risks outweigh the benefits? Why or why not?
2. Many of the modern DNA technologies are seen by many as invasion of privacy. Do you think risks outweigh the benefits? Why or why not?

# 12

# PROCESSES OF EVOLUTION

## INTRODUCTION
Chapter 12 illustrates the many processes occurring in populations that lead to the evolution of the population.

## FOCAL POINTS
- Figure 12.1 [p.177] shows how similar adaptations can occur in separate populations living under similar conditions.
- Figure 12.8 [p.183] illustrates polymorphism in snails.
- Figures 12.10 and 12.13 [p.185, 187] graphically diagram various forms of selection.
- Figure 12.16 [p.188] shows an example of sexual dimorphism.
- Figure 12.18 [p.190] compares genetic drift in large and in small populations.

## Interactive Exercises

*Rise of the Super Rats* [p.176]

### 12.1. EARLY BELIEFS, CONFOUNDING DISCOVERIES [pp.177–180]
### 12.2. THE NATURE OF ADAPTATION [pp.181–182]

**Selected Words:** *Rattus* [p.176], warfarin [p.176], Charles Darwin [p.178], the *Beagle* [p.178], *descent with modification* [p.179], Alfred Wallace [p.180], *On the Origin of Species* [p.180], *short-term* adaptation [p.181], long-term adaptation [p.181]

### *Boldfaced Terms*

[p.176] evolution _____

_____

[p.177] species _____

_____

[p.177] biogeography _____

_____

106    Chapter Twelve

[p.177] comparative morphology _____

[p.178] fossils _____

[p.178] catastrophism _____

[p.179] uniformity _____

[p.180] fitness _____

[p.180] natural selection _____

[p.181] adaptation _____

## Matching

1. _____ the *Beagle* [p.178]
2. _____ comparative morphology [p.177]
3. _____ uniformity [p.179]
4. _____ *Principles of Geology* [p.179]
5. _____ biogeography [p.177]
6. _____ natural selection [p.180]
7. _____ fossils [p.178]
8. _____ descent with modification [p.179]
9. _____ Great Chain of Being [p.177]
10. _____ *On the Origin of Species* [p.180]
11. _____ Alfred Wallace [p.180]
12. _____ Jean Lamarck [p.178]
13. _____ catastrophism [p.178]
14. _____ Thomas Malthus [pp.179–180]

a. Extended from the "lowest" forms of life to humans and on to spiritual beings
b. Sketched out the same ideas that Darwin had about evolution
c. Claimed that humans reproduce too much and would run out of food, living space, and other resources
d. Published in 1859, Darwin's detailed evidence to support the theory of natural selection
e. Theory of Georges Cuvier; he saw the Earth as a fixed stage for human drama and a series of divinely invoked catastrophes
f. The characteristics of organisms change through time
g. Darwin served as naturalist on this ship for a five-year voyage around the world
h. Studies processes that caused species distribution patterns
i. A theory of gradual, repetitive change on the Earth; challenged the idea that the Earth was only 6,000 years old
j. Stone-hard evidence of ancient life
k. Studies similarities and differences among groups of organisms
l. A book by Charles Lyell; in it he presented evidence that supported uniformity
m. One outcome of variation in the shared traits that affect which individuals of a population survive and reproduce in each generation; adaptation results
n. The environment modifies traits of individuals that are passed on to offspring

*Short Answer*

15. Describe why tomatoes and llamas are used as examples of adaptation. [pp.181–182]

_____
_____
_____
_____
_____
_____
_____
_____
_____

## 12.3. INDIVIDUALS DON'T EVOLVE, POPULATIONS DO [pp.182–183]
## 12.4. WHEN IS A POPULATION *NOT* EVOLVING? [p.184]

**Selected Words:** *morphological* traits [p.182], *physiological* traits [p.182], *behavioral* traits [p.182], *qualitative* differences [p.182], *quantitative* differences [p.182], *phenotype* [p.182], mutation [p.182], polymorphism [p.182], *natural selection* [p.183], *gene flow* [p.183], *genetic drift* [p.183], neutral mutation [p.183], *Hardy–Weinberg rule* [p.184]

**Boldfaced Terms**

[p.182] population _____
_____

[p.182] gene pool _____
_____

[p.182] alleles _____
_____

[p.183] allele frequencies _____
_____

[p.183] genetic equilibrium _____
_____

[p.183] microevolution _____
_____

[p.183] lethal mutation _____
_____

## Identification

Identify each of these traits as morphological (M), physiological (P), or behavioral (B). [p.182]

1. _____ Frogs have a three-chambered heart.
2. _____ The active transport of $Na^+$ and $K^+$ ions is unequal.
3. _____ Jays have two eyes, two wings, feathers, three toes forward and one back.
4. _____ Human babies instinctively imitate adult facial expressions.
5. _____ Cells and body parts of an individual work much the same way in metabolism, growth, and reproduction.

## Fill in the Blanks

A (6) _____ [p.182] is a group of individuals of the same species occupying a given area. Traits such as those in Mendel's pea plants show (7) _____ [p.182] differences such as purple or white flowers. Qualitative variation is called (8) _____ [p.182]. Height and other traits show (9) _____ [p.182] differences. All of the genes in a population make up the gene (10) _____ [p.182]. Most often, each kind of gene in the (10) is present in two or more slightly different molecular forms, or (11) _____ [p.182]. When individuals have different combinations of (11), this leads to variations in (12) _____ [p.182], or differences in details of traits. Researchers track allele (13) _____ [p.183], or the abundance of certain alleles in a population. They start from a theoretical reference point known as (14) _____ _____ [p.183]. At that point, a population is not (15) _____ [p.183] with respect to the allele frequencies being studied.

(16) _____ [p.183] refers to the small-scale changes in allele frequencies that arise from mutation, natural selection, gene flow, and genetic drift. Each gene has its own (17) _____ [p.183] rate, the probability of its mutating during or between DNA replications. A mutation that severely changes phenotype usually causes death; it is a (18) _____ [p.183] mutation. A (19) _____ [p.183] mutation doesn't help or hurt. (20) _____ [p.183] selection neither increases nor decreases its frequency in a population, because it won't have a discernible effect on whether an individual survives and reproduces.

## Short Answer

21. List the conditions (in any order) that must be met before genetic equilibrium (or nonevolution) will occur. [p.184] _____

## Problems

Following careful study of the Hardy–Weinberg rule, use the rule to determine the values requested. [p.184]

22. In a population of 200 individuals, determine the following for a particular locus if $p = 0.80$.
    a. the number of homozygous dominant individuals _____
    b. the number of homozygous recessive individuals _____
    c. the number of heterozygous individuals _____

23. If the prevalence of gene *D* is 70 percent in a gene pool, find the percentage of gene *d*.

24. If the frequency of gene *R* in a population is 0.60, what percentage of the individuals are heterozygous *Rr*?

## 12.5. NATURAL SELECTION REVISITED [pp.185–188]
## 12.6. MAINTAINING VARIATION IN A POPULATION [pp.188–189]

*Selected Words:* pesticide resistance [p.186], *pest resurgence* [p.186], antibiotic resistance [p.186], *balancing selection* [p.189], *sickle-cell anemia* [p.189], *Hb$^A$, Hb$^S$* [p.189], *malaria* [p.189], *Plasmodium* [p.189]

### Boldfaced Terms

[p.185] directional selection _____

[p.186] antibiotics _____

[p.186] stabilizing selection _____

[p.187] disruptive selection _____

[p.188] sexual dimorphism _____

[p.188] sexual selection _____

[p.189] balanced polymorphism _____

### Identification

1. Identify the three curves below as stabilizing selection, directional selection, or disruptive selection. [pp.185–187]

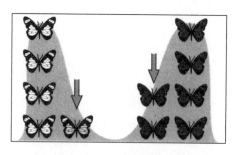

a. _____    b. _____    c. _____

*Choice*

Choose from these categories of natural selection:

      a. directional selection    b. stabilizing selection    c. disruptive selection

2. \_\_\_\_ Phenotypic forms at both ends of the variation range are favored and intermediate forms are selected against [p.187]
3. \_\_\_\_ Coat colors in rock pocket mice [p.185]
4. \_\_\_\_ Allele frequencies that give rise to a range of phenotype variation that tend to shift in a constant direction; this may be due to a change in the environment or a new mutation [p.185]
5. \_\_\_\_ Prospects are not good for human babies who weigh far more or far less than average at birth [p.186]
6. \_\_\_\_ Intermediate forms of a trait in a population are favored and alleles for the extreme forms are not [p.186]
7. \_\_\_\_ Black-bellied seedcrackers have two bill sizes, either large or small; this is found throughout the geographic range of the species [p.187]

*Choice*

Choose from these categories of natural selection. (In some cases, two letters may be correct.)

      a. balanced polymorphism    b. sexual selection

8. \_\_\_\_ Striking sexual dimorphism among many mammals and birds [p.185]
9. \_\_\_\_ Selection maintains two or more alleles for the same trait in a population in steady fashion over a long period of time [p.189]
10. \_\_\_\_ The maintenance of the $Hb^A/Hb^S$ combination of alleles at high frequency [p.189]
11. \_\_\_\_ Females of some species acting as selection agents [p.188]
12. \_\_\_\_ Malaria has been the selecting force [p.189]

## 12.7. GENETIC DRIFT—THE CHANCE CHANGES [pp.190–191]
## 12.8. GENE FLOW—KEEPING POPULATIONS ALIKE [pp.191–192]

**Selected Words:** *probability* [p.190], *Ellis–van Creveld syndrome* [p.191], *emigration* [p.191], *immigration* [p.191]

**Boldfaced Terms**

[p.190] genetic drift _____

[p.190] fixation _____

[p.191] bottleneck _____

[p.191] founder effect _____

[p.191] inbreeding _____

[p.191] gene flow _____

## Fill in the Blanks

(1) _____ _____ [p.190] is a random change in allele frequencies over time, brought about by chance alone. This is significant in (2) _____ [p.190] populations. (3) _____ [p.190] means only one kind of allele remains at a particular locus in a population. All individuals have become (4) _____ [p.190] for it. Genetic drift is pronounced when a few individuals rebuild a population or start a new one. This happens after a (5) _____ [p.191], a drastic reduction in population size brought about by a severe pressure or calamity. When a few individuals leave a population and establish a new one elsewhere, this form of bottlenecking is called a (6) _____ [p.191] effect.

(7) _____ [p.191] is a nonrandom mating among very close relatives, which share many identical alleles. It leads to the homozygous condition and lowers (8) _____ _____ [p.191]. For example, the Old Order Amish of Pennsylvania are a moderately inbred group. One outcome of their inbreeding is a high frequency of the recessive allele that causes (9) _____-_____ _____ [p.191] syndrome.

A population loses alleles whenever an individual leaves it, an act called (10) _____ [p.191]. A population gains alleles when new individuals permanently move in, an act called (11) _____ [p.191]. This (12) _____ _____ [p.191], a physical flow of alleles between populations, tends to counter genetic differences expected to develop through mutation, natural selection, and genetic drift.

## Self-Quiz

____ 1. Studies of the distribution of organisms on the Earth are known as _____ . [p.177]
   a. the Great Chain of Being
   b. stratification
   c. geological evolution
   d. comparative morphology
   e. biogeography

____ 2. The most direct evidence available to demonstrate life has changed as time passes are _____ . [p.178]
   a. old books written by old and very wise scientists
   b. mutations
   c. fossils
   d. endangered species
   e. living fossils

____ 3. In his book, *Principles of Geology,* Lyell suggested that _____ . [p.179]
   a. tailbones in a human body have no place in a perfectly designed body
   b. perhaps species originated in more than one place and have been modified over time
   c. the force for change in organisms was a built-in drive for perfection, up the Chain of Being
   d. gradual processes now molding the Earth's surface had also been at work in the past
   e. the environment determines inheritance of traits

*112* Chapter Twelve

_____ 4. The two scientists most closely associated with the concept of evolution are _____ . [pp.179–180]
   a. Lyell and Malthus
   b. Henslow and Cuvier
   c. Darwin and Wallace
   d. Henslow and Malthus
   e. Lamarck and Darwin

_____ 5. A(n) _____ is some heritable aspect of form, function, behavior, or development that improves the odds for surviving and reproducing in a certain environment. [p.181]
   a. polymorphism
   b. neutral mutation
   c. adaptation
   d. selective force
   e. uniformity

_____ 6. _____ refers to small-scale changes in allele frequencies of a population brought about by mutation, natural selection, gene flow, and genetic drift. [p.183]
   a. Genetic equilibrium
   b. Microevolution
   c. Natural selection
   d. Directional selection
   e. Fixation

_____ 7. The only microevolutionary process that adds heritable changes to the gene pool of a population is _____ . [p.183]
   a. fitness
   b. natural selection
   c. gene flow
   d. genetic drift
   e. mutation

_____ 8. When traits do not appear in proportions predicted from the Hardy–Weinberg rule, this tells you that _____ . [p.184]
   a. mutation might be occurring
   b. gene flow might be occurring
   c. natural selection might be occurring
   d. genetic drift might be occurring
   e. any or all of the above might be occurring

_____ 9. In cases of _____ , allele frequencies give rise to a range of variation in phenotypes that tends to shift in a consistent direction. [p.185]
   a. stabilizing selection
   b. directional selection
   c. sexual selection
   d. disruptive selection
   e. balanced polymorphism

_____ 10. When a few individuals rebuild a population or start a new one, it occurs after _____ . [p.191]
   a. fixation
   b. inbreeding
   c. a bottleneck
   d. a sampling error occurs
   e. disruptive selection

## Chapter Objectives/Review Questions

1. Explain why the effectiveness of warfarin on rats is an example of life evolving. [p.176]
2. Be able to tell why the following lines of inquiry provide evidence for a change in life forms with time and give examples of each: biogeography, comparative morphology, and fossils. [pp.177–178]
3. Cite the significant information about the following as related to evolutionary studies: Cuvier and catastrophism; Lamarck and environmental influences on traits; the *Beagle*; *Principles of Geology* by Charles Lyell; uniformity; *descent with modification*; Thomas Malthus, a clergyman and economist; Galápagos finch species; *On the Origin of Species*. [pp.178–180]
4. _____ _____ is an outcome of variation in the shared traits that affect which individuals of a population survive and reproduce in each generation. [p.180]

5. Define *adaptation* and explain the examples of salt-tolerant tomatoes and llamas. [p.181]
6. Individuals don't evolve, _____ do. [p.182]
7. Define *population, morphological traits, physiological traits, behavioral traits, qualitative* and *quantitative* differences and *polymorphism*. [p.182]
8. Define *allele frequencies* and *genetic equilibrium*. [p.183]
9. Distinguish between mutation rate, lethal mutation, and neutral mutation. [p.183]
10. Be able to solve simple allele frequency and genotype frequency problems related to the Hardy–Weinberg rule. [p.184]
11. Provide definitions for *directional, stabilizing*, and *disruptive* selection and cite one example for each. [pp.185–188]
12. Relate sexual dimorphism to sexual selection. [p.188]
13. The maintenance of $Hb^A/Hb^S$ in human populations is a case of _____ _____ . [p.189]
14. Define *genetic drift, sampling error, fixation, bottleneck, founder effect, inbreeding, emigration, immigration,* and *gene flow*. [pp.190–192]

## Integrating and Applying Key Concepts

If a nuclear holocaust were to sweep the entire Earth, do you think all humans and other organisms would be equally susceptible to the radiation and die? Be able to suggest reasons for your answer in terms of what you have learned about evolution.

# 13

# EVOLUTIONARY PATTERNS, RATES, AND TRENDS

## INTRODUCTION

This chapter looks at evidence for evolution, patterns of evolution, and how evidence and patterns are organized into a coherent whole.

## FOCAL POINTS

- Figure 13.3 [p.197] describes radiometric dating.
- Figure 13.4 [p.198] ties the geologic time scale to major biological events.
- Figures 13.6 and 13.7 [p.199, 200] look at crustal movements.
- Figure 13.8 [p.201] illustrates morphological divergence.
- Figure 13.9 [p.202] shows morphological convergence.
- Figure 13.12 [pp.204–205] compares the amino acid sequences of cytochrome C in three organisms.
- Figures 13.17 and 13.18 [pp.208, 209] illustrate allopatric speciation.
- Figure 13.20 [p.210] looks at sympatric speciation via polyploidy.
- Figure 13.27 [p.215] is a representation of an evolutionary tree of life.

## Interactive Exercises

*Measuring Time* [p.194]

### 13.1. FOSSILS—EVIDENCE OF ANCIENT LIFE [pp.195–196]
### 13.2. DATING PIECES OF THE PUZZLE [pp.196–198]

***Selected Words:*** asteroid impact [p.194], *fossil record* [p.195], *trace* fossils [p.195], *relative* ages [p.196], radioisotope decay [p.197]

***Boldfaced Terms***

[p.195] fossils _____

_____

[p.195] fossilization _____

[p.196] stratification _____

[p.196] lineage _____

[p.197] radiometric dating _____

[p.197] half-life _____

[p.197] geologic time scale _____

[p.198] macroevolution _____

## Matching

Match each term with its description.

1. ____ fossilization [p.195]
2. ____ half-life [p.197]
3. ____ stratification [p.196]
4. ____ trace fossils [p.195]
5. ____ geologic time scale [pp.197–198]
6. ____ lineage [p.196]
7. ____ macroevolution [p.198]
8. ____ fossils [p.195]
9. ____ fossil record [p.195]
10. ____ radiometric dating [p.197]
11. ____ formation of sedimentary rock layers [p.196]
12. ____ rupture or tilt of sedimentary layers [p.196]

a. A line of descent
b. A way to measure the proportions of a daughter isotope in a rock and the parent radioisotope of some element that had been trapped inside the rock when the rock formed
c. A chronology of Earth's history
d. Usually evidence of crustal movements or upheavals much later in time
e. Buried remains and impressions of organisms that had lived in the past
f. The older the rock layer, the older the fossils
g. A very slow process that starts when an organism, or traces of it, becomes buried in volcanic ash or sediments
h. Examples are imprints of leaves, stems, tracks, and burrows
i. The formation of sedimentary rock layers
j. The ordered array of fossils within the rock layers
k. The time it takes for half of a given quantity of a radioisotope to decay into a less unstable, daughter isotope
l. Major patterns, trends, and rates of change among lineages

## Sequence

Earth history has been divided into four great eras that are based on four abrupt transitions in the fossil record. The oldest era has been subdivided. Arrange the eras in correct chronological sequence from the oldest to the youngest. [p.198]

13. _____
14. _____
15. _____
16. _____
17. _____

a. Mesozoic: 248 to 65 million years ago
b. Archean: 3.8 billion to 2.5 billion years ago
c. Paleozoic: 544 to 248 million years ago
d. Cenozoic: 65 million years ago to the present
e. Proterozoic: 2.5 billion to 544 million years ago

*Choice*

Choose from the following: [p.198]

      a. Archean    b. Cenozoic    c. Mesozoic    d. Paleozoic    e. Proterozoic

18. _____ Adaptive radiations of marine vertebrates, fishes, insects, and dinosaurs; origin of flowering plants

19. _____ Adaptive radiations of flowering plants, insects, birds, and mammals; major glaciations; modern humans evolve

20. _____ Aerobic metabolism, oxygen accumulates in the atmosphere, and eukaryotic cells have their origin and divergences

21. _____ Supercontinent breaks up; adaptive radiations of marine invertebrates and early fishes; adaptive radiations of insects and amphibians; spore-bearing plants dominate; adaptive radiations of reptiles and gymnosperms

22. _____ Origin of photosynthetic, prokaryotic cells; origin of Earth's crust; first atmosphere, first seas; chemical, molecular evolution leads to origin of life

23. _____ It ended with the greatest of all extinctions, the Permian extinction.

24. _____ Adaptive radiation of fishes; origin of vascular plants and the origin of amphibians

## 13.3. EVIDENCE FROM BIOGEOGRAPHY [pp.199–200]
## 13.4. MORE EVIDENCE FROM COMPARATIVE MORPHOLOGY [pp.201–202]

*Selected Words:* theory of uniformity [p.199], continental drift theory [p.199], divergence [p.201], "stem reptile" [p.201]

*Boldfaced Terms*

[p.199] Pangea _____

[p.200] plate tectonics theory _____

[p.200] Gondwana _____

[p.201] comparative morphology _____

[p.201] homologous structures _____

[p.201] morphological divergence _____

[p.202] morphological convergence _____

[p.202] analogous structures _____

## Fill in the Blanks

The theory of (1) _____ [p.199] held that mountain building and erosion had repeatedly worked over the Earth's surface in precisely the same ways through (2) _____ [p.199]. Geologists also came to realize that the Atlantic coasts of South America and Africa seemed to "fit" like (3) _____ pieces [p.199]. Perhaps all continents were once part of a bigger one that split into fragments that drifted apart. (4) _____ [p.199] was a proposed supercontinent. Evidence kept piling up to support the idea. Scientists made a magnetic (5) _____–_____ [p.199] alignment work by joining North America and western Europe. Deep-sea probes showed that the seafloor is spreading away from mid-oceanic ridges. This spreading new (6) _____ [p.199] forces older (6) into deep trenches elsewhere in the seafloor. All the ridges and trenches form the thin edges of enormous (7) _____ [p.199] that move with almost imperceptible slowness. These findings have been put into a broader explanation of crustal movements, now called a (8) _____ _____ [p.200] theory. As evidence that continents were once joined, two (9) _____ [p.200], a particular seed fern and a particular mammal-like reptile have been found in Africa, India, Australia, South America, and more recently, Antarctica. The plant's seeds were too (10) _____ [p.200] and the animal too (11) _____ [p.200] for dispersal across the ocean. But what if both had evolved on (12) _____ [p.200], a supercontinent that preceded Pangea? All changes in the land, oceans, and atmosphere have influenced life's (13) _____ [p.200].

## Labeling

Label each sketch as either morphological divergence or morphological convergence to test your understanding of the principles being illustrated. [pp.201–202]

a          b          c

14. Morphological _____

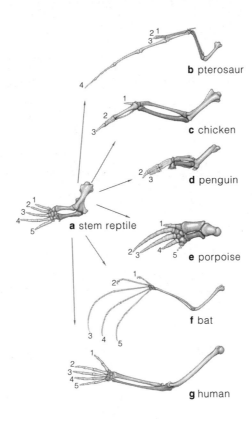

15. Morphological _____

*Choice*

Choose from the following:

      a. analogous structures    b. homologous structures    c. examples of lineages
      d. a specific example of morphological convergence
      e. a specific example of morphological divergence

16. _____ Insect wings compared to bird wings [p.202]
17. _____ Birds, bats, and insects [p.202]
18. _____ May have evolved independently in lineages that aren't closely related but that faced similar environmental pressures [p.202]
19. _____ The human arm and hand, the porpoise's front flipper, and the bird's wing [p.201]
20. _____ Similarity of body parts of different kinds of organisms that may have different functions and may be attributed to descent from a common ancestor [p.201]

## 13.5. EVIDENCE FROM PATTERNS OF DEVELOPMENT [pp.203–204]
## 13.6. EVIDENCE FROM DNA, RNA, AND PROTEINS [pp.204–205]

*Selected Words:* transposons [p.203], *Alu* elements [p.204], conserved genes [p.204], mitochondrial DNA (mtDNA) [p.204], DNA sequencing [p.204], nucleic acid hybridization [p.204], cytochrome *c* [p.204], restriction fragments [p.205]

*Boldfaced Term*

[p.205] molecular clock _____

_____

*Identification*

Identify each statement as evidence coming from developmental patterns (D) or comparative biochemistry (B).

1. _____ Comparison of the developmental rates of changes in a chimpanzee skull and a human skull [p.203]
2. _____ Using accumulated neutral mutations to date the divergence of two species from a common ancestor, a kind of "molecular clock" [p.205]
3. _____ The entire amino acid sequence in the cytochrome *c* of humans precisely matches the sequence in the cytochrome *c* in chimpanzees. [pp.204–205]
4. _____ Establishing a rough measure of evolutionary distance between two organisms by use of nucleic acid hybridization studies that demonstrate the extent to which the DNA from one species base-pairs with another [pp.205–206]
5. _____ Ancestors of humans and chimpanzees diverged from the same ancestral stock somewhere between 6 million and 4 million years ago; approximately 99 percent of human and chimp DNA is identical. [p.204]
6. _____ The same gene controls the fate of cells that give rise to crab legs, beetle legs, sea star arms, butterfly wings, fish fins, and mouse feet. [p.203]
7. _____ Mitochondrial DNA (mtDNA) is especially useful for studying eukaryotes. [p.204]

## 13.7. REPRODUCTIVE ISOLATION, MAYBE NEW SPECIES [pp.205–207]
## 13.8. INTERPRETING THE EVIDENCE: MODELS FOR SPECIATION [pp.208–211]

**Selected Words:** *reproduction* [p.205], Ernst Mayr [p.205], reproductive isolation [p.205], *mechanical isolation* [p.206], *temporal isolation* [p.207], *behavioral isolation* [p.207], *ecological isolation* [p.207], *gamete mortality* [p.207], *postzygote* isolation mechanisms [p.207], *physical separation* [p.208], *ecological* separation [p.210], the hybrid zone [p.211]

*Boldfaced Terms*

[p.205] biological species concept _____

_____

[p.206] gene flow _____

_____

[p.206] genetic divergence _____

_____

[p.206] speciation _____

_____

[p.206] reproductive isolating mechanisms _____

_____

[p.208] allopatric speciation _____

_____

[p.210] sympatric speciation _____

_____

[p.210] polyploidy _____

_____

[p.211] parapatric speciation _____

_____

## Choice

Choose from these isolating mechanisms: [pp.206–207]

      a. temporal    b. behavioral    c. mechanical    d. gamete mortality
      e. ecological    f. postzygotic

1. _____ Two sage species keep their pollen to themselves with different-sized floral landing platforms that attract different pollinators.
2. _____ Comparing two species of periodical cicadas, one matures, emerges, and reproduces every thirteen years, its "sibling" species every seventeen years.
3. _____ Two different populations of manzanita have different built-in physiological mechanisms that enhance water conservation during the dry season.
4. _____ Molecularly mismatched gametes of different species
5. _____ Prior to copulation, male and female birds engage in complex courtship displays recognized only by birds of their own species.
6. _____ Sterile mules

## Matching

Match each term with its description.

7. _____ allopatric speciation [p.208]
8. _____ archipelago [pp.208–209]
9. _____ sympatric speciation [pp.209–210]
10. _____ speciation via polyploidy [pp.210–211]
11. _____ parapatric speciation [p.211]

a. Species may form *within* the home range of an existing species in the absence of a physical barrier; cichlid speciation in Africa is an example
b. Populations that maintain contact along a common border become distinct species; hybrids form in the hybrid zone and are selected against; Australian wingless grasshoppers are an example
c. Rapid speciation for flowering plants that self-fertilize or hybridize; bread wheat may have arisen this way
d. Genetic changes leading to a new species require physical separation between populations; honey creeper speciation on the Hawaiian Archipelago and finches on the Galápagos Islands are good examples
e. An island chain some distance away from a continent; gene flow is impeded

# 13.9. PATTERNS OF SPECIATION AND EXTINCTIONS [pp.211–212]
# 13.10. ORGANIZING INFORMATION ABOUT SPECIES [pp.213–215]

*Selected Words:* branch [p.211], branch point [p.211], *niche* [p.212], specific epithet [p.213], phylogenic [p.213]

*Boldfaced Terms*

[p.211] cladogenesis _____

[p.211] anagenesis _____

[p.211] evolutionary trees _____

[p.211] gradual model of speciation _____

[p.212] punctuation model of speciation _____

[p.212] adaptive radiation _____

[p.212] adaptive zones _____

[p.212] key innovation _____

[p.212] extinction _____

[p.212] mass extinctions _____

[p.213] genus _____

[p.213] classification systems _____

[p.213] higher taxa (singular, taxon) _____

[p.213] six-kingdom classification system _____

[p.213] three-domain system _____

[p.214] monophyletic group _____

[p.214] clade _____

[p.214] derived trait _____

[p.214] cladograms _____

## Matching

Match each term with its description.

1. _____ cladogenesis [p.211]
2. _____ anagenesis [p.211]
3. _____ evolutionary trees [p.211]
4. _____ gradual model of speciation [p.211]
5. _____ punctuation model of speciation [p.212]
6. _____ adaptive radiation [p.212]
7. _____ adaptive zones [p.212]
8. _____ key innovations [p.212]
9. _____ extinction [p.212]
10. _____ mass extinction [p.212]

a. Modification of some structure or function that permits a lineage to exploit the environment in new or more efficient ways
b. In this speciation pattern, most morphological changes are said to occur only in a brief period when populations are starting to diverge—within hundreds or thousands of years only; bottlenecks, founder effects, directional selection, or a combination of these foster rapid speciation
c. Ways of life such as "burrowing in the seafloor sediments"
d. Holds that species originate by small morphological changes over long time spans; fits well with many fossil sequences
e. Large, catastrophic losses of entire families or other major groups; differ in size
f. Changes in allele frequencies and morphology accumulate in one line of descent; gene flow never stops between populations, so directionally changes are confined to that lineage alone
g. A method of summarizing information about the continuity of relationships among groups
h. A branching speciation pattern; genetically isolated populations diverge
i. A burst of divergences from a single lineage that gives rise to many new species, adapted to an unoccupied or a new habitat or to using novel resources
j. An irrevocable loss of a species

## Short Answer

11. Arrange the following jumbled taxa and their categories in proper order, the most inclusive first: genus: *Archibaccharis*; kingdom: Plantae; order: Asterales; specific epithet: *lineariloba*; class: Dicotyledonae; Phylum: Anthophyta; family: Asteraceae. [p.213]

_____

_____

_____

Use this simple cladogram to answer the following questions. [p.214]

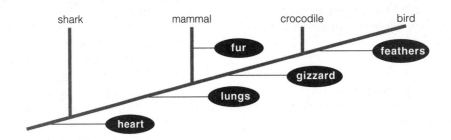

12. Which animals have a heart as a common trait? _____

13. Which animals have lungs as a common trait? _____

14. Which animals have a gizzard as a common trait? _____

15. Which animals have feathers as a common trait? _____

16. Are birds more closely related to crocodiles or mammals? _____

## Self-Quiz

For questions 1–6, choose from the following:

    a. fossilization
    b. half-life
    c. geologic time scale
    d. plate tectonics theory
    e. morphological convergence
    f. macroevolution

_____ 1. _____ refers to dissimilar body parts that become similar in independent, evolutionarily distant lineages. [p.202]

_____ 2. A _____ is a constructed chronology of Earth's history. [p.197]

_____ 3. _____ is a slow process that starts when an organism or traces of it are buried in volcanic ash or sediments. [p.195]

_____ 4. The time it takes for fifty percent of a given quantity of a radioisotope to decay into a less unstable, daughter isotope is its _____ . [p.197]

_____ 5. _____ refers to the major patterns, trends, and rates of change among lineages. [p.198]

_____ 6. A broad explanation of crustal movements is now called a _____ . [p.200]

For questions 7–12, choose from the following:

    a. homologous structures
    b. biological species
    c. temporal isolation
    d. allopatric speciation
    e. cladogenesis
    f. phylogeny

_____ 7. Consideration of evolutionary relationships among species, from the most ancestral species through divergences that led to all descendant species, defines _____. [p.213]

_____ 8. In _____, a physical barrier arises and stops gene flow between two populations or subpopulations of a species. [p.208]

_____ 9. A population of interbreeding leopard frogs living in a pond represents an example of _____. [p.205]

_____ 10. A branching pattern of evolutionary change is referred to as _____. [p.211]

_____ 11. 17-year-cycle cicadas and 13-year-cycle cicadas that do not get together to interbreed represent an example of _____. [p.207]

_____ 12. Comparisons of the forelimb of a human and the forelimb of a bat represent _____. [p.201]

## Chapter Objectives/Review Questions

1. State the possible significance of asteroid impacts to the evolutionary process. [p.194]
2. Define *fossil* and describe the fossilization process. [p.195]
3. The age of rocks and the fossils within them are carefully measured with _____ dating techniques. [p.197]
4. Explain why the geologic time scale is so useful. [pp.197–198]
5. Pangea and Gondwana fit into a broad explanation of crustal movements called a _____ _____ theory. [pp.199–200]
6. Define *comparative morphology, homologous structures, morphological divergence, morphological convergence,* and *analogous structures.* [pp.201–202]
7. List biochemical techniques that have been used to show similarities between species. [pp.204–205]
8. Be able to state and explain the biological species concept. [p.205]
9. Give one example each for temporal isolation, mechanical isolation, behavioral isolation, ecological isolation, gamete mortality, and postzygote isolating mechanisms. [pp.206–207]
10. Compare and contrast the following models for speciation: allopatric speciation, sympatric speciation, polyploidy, and parapatric speciation. [pp.208–211]
11. Define and contrast the *gradual model of speciation* with the *punctuation model of speciation.* [pp.211–212]
12. Define *adaptive radiation, adaptive zones, key innovations, extinction, mass extinctions.* [p.212]
13. Which is currently in favor, the six-kingdom system or the three-domain system? Why? [pp.213–214]
14. Provide definitions for *derived trait, cladograms, monophyletic group,* and *clade.* [p.214]

## Integrating and Applying Key Concepts

*Systematics* is defined as the practice of describing, naming, and classifying living things; this includes the comparative study of organisms and all relationships among them. Plant systematists who work with flowering plants readily accept Ernst Mayr's definition of a "biological species" as described in this chapter. However, in real practice, systematists often must also work with a concept known as the "morphological species." For example, the statement is sometimes made that "two plant specimens belong to the same morphological species but not to the same biological species." Explain the meaning of that statement. What type of experimental evidence would be necessary as a basis for this statement? From your study of this chapter, can you suggest a reason that application of the term "morphological species" is sometimes necessary? What difficulties and inaccuracies, in terms of identifying a species, might the use of both species' concepts present?

# 14

# EARLY LIFE

## INTRODUCTION

Chapter 14 discusses the origin of life. In addition it surveys bacteria, Archaea, protists, fungi, viruses, viroids, and prions.

## FOCAL POINTS

- Figure 14.4 [pp.222–223] diagrams the evolutionary history of living organisms.
- Figure 14.5 [p.224] looks at a typical bacterium.
- Figures 14.6, 14.9, and 14.12 [pp.224, 227, 228] illustrate several different bacteria.
- Figure 14.13 [p.228] is an evolutionary tree of protists.
- Figures 14.14, 14.15, 14.16, 14.17, 14.18, 14.20, 14.21, 14.23, and 14.24 [pp.229–234] show protist diversity.
- Figure 14.19 [p.231] diagrams the life cycle of *Plasmodium*.
- Figure 14.25 [p.234] is an evolutionary tree of fungi.
- Figure 14.26 [p.235] illustrates a typical fungal life cycle.
- Figures 14.27, 14.28, and 14.29 [pp.235, 236] show typical fungi.
- Figure 14.30 [p.237] illustrates the variations seen in lichens.
- Figure 14.33 [p.239] shows typical viruses.
- Figure 14.34 [p.240] diagrams viral life cycles.

## Interactive Exercises

*Looking for Life in All the Odd Places* [p.218]

### 14.1. ORIGIN OF THE FIRST LIVING CELLS [pp.219–223]

**Selected Words**: *bioprospecting* [p.218], nanobes [p.218], Stanley Miller [p.219], *metabolism* [p.220]

**Boldfaced Terms**

[p.220] RNA world _____

126  Chapter Fourteen

[p.221] proto-cells _____

[p.222] Archean _____

[p.222] prokaryotic cells _____

[p.222] bacteria _____

[p.222] archaea _____

[p.222] eukaryotic cells _____

[p.222] stromatolites _____

## Fill in the Blanks

When Thomas Brock looked at (1) _____ [p.218] springs and pools at (2) _____ [p.218] National Park, he found bacteria living in water that was (3) _____ [p.218] °C ((4) _____ [p.218] °F). But life has been found at even higher temperatures. (5) _____ [p.218] live at hydrothermal vents where temperatures reach (6) _____ [p.218] °C ((7) _____ [p.218] °F). The champions, however, are the (8) _____ [p.218] that live beneath the earth's surface in rocks that reach (9) _____ [p.218] °C ((10) _____ [p.218] °F). These latter are so small that they can only be visualized with an (11) _____ [p.218] microscope.

These organisms may be similar to early life forms on this planet that had to exist in similar conditions. About (12) _____ [p.219] billion years ago, gasses like (13) _____ [p.219], (14) _____ [p.219], (15) _____ _____ [p.219], and (16) _____ _____ [p.219] blanketed the Earth, forming its first atmosphere. After the Earth's crust cooled, (17) _____ _____ [p.219] collected on the surface. The first living cells appeared less than (18) _____ [p.219] million years after the crust of the Earth solidified.

Early Life 127

*Matching*

Match each phrase with a number from the chart. [pp.222–223]

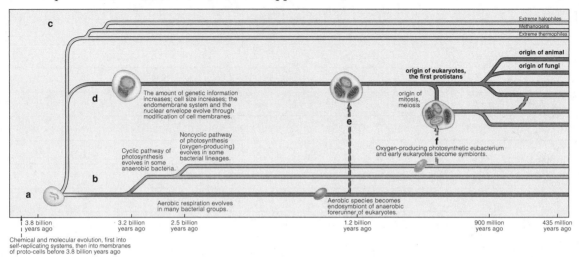

19. _____ origin of prokaryotes
20. _____ origin of chloroplasts
21. _____ origin of mitochondria
22. _____ Archaean lineage
23. _____ bacterial lineage
24. _____ eukaryotic lineage

*Fill in the Blanks*

The first cells arose in the (25) _____ [p.222] eon. The first major divergence gave rise to the ancestors of the (26) _____ [p.222] and to the ancestors of the (27) _____ [p.222] and (28) _____ [p.222]. The first photosynthetic pathway that evolved was the (29) _____ [p.222] pathway. Mats of these early photosynthetic bacteria and their descendants have become fossilized as (30) _____ [p.222]. When the (31) _____ [p.222] pathway evolved, its byproduct, (32) _____ [p.222], drastically changed the atmosphere. This byproduct was used in (33) _____ _____ [p.223] which soon became the dominant energy-releasing pathway, and the byproduct's presence stopped the further (34) _____ _____ [pp.222–223] of living cells.

## 14.2. WHAT ARE EXISTING PROKARYOTES LIKE? [pp.224–228]

**Selected Words:** *prokaryotic* [p.224], capsule [p.224], layer of slime [p.224], flagella [p.224], pili (pilus) [p.224], *photoautotrophs* [p.224], *chemoautotrophs* [p.224], *photoheterotrophs* [p.224], *chemoheterotrophs* [p.224], prokaryotic fission [p.225], cyanobacteria [p.226], heterocysts [p.227], *Lactobacillus* [p.227], *Clostridium* [p.227], *Bacillus* [p.227], endospore [p.227], *anthrax* [p.227], *botulism* [p.227], cholera [p.227], gonorrhea [p.227], *Rhizobium* [p.227], *Borrelia burgdorferi* [p.228], Lyme disease [p.228]

*Boldfaced Terms*

[p.224] prokaryotic conjugation _____
_____

[p.224] plasmid _____
_____

[p.224] pathogens _____

[p.225] lateral gene transfer _____

[p.226] extreme halophiles _____

[p.226] methanogens _____

[p.226] extreme thermophiles _____

## Matching

Match the types of metabolism with the proper definition. [p.224]

1. _____ chemoautotrophs
2. _____ chemoheterotrophs
3. _____ photoautotrophs
4. _____ photoheterotrophs

a. Self-feeding. Use light energy to make food.
b. Self-feeding. Use light chemical to make food.
c. Not self-feeding but can use light energy to make ATP
d. Get nutrients and energy from other organisms

## Identification

Identify each prokaryotic cell part on the figure. Then note its property in the parentheses using the list given. [p.224]

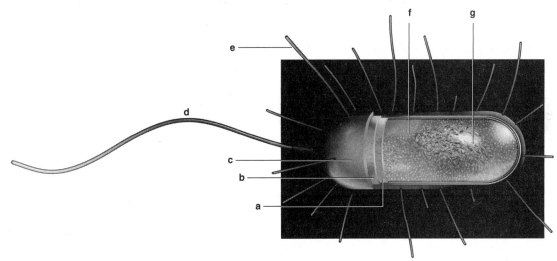

5. _____ capsule ( )
6. _____ cell wall ( )
7. _____ cytoplasm ( )
8. _____ flagellum ( )
9. _____ nucleoid ( )
10. _____ pilus ( )
11. _____ plasma membrane ( )

a. Semi-rigid; helps cell maintain shape
b. Helps cell hold on to surfaces and avoid phagocytosis
c. Allows cell to move
d. Helps in attachment and in DNA transfer
e. Major portion of cell; protein synthesis occurs here
f. Protects the DNA from damage
g. Serves as a selectively permeable separation between the cell and the external environment

*Complete the Table*

12. Complete the table by entering the missing names, habitats, and other information. [pp.226–228]

| Name | Habitat | Metabolism/Other Information |
|---|---|---|
| methanogen | (a) | (b) |
| (c) | Dead Sea, Great Salt Lake, etc. | (d) |
| (e) | (f) | Uses hydrogen sulfide as electron source |
| cyanobacteria | (g) | (h) |
| (i) | (j) | Fixes nitrogen when symbiotic with legumes |
| (k) | Skin, vaginal lining, fermented dairy products | Ferment, forming lactose |
| (l) | (m) | Causes Lyme disease |
| *Clostridium botulinum* | (n) | (o) |

## 14.3. THE CURIOUSLY CLASSIFIED PROTISTS [pp.228–234]

**Selected Words:** protozoans [p.228], *giardiasis* [p.228], *trichomoniasis* [p.228], trypanosomes [p.229], *African sleeping sickness* [p.229], *Chagas disease* [p.229], euglenoids [p.229], radiolarians [p.229], foraminiferans [p.229], alveolates [p.230], *Paramecium* [p.230], red tides [p.230], *Plasmodium* [p.231], malaria [p.231], *Anopheles* mosquitoes [p.231], diatoms [p.231], brown algae [p.231], stramenopiles [p.231], oomycote [p.231], *late blight* [p.231], *sudden oak death* [p.232], giant kelps [p.232], red algae [p.232], agar [p.232], carrageenan [p.232], *nori* [p.232], green algae [p.232], amoebozoans [p.234], amoebas [p.234], *amoebic dysentery* [p.234], slime molds [p.234]

**Boldfaced Terms**

[p.228] protists _____

[p.228] flagellated protozoans _____

[p.228] cyst _____

[p.229] pellicle _____

[p.229] plankton _____

[p.230] ciliates _____

[p.230] dinoflagellates _____

[p.230] algal bloom _____

[p.230] apicomplexans _____

## Matching

Match each of the protists with all of their characteristics from the lettered list.

1. _____ euglenoids [p.229]
2. _____ *Trypanosoma* [p.229]
3. _____ *Giardia lamblia* [p.228]
4. _____ *Amoeba* [p.234]
5. _____ foraminiferans [p.229]
6. _____ *Paramecium* [p.230]
7. _____ *Plasmodium* [p.231]
8. _____ dinoflagellates [p.230]
9. _____ diatoms [p.232]

a. photosynthetic
b. flagella
c. cilia
d. pseudopods
e. hard shell
f. pathogenic/toxic
g. transmitted by insects
h. bioluminescent

## Choice

10. _____ Protists usually have _____. [p.228]

    a. nucleus    b. mitochondria    c. both a. and b.    d. neither a. nor b.

11. _____ Which of these protist groups has/have both photosynthetic and heterotrophic members? [p.229]

    a. diatoms    b. dinoflagellates    c. euglenoids    d. red algae

12. _____ Trypanosomes cause _____. [p.229]

    a. African sleeping sickness    b. Chagas disease    c. both a. and b.    d. neither a. nor b.

13. _____ Which of these parasites is/are usually transmitted sexually? [pp.228–229]

    a. *Giardia lamblia*    b. *Trichomonas vaginalis*    c. both a. and b.    d. neither a. nor b.

14. _____ *Paramecium* _____. [p.230]

    a. is an autotroph    b. is a heterotroph    c. may be either an autotroph or a heterotroph

15. _____ Gametes of the malaria parasite mature in _____. [p.231]

    a. the human liver    b. human red blood cells    c. human white blood cells    d. mosquitoes

Early Life    131

16. _____ Red tides are caused by blooms of _____. [p.230]

   a. diatoms   b. dinoflagellates   c. oomycotes   d. red algae

17. _____ The Irish potato famine was caused by a pathogenic _____. [pp.231–232]

   a. amoeba   b. ciliate   c. flagellate   d. oomycote

18. _____ The largest protists, the giant kelps, are classified as _____. [p.232]

   a. brown algae   b. chrysophytes   c. green algae   d. red algae

19. _____ Which algae is most like plants in biochemistry? [p.232]

   a. brown algae   b. chrysophytes   c. green algae   d. red algae

20. _____ Most red algae are found in _____. [p.232]

   a. lakes   b. soil   c. swamps   d. tropical seas

21. _____ Nori, used to wrap sushi, is derived from _____. [p.232]

   a. brown algae   b. chrysophytes   c. green algae   d. red algae

## 14.4. THE FABULOUS FUNGI [pp.234–238]

*Selected Words:* chitin [p.234], saprobic [p.234], spore [p.234], mushroom [p.235], gills [p.235], rusts [p.235], smuts [p.235], *Agaricus bisporus* [p.235], black bread mold [p.236], asci (ascus) [p.236], truffles [p.236], yeasts [p.237], *histoplasmosis* [p.238], molds [p.238], *ergotism* [p.238]

### Boldfaced Terms

[p.234] saprobes _____

_____

[p.234] mutualism _____

_____

[p.234] mycelium _____

_____

[p.234] hypha _____

_____

[p.235] club fungi _____

_____

[p.235] sac fungi _____

_____

[p.235] zygomycetes _____

_____

[p.237] symbiosis _____

[p.237] lichen _____

[p.237] mycorrhiza _____

## Choice

1. _____ How do fungi get their necessary nutrients? [p.234]

    a. photosynthesis    b. as saprobes    c. both a. and b.    d. neither a. nor b.

2. _____ A single fungal filament is called a _____. [p.234]

    a. hypha    b. mycelium    c. spore    d. yeast

3. _____ Mushrooms are reproductive structures formed by _____. [p.235]

    a. club fungi    b. imperfect fungi    c. sac fungi    d. zygomycetes

4. _____ For reproduction, all fungi produce _____. [p.236]

    a. seeds    b. spores    c. both a. and b.    d. neither a. nor b.

5. _____ Truffles are _____. [p.236]

    a. club fungi    b. imperfect fungi    c. sac fungi    d. zygomycetes

6. _____ Bread yeast is a _____. [p.236]

    a. club fungus    b. imperfect fungus    c. sac fungus    d. zygomycete

7. _____ The photosynthetic partner in a lichen can be a _____. [p.237]

    a. green algae    b. red algae    c. both a. and b.    d. neither a. nor b.

8. _____ Which of the following is able to grow on hostile sites like bare rocks and gravestones? [p.237]

    a. lichen    b. mycorrhiza    c. sac fungus    d. yeast

9. _____ Air pollution causes a decline in _____. [pp.237–238]

    a. lichens    b. mushrooms    c. both a. and b.    d. neither a. nor b.

10. _____ Which of the following is caused by a fungus? [p.238]

    a. athlete's foot    b. malaria    c. both a. and b.    d. neither a. nor b.

11. _____ Which fungus is responsible for ergotism? [p.238]

    a. *Ajellomyces*    b. *Claviceps*    c. *Epidermophyton*    d. *Histoplasma*

Early Life 133

## 14.5. VIRUSES, VIROIDS, AND PRIONS [pp.239–241]
## 14.6. EVOLUTION AND INFECTIOUS DISEASES [p.241]

*Selected Words:* protein coat [p.239], parvoviruses [p.239], poxviruses [p.239], HIV [p.239], *AIDS* [p.239], *attachment* [p.240], *penetration* [p.240], *replication and protein synthesis* [p.240], *assembly* [p.240], *release* [p.240], *lytic* pathway [p.240], *lysogenic* pathway [p.240], *cold sores* [p.240], retrovirus [p.240], "mad cow disease" [p.241], bovine spongiform encephalopathy (BSE) [p.241], variant Creutzfeldt-Jacob disease (vCJD) [p.241], *sporadic* diseases [p.241], *endemic* diseases [p.241], *epidemic* [p.241], *pandemic* [p.241], *SARS* [p.241], coronavirus [p.241]

### Boldfaced Terms

[p.239] virus _____

[p.239] bacteriophages _____

[p.240] lysis _____

[p.240] viroids _____

[p.241] prion _____

[p.241] infection _____

[p.241] disease _____

[p.241] antibiotics _____

### True–False

If the statement is true, write a T in the blank. If the question is false, correct it by changing the underlined word(s) and writing the correct word(s) in the answer blank.

_____ 1. The two major components of viruses are <u>proteins and carbohydrates</u>. [p.239]

_____ 2. Viruses can <u>only replicate inside living cells</u>. [p.239]

_____ 3. Most viruses can infect <u>many different</u> host species. [p.239]

_____ 4. In the <u>lytic pathway</u>, the viral DNA becomes integrated into the host's DNA. [p.240]

_____ 5. Cold sores are caused by <u>bacteriophages</u>. [p.240]

_____ 6. Prions are infectious <u>RNAs</u>. [p.241]

## Identification

Identify each stage of the lytic cycle with a letter on the diagram. [p.240]

7. _____ Assembly of new virus particles
8. _____ Attachment of the virus to a bacterium
9. _____ Host cell lyses releasing new viral particles
10. _____ Penetration of viral DNA into host cell
11. _____ Viral DNA directs cell to make viral proteins and replicate viral DNA

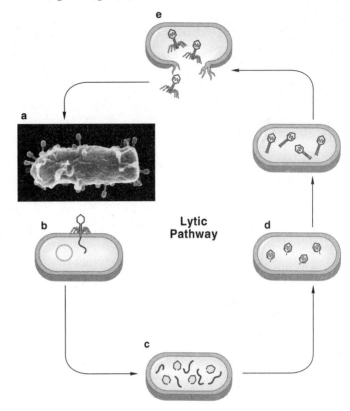

## Choice

12. _____ Viroids contain _____ . [pp.240–241]

   a. protein     b. RNA     c. both a. and b.     d. neither a. nor b.

13. _____ Which of the following is a virally caused pandemic? [p.241]

   a. AIDS     b. SARS     c. both a. and b.     d. neither a. nor b.

14. _____ Tuberculosis is considered to be a/an _____ disease. [p.241]

   a. endemic     b. epidemic     c. pandemic     d. sporadic

## Identification

Identify the main causative agent(s) of each of the following (some may have more than one causative agent). [p.241]

a. bacteria     b. protists     c. viruses

15. _____ acute respiratory infections
16. _____ diarrhea

a. bacteria   b. protists   c. viruses

17. _____ tuberculosis
18. _____ malaria
19. _____ AIDS
20. _____ measles
21. _____ hepatitis B
22. _____ tetanus

## Self-Quiz

_____ 1. Which of these gasses was probably not part of Earth's earliest atmosphere? [p.219]
   a. carbon dioxide
   b. hydrogen
   c. nitrogen
   d. oxygen

_____ 2. The initial genetic material was probably _____. [p.220]
   a. DNA
   b. lipid
   c. protein
   d. RNA

_____ 3. Which of the following arose endosymbiotically? [p.223]
   a. chloroplasts
   b. mitochondria
   c. both a. and b.
   d. neither a. nor b.

_____ 4. An organism that can make its own food through photosynthesis is called a _____. [p.224]
   a. chemoautotroph
   b. chemoheterotroph
   c. photoautotroph
   d. photoheterotroph

_____ 5. Which of these can help a bacterium adhere to surfaces like teeth and intestinal epithelium? [p.224]
   a. capsule
   b. pilus
   c. both a. and b.
   d. neither a. nor b.

_____ 6. Which type of archaebacterium is common in the guts of cows and termites? [p.226]
   a. acidophiles
   b. halophiles
   c. methanogens
   d. thermophiles

_____ 7. Which of the following is not a chemoheterotroph? [p.226]
   a. actinomycetes
   b. cyanobacteria
   c. *Lactobaclus*
   d. *Rhizobium*

_____ 8. Which of the following is caused by a *Clostridium* species? [p.227]
   a. botulism
   b. tetanus
   c. both a. and b.
   d. neither a. nor b.

_____ 9. Which of the following can cause intestinal upsets in humans? [pp.229, 234]
   a. amoeba
   b. foraminiferan
   c. both a. and b.
   d. neither a. nor b.

_____ 10. Ciliates use their cilia for _____. [p.230]
   a. feeding
   b. locomotion
   c. both a. and b.
   d. neither a. nor b.

*136* Chapter Fourteen

____ 11. Oomycotes can cause _____ . [pp.231–232]
   a. malaria
   b. red tide
   c. sudden oak death
   d. all of the preceding

____ 12. Which of the following algae have chloroplasts that are like those in plants? [pp.232–233]
   a. green algae
   b. red algae
   c. both a. and b.
   d. neither a. nor b.

____ 13. The majority of fungi are _____ . [p.234]
   a. parasites
   b. photoautotrophs
   c. photoheterotrophs
   d. saprobes

____ 14. The largest known organism is most probably a/an _____ . [p.235]
   a. alga
   b. club fungus
   c. lichen
   d. sac fungus

____ 15. Viruses are able to infect _____ . [p.239]
   a. animals
   b. bacteria
   c. plants
   d. all of the preceding

____ 16. For its genetic information, a retrovirus contains _____ . [p.240]
   a. DNA
   b. lipid
   c. protein
   d. RNA

## Chapter Objectives/Review Questions

1. Describe Stanley Miller's experiments. What were the results? [p.219]
2. How did eukaryotic cells form? [pp.222–223]
3. Describe the different ways bacteria get the energy and building blocks they need. [p.224]
4. Describe prokaryotic fission. [p.225]
5. What causes algal blooms? How can they adversely affect aquatic environments? [p.230]
6. Describe the life cycle of a slime mold. [p.234]
7. What are the four main groups of fungi? What are the characteristics of each group? [pp.235–236]
8. What mechanisms do lichens have to survive hostile habitats? [p.237]
9. Why are mycorrhizae important to plants? [pp.237–238]
10. Compare and contrast the lytic and lysogenic pathways in bacteriophages. [p.240]
11. Compare and contrast viruses, viroids, and prions. [pp.239–241]

12. Match each organism with its photograph or illustration. [pp.224, 229, 232–234, 236, 237, 239]

   a. _____ amoeba
   b. _____ bacteria
   c. _____ brown algae
   d. _____ club fungus
   e. _____ diatom
   f. _____ green algae
   g. _____ lichen
   h. _____ *Paramecium*
   i. _____ red algae
   j. _____ sac fungus
   k. _____ trypanosome
   l. _____ virus

## Integrating and Applying Key Concepts

1. How have bacteria, protists, and viruses had major effects on human history?
2. What similarities do viruses share with living organisms? How might viruses have arisen?

# 15

# PLANT EVOLUTION

## INTRODUCTION
Chapter 15 describes the evolution of land plants and includes a survey of the various plant groups.

## FOCAL POINTS

- Figure 15.2 [p.245] diagrams the generalized life cycle of plants.
- Figures 15.3 and 15.4 [p.246] illustrate plant evolution.
- Figure 15.6 [p.248] shows the moss life cycle.
- Figures 15.7, 15.8, and 15.10 [pp.249, 250] illustrate diversity of seedless vascular plants.
- Figure 15.9 [p.250] diagrams the fern life cycle.
- Figure 15.12 [p.252] shows gymnosperm diversity.
- Figure 15.13 [p.253] illustrates the gymnosperm life cycle.
- Figure 15.14 [p.254] diagrams the parts of a typical flower.
- Figure 15.16 [p.255] shows the angiosperm life cycle.
- Tables 15.1 and 15.2 [p.257] summarize the characteristics of plants.

## Interactive Exercises

*Beginnings, and Endings* [p.244]

### 15.1. PIONEERS IN A NEW WORLD [pp.245–247]

**Selected Words:** *Cooksonia* [p.245], lignin [p.245], xylem [p.245], phloem [p.245], *nonvascular* plants [p.246], *vascular* types [p.246], *seedless* vascular plants [p.246], *seed-bearing* vascular plants [p.246], megaspores [p.246], microspores [p.246]

### Boldfaced Terms

[p.244] deforestation _____

_____

[p.245] cuticle _____

_____

[p.245] stomata _____

[p.245] gametophyte _____

[p.245] sporophyte _____

[p.245] spore _____

[p.246] pollen grains _____

[p.247] seed _____

## Matching

1. _____ *Cooksonia* [p.245]
2. _____ root systems [p.245]
3. _____ shoot systems [p.245]
4. _____ xylem and phloem [p.245]
5. _____ cuticle [p.245]
6. _____ gametophytes [p.245]
7. _____ sporophyte [p.245]
8. _____ spores [p.245]
9. _____ pollen grains [p.246]
10. _____ seed [p.247]

a. A waxy coat that conserves water in shoots on hot, dry days
b. Structures composed of haploid cells, some of which develop into haploid gametes
c. One distributes water and mineral ions, the other distributes sugars and other photosynthetic products through the plant
d. Develops from a diploid zygote
e. Following dispersal, they germinate and develop into haploid gametophytes
f. An embryo sporophyte with tissues that nourish it, and an outer coat
g. Simple, branched, vascular plant without leaves
h. Stems and leaves absorb energy from the sun and carbon dioxide from the air; became erect, taller, and branched with the evolution of lignin
i. Possess many underground absorptive structures that, taken together, have a large surface area
j. Formed from smaller microspores in gymnosperms and angiosperms

## Fill in the Blanks

Only 24,000 of the 295,000 existing plant species are (11) _____ [p.246] plants, or bryophytes. (12) _____ [p.246] plants have internal tissue systems that conduct water and solutes through roots, stems, and leaves. The lycophytes, horsetails, and ferns are among the (13) _____ vascular plants [p.246]. Cycads, ginkgos, gnetophytes, and conifers are (14) _____ [p.246], a group of (15) _____-_____ [p.246] vascular plants. Another group of vascular plants, the (16) _____ [p.246], bear flowers and seeds. As with some seedless species, all seed-bearing plants produce two types of (17) _____ [p.246]. Gymnosperms and angiosperms form (18) _____ [p.246] that give rise to female (19) _____ [p.246]. They also form smaller (20) _____ [p.246], which develop into (21) _____ _____ [p.246], the male (19).

140   Chapter Fifteen

# 15.2. THE BRYOPHYTES—NO VASCULAR TISSUES [pp.247–248]

*Selected Words:* mosses [p.247], *Sphagnum* [p.247], peat [p.247], hornworts [p.247], liverworts [p.247]

## Boldfaced Terms

[p.247] bryophytes _____

_____

[p.247] rhizoids _____

_____

## True–False

If the statement is true, write a T in the blank. If the statement is false, correct it by writing the correct word(s) for the underlined word(s) in the answer blank.

_____ 1. Mosses particularly are sensitive to <u>air</u> pollution. [p.247]

_____ 2. Bryophytes <u>have</u> true leaves, stems, and roots. [p.247]

_____ 3. Most bryophytes have <u>rhizomes</u> that absorb and attach gametophytes to the soil. [p.247]

_____ 4. In mosses, eggs and sperms form by mitosis at <u>gametophyte</u> shoot tips. [p.247]

_____ 5. The moss <u>gametophyte</u> is a stalk and a jacketed structure in which haploid spores develop by meiosis. [p.247]

_____ 6. The metabolic products of bryophytes have a(n) <u>basic</u> pH and restrict the growth of bacterial and fungal decomposers. [p.247]

_____ 7. Because of its high absorbency and antiseptic properties, doctors used <u>peat</u> as an emergency poultice on wounded soldiers during World War I. [p.247]

## Dichotomous Choice

For each statement on page 142, refer to the corresponding numbers on the sketches of the moss life cycle (also see Figure 15.6, p.248 in the main text). Circle the words in parentheses that make each statement correct. [p.248]

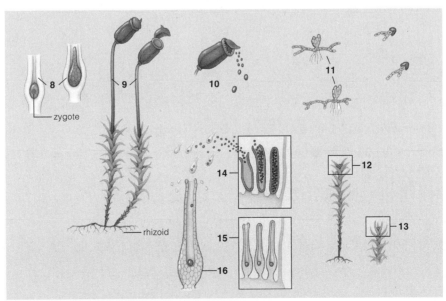

Plant Evolution 141

8. The zygote shown is (haploid/diploid) and represents the future (gametophyte/sporophyte) plant.
9. The moss plants shown above the line represent (gametophyte/sporophyte) plants and their tissues are (haploid/diploid); the moss plants shown below the dotted line represent (gametophyte/sporophyte) plants and their tissues are (haploid/diploid).
10. (Mitosis/Meiosis) is occurring in the capsule-shaped structure and the cells being released from it are (haploid gametes/haploid spores).
11. The two structures shown represent young (gametophyte/sporophyte) plants and their tissues are (haploid/diploid).
12. The moss plant shown here represents a (gametophyte/sporophyte) plant.
13. The moss plant shown here represents a (gametophyte/sporophyte) plant.
14. The cells being released from this structure are (haploid sperms/diploid spores).
15. The cells at the base of the flask-shaped structures are (haploid spores/haploid eggs).
16. The flask-shaped structure is actually found growing at the tip of moss plant (14/15).

## 15.3. SEEDLESS VASCULAR PLANTS [pp.248–250]

**Selected Words:** club mosses [p.248], *Lycopodium* [p.249], sphenophytes [p.249], *Equisetum* [p.249], *epiphyte* [p.250], sori (singular, sorus) [p.250]

### Boldfaced Terms

[p.248] lycophytes _____

_____

[p.248] horsetails _____

_____

[p.248] ferns _____

_____

[p.249] rhizomes _____

_____

[p.249] strobilus (plural, strobili) _____

_____

### Choice

Choose from the following:

      a. lycophytes    b. horsetails    c. ferns    d. applies to a., b., and c.

1. _____ A group in which only the genus *Equisetum* survives. [p.249]
2. _____ Club mosses, the most familiar seedless vascular plants, live in the Arctic, tropics, and regions in between. [p.249]
3. _____ Seedless vascular plants [p.248]
4. _____ Rust-colored patches, the sori, are on the lower surface of their fronds. [p.250]
5. _____ Some tropical species are the size of trees and some are epiphytes. [p.250]

6. _____ Ancestral plants that lived in the Carboniferous became peat and coal through time, heat, and pressure. [p.250]
7. _____ Stems were used by pioneers of the American West to scrub cooking pots. [p.250]
8. _____ Mature leaves (fronds) are usually divided into leaflets. [p.250]
9. _____ When the spore chamber snaps open, spores catapult through the air. [p.250]
10. _____ Grow along streambanks and in disrupted habitats, such as roadsides and vacant lots [p.249]
11. _____ Sporophytes usually have rhizomes and hollow, silica-reinforced, photosynthetic stems. [p.249]
12. _____ Their sporophytes are adapted to conditions on land. [pp.248–250]
13. _____ *Lycopodium* [p.249]
14. _____ The young leaves are coiled into the shape of a fiddlehead. [p.250]
15. _____ A germinating spore develops into a small, green, heart-shaped gametophyte. [p.250]
16. _____ Films of water must be available at the time of sexual reproduction if their motile sperm are to reach the eggs. [pp.248–250]

*Dichotomous Choice*

For each statement below, refer to the corresponding numbers on the sketches of the fern life cycle (also see Figure 15.9, p.250 in the main text). Circle the words in parentheses that make each statement correct. [p. 250]

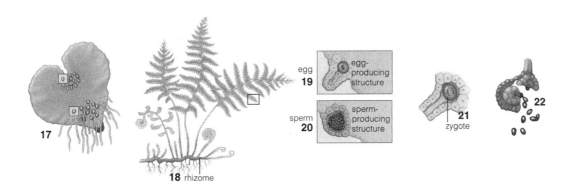

17. This structure represents the (gametophyte/sporophyte) plant; its tissues are (haploid/diploid).
18. The entire structure represents the (gametophyte/sporophyte) plant; its tissues are (haploid/diploid).
19. The egg shown is (haploid/diploid); (mitosis/meiosis) gave rise to the egg.
20. The sperm-producing structure is located on the (gametophyte/sporophyte) plant; its tissues are (haploid/diploid).
21. The zygote is (haploid/diploid) and is located within the egg-producing structure found on the (gametophyte/sporophyte) plant.
22. This structure functions in (sperm production/spore production) and the cells being released from it are (haploid/diploid); the structure itself is located on the (gametophyte/sporophyte) plant.

## 15.4. THE RISE OF SEED-BEARING PLANTS [p.251]
## 15.5. GYMNOSPERMS—PLANTS WITH "NAKED" SEEDS [pp.252–253]

*Selected Words:* "seed ferns" [p.251], ovules [p.251], *Ephedra* [p.252], *Welwitschia* [p.252], evergreen [p.252], deciduous [p.252]

## Boldfaced Terms

[p.251] microspores _____

_____

[p.251] pollination _____

_____

[p.251] megaspores _____

_____

[p.252] cycads _____

_____

[p.252] gnetophytes _____

_____

[p.252] conifers _____

_____

[p.252] cones _____

_____

## Matching

Match each term with its description. [p. 251]

1. _____ seed ferns
2. _____ pollination
3. _____ ovules
4. _____ megaspores
5. _____ microspores

a. The arrival of pollen on female reproductive parts
b. Produces the female gametophyte
c. Earliest seed plants
d. Precursor of pollen
e. Female reproductive structures that mature into seeds

## Choice

Choose from the following: [pp.252–253]

      a. cycads    b. ginkgos    c. gnetophytes
      d. conifers    e. gymnosperms (includes a., b., c., d.)

6. _____ Crushed, fleshy seeds of female trees produce quite a stench.
7. _____ Includes pines and redwoods
8. _____ Seeds and trunk are ground into an edible flour.
9. _____ Only a single species survives, the maidenhair tree.
10. _____ Includes *Welwitschia* and *Ephedra*
11. _____ Seeds are mature ovules

12. _____ The favored male trees are now widely planted; they have attractive, fan-shaped leaves and are resistant to insects, disease, and air pollutants.
13. _____ Their ovules and seeds are not covered; they are borne on surfaces of spore-producing reproductive structures.
14. _____ Some sporophyte plants in this group mainly have a deep-reaching taproot on a woody stem bearing cones and one or two strap-shaped leaves that split lengthwise repeatedly as the plant ages.
15. _____ Most species are "evergreen" trees and shrubs with needlelike or scalelike leaves; a few are deciduous.
16. _____ Includes conifers, cycads, ginkgos, and gnetophytes

## Dichotomous Choice

For each statement below, refer to the corresponding numbers on the sketches of the pine life cycle (also see Figure 15.13, p.253 in the main text). Circle the words in parentheses that make each statement correct. [p. 253]

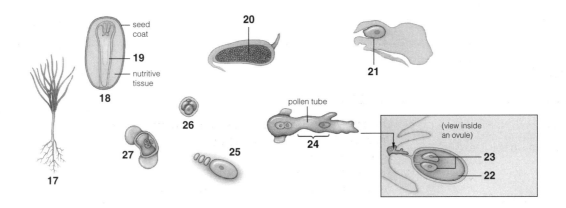

17. This plant is a (gametophyte/sporophyte) and its tissues are (haploid/diploid).
18. The entire structure shown is a (seed/fruit).
19. This structure represents a young (gametophyte/sporophyte) plant; its tissues are (haploid/diploid).
20. The entire structure shown is a(n) (egg sac/pollen sac) and may be found as part of the pine (flower/cone).
21. This structure is the (egg/ovule) and it is (haploid/diploid).
22. The inside cellular mass of this structure represents the (male gametophyte/female gametophyte) plant; its cells are (haploid/diploid).
23. These structures are (eggs/sperms) and they are (haploid/diploid); they were produced by (mitosis/meiosis).
24. The cellular portion of this structure is known as the (male gametophyte/female gametophyte) plant; it is (haploid/diploid).
25. The largest cell shown is a (microspore/megaspore); it was produced by (mitosis/meiosis) and is (haploid/diploid).
26. The quartet of cells shown are (microspores/megaspores); each is (haploid/diploid) and will develop into a(n) (ovule/pollen grain).
27. This structure is a (megaspore/pollen grain) and through pollination, it will be moved to a (male cone/female cone).

Plant Evolution 145

## 15.6. ANGIOSPERMS—THE FLOWERING PLANTS [pp.254–255]
## 15.7. DEFORESTATION IN THE TROPICS [p.256]

***Selected Words:*** erosion [p.256], leaching of nutrients [p.256], mechanized logging [p.256], shifting cultivation [p.256]

### Boldfaced Terms

[p.254] flower _____

_____

[p.254] pollinator _____

[p.254] coevolution _____

[p.254] magnoliids _____

[p.254] eudicots _____

[p.254] monocots _____

[p.256] deforestation _____

_____

### Matching

Match each term with its description.

1. _____ flower [p.254]
2. _____ coevolution [p.254]
3. _____ pollinator [p.254]
4. _____ ovary [p.254]
5. _____ examples of magnoliids [p.254]
6. _____ examples of eudicots [p.255]
7. _____ examples of monocots [p.255]

a. Cabbages, daisies, most flowering shrubs and trees, and cacti
b. Magnolias, avocados, nutmeg, and pepper plants
c. Any agent that transfers pollen to female floral parts of the same plant species; examples are insects, bats, birds, and other animals
d. A specialized reproductive shoot found only in angiosperms
e. Orchids, palms, lilies, and grasses, such as rye, sugarcane, corn, rice, wheat, and many other highly valued plants
f. Where ovules and seeds develop; also usually an immature fruit
g. Refers to two or more species evolving jointly by close ecological interactions

*Dichotomous Choice*

For each statement below, refer to the corresponding numbers on the sketches of the flowering plant life cycle (monocot) (also see Figure 15.16, p.255 in the main text). Circle the words in parentheses that make each statement correct. [p. 255]

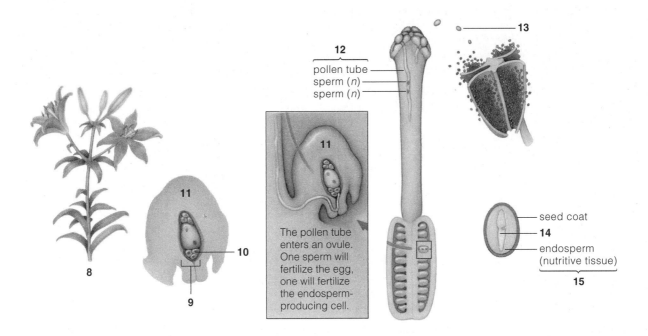

8. The plant is a (gametophyte/sporophyte); its tissues are (haploid/diploid).
9. The reduced plant is a (female gametophyte/female sporophyte); its cells are (haploid/diploid).
10. The single cell is a(n) (egg/spore); it is (haploid/diploid) and it was produced by (mitosis/meiosis).
11. This structure is called an (ovule/ovary); its cells are (haploid/diploid) and it later can develop into a(n) (fruit/seed).
12. The reduced plant is a (male gametophyte/male sporophyte); its cells are (haploid/diploid) and they were produced by (mitosis/meiosis).
13. The small cell is a (released spore/released pollen grain) and it is (haploid/diploid).
14. This structure is a(n) (embryo gametophyte plant/embryo sporophyte plant); its cells are (haploid/diploid) and it was produced by (mitosis/meiosis) of a (haploid zygote/diploid zygote).
15. The entire structure is called a (seed/fruit); it is a matured (ovary/ovule).

*Short Answer*

16. List the benefits for allowing forested watersheds to remain. [p.256]

17. Why is it difficult to develop nutrient-rich topsoils in the tropics? [p.256]

18. Describe advantages to the practice of shifting cultivation. [p.256]

Plant Evolution  **147**

## Self-Quiz

___ 1. The _____ is *not* a trend in the evolution of plants. [pp.245–246]
   a. evolution of roots, stems, and leaves
   b. shift from diploid to haploid dominance
   c. development of xylem and phloem
   d. development of cuticles and stomata

___ 2. Plants possessing xylem and phloem are called _____ plants. [p.245]
   a. gametophytes
   b. nonvascular
   c. vascular
   d. sporophytes
   e. seedless

___ 3. Existing nonvascular plants do *not* include _____ . [p.247]
   a. horsetails
   b. mosses
   c. liverworts
   d. hornworts

___ 4. _____ are *not* seedless vascular plants. [pp.248–250]
   a. Lycophytes
   b. Gymnosperms
   c. Horsetails
   d. *Equisetum*
   e. Ferns

___ 5. In horsetails, lycopods, and ferns, _____ . [pp.248–250]
   a. spores give rise to gametophytes
   b. the dominant plant is a gametophyte
   c. the sporophyte bears sperm- and egg-producing structures
   d. the gametophyte bears sperm- and egg-producing structures
   e. both a. and d. are correct

___ 6. _____ are seed plants. [pp.252–254]
   a. Cycads
   b. Ginkgos
   c. Conifers
   d. Angiosperms
   e. All of the above

___ 7. The diploid stage progresses through this sequence: _____ . [pp.248, 250, 253, 255]
   a. gametophyte → male and female gametes
   b. spores → sporophyte
   c. zygote → sporophyte
   d. zygote → gametophyte
   e. spores → gametes

___ 8. In which of the following is the gametophyte *not* the dominant plant? [pp.248, 250, 253, 255]
   a. lycophytes
   b. gymnosperms
   c. bryophytes
   d. angiosperms
   e. ferns

___ 9. Microspores give rise to _____ . [p.251]
   a. megaspores
   b. female gametophytes
   c. male cones
   d. pollen grains
   e. embryos

___ 10. Magnoliids, eudicots, and monocots are groups of _____ . [p.254]
   a. gymnosperms
   b. bryophytes
   c. club mosses
   d. lycophytes
   e. angiosperms

## Chapter Objectives/Review Questions

1. Define *deforestation* and cite the damage it has done to the world's forests. [pp.244, 256]
2. Be able to list the trends that have occurred in the evolution of plants. [pp.245–247]
3. List the groups included in the modern bryophytes and describe their major features. [pp. 247–248]
4. Be familiar with the major features of the moss life cycle. [p.248]

5. The main lineages of seedless vascular plants are lycophytes, horsetails, and _____ ; describe the habitats and major features of each. [pp.248–250]
6. Be familiar with the major features of the fern life cycle. [p.250]
7. Define the following as features that developed to allow the rise of seed-bearing plants: *microspores, pollination, megaspores,* and *ovules.* [p.251]
8. Be familiar with representatives and the features of conifers, cycads, ginkgos, and gnetophytes. [pp.252–253]
9. Be familiar with the major features of the pine life cycle. [p.253]
10. The outstanding feature of the angiosperms is the _____ , a specialized reproductive shoot. [p.254]
11. Define *coevolution, pollinator, magnoliids, eudicots,* and *monocots.* [pp.254–255]
12. Be familiar with the major features of the angiosperm life cycle. [p.255]

## Integrating and Applying Key Concepts

Prepare a table with lined columns having the following four titles at the top: "plant group, dominant plant, vascular tissue present?, and seeds present?" List the following plant groups in lined rows in the leftmost column: bryophytes, lycophytes, horsetails, ferns, gymnosperms, and angiosperms. Then fill in the rows of information for each plant group. Check your work with the main text. List evolutionary trends you can discern after looking over the information.

# 16

# ANIMAL EVOLUTION

## INTRODUCTION
This chapter combines animal evolution with animal diversity. It compares many groups of animals.

## FOCAL POINTS
- Figure 16.1 [p.260] shows an evolutionary tree of animals.
- Figure 16.3 [p.261] contrasts the various types of body cavities.
- Figure 16.4 [p.262] illustrates sponge anatomy.
- Figure 16.6 [p.263] looks at cnidarian diversity.
- Figures 16.7 and 16.8 [pp.263, 264] compare flatworm types.
- Figures 16.9 and 16.10 [p.265] show annelid diversity.
- Figure 16.11 [p.266] compares various mollusks.
- Figure 16.12 [p.267] shows nematodes and some problems caused by them.
- Figures 16.14, 16.15, 16.17, and 16.18 [pp.268–271] explore arthropod diversity.
- Figure 16.19 [p.272] shows some echinoderm diversity.
- Figure 16.20 [p.273] diagrams chordate evolution and illustrates a simple chordate, the lancelet.
- Figures 16.23, 16.24, and 16.25 [pp.275, 276] look at fish diversity.
- Figures 16.28 and 16.29 [pp.278, 279] show amniotes.
- Figures 16.30 and 16.31 [pp.280, 281] illustrate mammalian diversity.
- Figure 16.36 [p.285] diagrams the hominid family tree.

---

## Interactive Exercises

*Interpreting and Misinterpreting the Past* [p.259]

### 16.1. OVERVIEW OF THE ANIMAL KINGDOM [pp.260–261]

**Selected Words:** "missing links" [p.259], *Archaeopteryx* [p.259], *anterior* [p.260], *posterior* [p.260], *ventral* [p.260], *dorsal* [p.260], *incomplete* digestive system [p.261], *complete* digestive system [p.261], blastopore [p.261], peritoneum [p.261], pseudocoel [p.261]

## Boldfaced Terms

[p.260] animals _____

[p.260] ectoderm _____

[p.260] endoderm _____

[p.260] mesoderm _____

[p.260] vertebrates _____

[p.260] invertebrates _____

[p.260] Ediacarans _____

[p.260] radial symmetry _____

[p.260] bilateral symmetry _____

[p.260] cephalization _____

[p.261] gut _____

[p.261] protostomes _____

[p.261] deuterostomes _____

[p.261] coelom _____

[p.261] segmentation _____

## Fill in the Blanks

Workers in a limestone quarry in (1) _____ [p.259] unearthed a fossil with characteristics of both (2) _____ [p.259] and (3) _____ [p.259]. The animal was named (4) _____ [p.259]. Fossils of many other ancient organisms like (5) _____ [p.259] and (6) _____ [p.259] were also found in the same limestone. Radiometric dating methods showed that the organisms that left these fossil remains were alive (7) _____ _____ [p.259] years ago.

The animals that left these fossils like the animals of today are defined by several characteristics. Animals are both (8) _____ [p.260] and (9) _____ [p.260], the latter meaning that they feed on other organisms. Animals also reproduce (10) _____ [p.260]. In addition, most animals produce three tissue layers during development: (11) _____ [p.260], (12) _____ [p.260], and (13) _____ [p.260].

*True–False*

If the statement is true, write a T in the blank. If the question is false, correct it by changing the underlined word(s) and writing the correct word(s) in the answer blank.

_____ 14. Most animal species are <u>vertebrates</u>. [p.260]

_____ 15. An animal that can be cut into two symmetrical halves by a single plane is said to be <u>radially</u> symmetric. [p.260]

_____ 16. The <u>belly or underside</u> of an organism is the dorsal side. [p.260]

_____ 17. A complete digestive system has a <u>sac-like gut with only one opening</u>. [p.261]

*Matching*

Match each term with its corresponding letter on this diagram of the evolutionary tree of life [p.260]

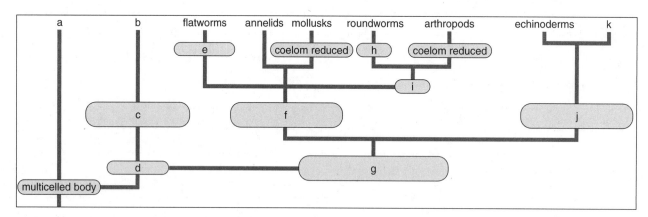

18. _____ bilateral coelomate ancestry, 3 germ layers
19. _____ mouth from blastopore
20. _____ anus from blastopore
21. _____ radial ancestry, 2 germ layers
22. _____ true tissues
23. _____ pseudocoel
24. _____ molting
25. _____ coelom lost
26. _____ chordates
27. _____ cnidarians
28. _____ sponges

## 16.2. GETTING ALONG WELL WITHOUT ORGANS [pp.262–263]
## 16.3. FLATWORMS—INTRODUCING ORGAN SYSTEMS [pp.263–264]

*Selected Terms:* spicules [p.262], collar cells [p.262], nematocysts [p.262], polyps [p.262], medusa [p.262], *epidermis* [p.262], *gastrodermis* [p.262], *nerve net* [p.262], *mesoglea* [p.263], *hydrostatic* skeleton [p.263], hermaphrodites [p.263], *definitive* host [p.264], *intermediate* host [p.264], *schistosomiasis* [p.264]

### Boldfaced Terms

[p.262] sponges _____

[p.262] larva _____

[p.262] cnidarians _____

[p.262] epithelium _____

[p.263] organ _____

[p.263] flatworms _____

### Complete the Table [pp.262–264]

| Animal | Tissues (Y/N) | Symmetry | Digestive System | Nervous System |
|---|---|---|---|---|
| Sponges | 1. | 4. | 7. | 10. |
| Cnidaria | 2. | 5. | 8. | 11. |
| Flatworms | 3. | 6. | 9. | 12. |

### Choice

Choose from the following: [pp.262–264]

a. cnidarian   b. flatworm   c. sponge

13. _____ They often have sharp, glassy spicules as skeletal elements.
14. _____ These can exist in either a polyp or a medusa form.
15. _____ Without a digestive system, these animals use collar cells to trap food.
16. _____ They use nematocysts to trap prey.
17. _____ Flukes are in this group.
18. _____ Most of these are hermaphrodites.
19. _____ The guts of these organisms have only one opening.
20. _____ Some of these are human parasites.

Animal Evolution   153

## 16.4. ANNELIDS—SEGMENTS GALORE [pp.264–265]
## 16.5. THE EVOLUTIONARILY PLIABLE MOLLUSKS [pp.266–267]

***Selected Words:*** *oligo–* [p.264], *poly–* [p.264], closed circulatory system [p.265], parapods [p.265], *molluscus* [p.266], gastropods [p.266], bivalves [p.266], cephalopods [p.266], *jet propulsion* [p.267]

### Boldfaced Terms

[p.264] annelids _____

_____

[p.265] brain _____

_____

[p.265] nerve cords _____

_____

[p.266] mollusks _____

_____

### Choice

For questions 1–10, choose from one of the following. [pp.264–267]

                    a. annelid    b. mollusk

_____ 1. clam                _____ 6. octopus

_____ 2. earthworm       _____ 7. oyster

_____ 3. leech              _____ 8. polychaete

_____ 4. nautilus          _____ 9. slug

_____ 5. nudibranch      _____ 10. snail

11. _____ Setae (bristles) are seen in _____. [p.264]

       a. leeches    b. oligochaetes    c. both a. and b.    d. neither a. nor b.

12. _____ Which of the following is/are usually marine? [p.264]

       a. oligochaete    b. polychaete    c. both a. and b.    d. neither a. nor b.

13. _____ The mantle is a characteristic of _____. [p.266]

       a. annelids    b. mollusks    c. both a. and b.    d. neither a. nor b.

14. _____ The vast majority of mollusks are _____. [p.266]

       a. bivalves    b. cephalopods    c. chitons    d. gastropods

15. _____ Which of the following is not a mollusk? [p.266]

       a. barnacle    b. octopus    c. scallop    d. squid

16. \_\_\_\_ The largest mollusk is found in which group? [p.266]

   a. bivalves   b. cephalopods   c. chitons   d. gastropods

17. \_\_\_\_ What kind of mollusk has a closed circulatory system? [p.267]

   a. bivalve   b. cephalopod   c. both a. and b.   d. neither a. nor b.

18. \_\_\_\_ A brain can be found in _____. [pp.264–266]

   a. mollusks   b. Cnidaria   c. both a. and b.   d. neither a. nor b.

## 16.6. AMAZINGLY ABUNDANT ROUNDWORMS [p.267]
## 16.7. ARTHROPODS—THE MOST SUCCESSFUL ANIMALS [pp.268–271]

***Selected Words:*** nematodes [p.267], false coelom [p.267], *trichinosis* [p.267], *elephantiasis* [p.267], edema [p.267], *joints* [p.268], *juvenile* [p.268], *Loxosceles* [p.269], *Borrelia burgdorferi* [p.269], "crust" [p.269], "shell" [p.270], copepods [p.270]

### Boldfaced Terms

[p.267] roundworms _____

_____

[p.267] molting _____

_____

[p.268] exoskeleton _____

_____

[p.268] metamorphosis _____

_____

### Fill in the Blanks

Another name for roundworm is (1) _____ [p.267]. Their body cavity is a (2) _____ _____ [p.267] and they (3) _____ [p.267], periodically shedding their outer covering. Most roundworms are (4) _____ [p.267] that live in the soil but others are (5) _____ [p.267]. (6) _____ [p.267] is caused by a roundworm that encysts in muscle tissue while (7) _____ [p.267] occurs when roundworms block lymph channels.

   Researchers have identified more than (8) _____ _____ [p.268] arthropods, the majority of which are (9) _____ [p.268]. Arthropods are characterized by their hard (10) _____ [p.268] and their (11) _____ [p.268] appendages. Aquatic arthropods usually use (12) _____ [p.268] for gas exchange while insects use a (13) _____ [p.268] system, tubes that bring oxygen directly to the tissues. Many arthropods go through (14) _____ [p.268] where the body changes, often completely, between juvenile and adult forms.

*Choice*

Choose from the following. [pp.269–271]

            a. chelicerates    b. crustaceans    c. insects

15. \_\_\_\_ barnacle    19. \_\_\_\_ louse    22. \_\_\_\_ shrimp

16. \_\_\_\_ beetle    20. \_\_\_\_ pill bug    23. \_\_\_\_ spider

17. \_\_\_\_ crab    21. \_\_\_\_ scorpion    24. \_\_\_\_ tick

18. \_\_\_\_ damselfly

*True–False*

If the statement is true, write a T in the blank. If the question is false, correct it by changing the underlined word(s) and writing the correct word(s) in the answer blank.

_____ 25. Bites of <u>both the brown recluse and black widow spiders</u> can be fatal to humans. [p.269]

_____ 26. Lyme disease, typhus, and tularemia are transmitted by <u>spiders</u>. [p.269]

_____ 27. Crustaceans usually have <u>three pairs</u> of walking legs. [p.270]

_____ 28. <u>Insects</u> are the most common arthropods in aquatic habitats. [p.270]

_____ 29. <u>Insects</u> are the only invertebrates with wings. [p.270]

## 16.8. THE PUZZLING ECHINODERMS [p.272]
## 16.9. EVOLUTIONARY TRENDS AMONG VERTEBRATES [pp.273–274]
## 16.10. MAJOR GROUPS OF JAWED FISHES [pp.275–276]

*Selected Words:* tube feet [p.272], notochord [p.273], gill slits [p.273], tunicates [p.273], lancelets [p.273], vertebrates [p.274], swim bladder [p.275], lung fishes [p.275], lobe-finned fishes [p.275], coelacanths [p.276]

*Boldfaced Terms*

[p.272] echinoderm _____

_____

[p.273] chordates _____

_____

[p.273] craniates _____

_____

[p.274] jaws _____

_____

[p.274] vertebrae (singular, vertebra) _____

[p.274] fins _____

[p.274] gills _____

[p.274] lungs _____

[p.275] cartilaginous fishes _____

[p.275] bony fishes _____

[p.275] ray-finned fishes _____

[p.276] tetrapods _____

## Fill in the Blanks

Nearly all echinoderms have an internal skeleton made of (1) _____ _____ [p.272] and a decentralized nervous system without a (2) _____ [p.272]. Echinoderms use their fluid-filled (3) _____ _____ [p.272] for movement, clinging to substrates and feeding. Some sea stars feed by extruding their (4) _____ [p.272] outside their mouths. Larval symmetry is (5) _____ [p.272] while adults usually display (6) _____ [p.272] symmetry.

All chordates share several features. They all have a (7) _____ [p.273], a stiffened rod that helps support the body; a (8) _____ [p.273] nerve cord that expands to form a (9) _____ [p.273] at the anterior end; (10) _____ _____ [p.273] in the wall of the pharynx; and a (11) _____ [p.273] that extends past the (12) _____ [p.273]. Early chordates used their (10) to become efficient (13) _____ [p.273] feeders. More than 530 million years ago, (14) _____ [p.273] arose, so named because they have a chamber covered in (15) _____ [p.273] or (16) _____ [p.273] to house their brains. Soon, (17) _____ [p.274], hinged bony feeding structures, evolved. (18) _____ [p.274] changed the course of evolution because they evolved a series of hardened bony structures called the (19) _____ _____ [p.274] that replaced the notochord.

Animal Evolution

*Choice*

Choose from the following:

    a. coelacanth    b. lamprey    c. lancelet    d. lung fish
    e. ostracoderm    f. perch    g. placoderm    h. shark

20. \_\_\_\_ A notochord is the only support structure; filter feeder [p.273]
21. \_\_\_\_ Modern jawless fish [p.273]
22. \_\_\_\_ Extinct jawless fish [p.274]
23. \_\_\_\_ Extinct jawed fish with armor plating [p.274]
24. \_\_\_\_ Cartilaginous jawed fish [p.275]
25. \_\_\_\_ Bony fish with ray fins [p.275]
26. \_\_\_\_ Which organism represents the most diverse and abundant vertebrate group? [p.275]
27. \_\_\_\_ Bony fish with lobe fins [p.276]
28. \_\_\_\_ Has both gills and modified gut outpocketings for gas exchange [p.276]
29. \_\_\_\_ Which of the two groups above have most in common with terrestrial tetrapods? [p.276]

## 16.11. EARLY AMPHIBIOUS TETRAPODS [pp.276–277]
## 16.12. THE RISE OF AMNIOTES [pp.278–280]

*Selected Words:* synapsids [p.278], sauropsids [p.278], turtles [p.279], lizards [p.279], snakes [p.279], crocodilians [p.279], therapod [p.279], *therapsids* [p.280]

*Boldfaced Terms*

[p.276] amphibians _____

[p.278] amniotes _____

[p.278] "reptiles" _____

[p.278] dinosaurs _____

[p.279] K–T asteroid impact theory _____

[p.279] birds _____

[p.280] mammals _____

[p.280] monotremes _____

[p.280] marsupials _____

[p.280] eutherians _____

## Fill in the Blanks

(1) _____ [p.276] were the first tetrapods on land. Land life presented new challenges and (2) _____ , _____ , and _____ [p.277] became the most advantageous senses. Modern (1) include the tailed (3) _____ [p.277] and the tailless (4) _____ [p.277] and (5) _____ [p.277]. These have not escaped water entirely since they must keep their (6) _____ [p.277] moist for respiration and most return to water for (7) _____ [p.277].

(8) _____ [p.278] were even better adapted to life on land. They were the first (9) _____ [p.278], animals that are able to protect embryos with four specialized protective layers. They also had skin rich in (10) _____ [p.278] and efficient (11) _____ [p.278] to help conserve water. They are subdivided into two groups: the (12) _____ [p.278] that were ancestral to mammals and the (13) _____ [p.278] that were ancestral to modern reptiles and birds. Huge reptiles, the (14) _____ [p.278], ruled the Earth for (15) _____ [p.279] million years until a large (16) _____ [p.279] struck the Earth and led to their extinction. Feathers first evolved in reptiles. Today they are seen in reptile descendants, the (17) _____ [p.279].

## Choice

18. _____ A four-chambered heart is seen in _____ . [pp.279–280]

    a. birds   b. mammals   c. both a. and b.   d. neither a. nor b.

19. _____ Hair is a characteristic seen in _____ . [p.280]

    a. birds   b. mammals   c. both a. and b.   d. neither a. nor b.

20. _____ Milk production is seen in _____ mammals. [p.280]

    a. egg-laying   b. pouched   c. both a. and b.   d. neither a. nor b.

21. _____ Mammals differ from other amniotes in the size, shape, and number of their _____ . [p.280]

    a. digits   b. ears   c. legs   d. teeth

Animal Evolution   159

## 16.13. FROM EARLY PRIMATES TO HUMANS [pp.281–286]

**Selected Words:** tarsiers [p.281], prosimians [p.281], anthropoids [p.281], hominoids [p.281], *prehensile* movements [p.282], *opposable* movements [p.282], *Sahelanthropus tchadensis* [p.283], *Australopithecus* [p.283], *Homo* [p.284], *H. habilis* [p.284], *H. erectus* [p.284], *H. sapiens* [p.284], Neandertals [p.284], *multiregional* model [p.286], *African emergence* model [p.286]

### Boldfaced Terms

[p.281] primates _____

[p.281] hominids _____

[p.281] bipedalism _____

[p.281] culture _____

[p.283] australopiths _____

[p.284] humans _____

### Fill in the Blanks

Primates include the (1) _____ [p.281], (2) _____ [p.281], and (3) _____ [p.281], the latter group including monkeys, apes, and (4) _____ [p.281]. Humans and their extinct human-like ancestors are called (5) _____ [p.281]. Five trends are seen in this line. First, less reliance on smell and more on (6) _____ _____ [p.281]; second, (7) _____ [p.281] or upright walking; third, hand changes that led to both (8) _____ [p.282] and (9) _____ [p.282] grips; fourth, less specialization in the (10) _____ [p.282]; and lastly, the development of (11) _____ [p.282], the sum of behavior patterns in a social group. Along with (11), the need to communicate became more important and the capacity for (12) _____ [p.282] appeared among the hominids.

The earliest primates looked like (13) _____ [p.282] or tree (14) _____ [p.282]. From them the first anthropoids evolved (15) _____ [p.282] million years ago. Early hominids like *Sahelanthropus tchadensis* first appeared about (16) _____ [p.283] million years ago. The footprints of the bipedal hominid, (17) _____ _____ [p.283], were preserved in volcanic ash 3.7 million years ago. The first members of our genus, *Homo*, lived in southern and eastern (18) _____ [p.284]. Many (19) _____ [p.284] tools have been dated to the time of (20) _____ _____ [p.284]. The first hominid to leave Africa was probably

(21) _____ _____ [p.284] who migrated to the Middle East about (22) _____ [p.284] million years ago. Our species, (23) _____ _____ [p.285], had evolved by (24) _____ [p.285] years ago. Scientists propose two models that explain the origin of *Homo sapiens*. The (25) _____ _____ [p.286] model says that the species arose in Africa and then spread out, replacing archaic forms, while the (26) _____ [p.286] model says that in several different places around the world, archaic forms evolved into *Homo sapiens*.

## Self-Quiz

_____ 1. A fossil with both avian and reptilian characteristics found in Germany in the mid-nineteenth century was named _____ . [p.259]
  a. *Archaeopteryx*
  b. *Australopithecus*
  c. ediacara
  d. pteranodon

_____ 2. Humans are classified as _____ . [p.261]
  a. bilateral acoelomates
  b. bilateral coelomates
  c. radial acoelomates
  d. radial coelomates

_____ 3. A tubular gut with two openings is seen in _____ . [p.261]
  a. flatworms
  b. roundworms
  c. both a. and b.
  d. neither a. nor b.

_____ 4. Which of the following is not classified as a Cnidarian? [p.262]
  a. coral
  b. jellyfish
  c. sea anemone
  d. sponge

_____ 5. Which of the following is not an annelid? [pp.264–265]
  a. earthworm
  b. fluke
  c. leech
  d. polychaete

_____ 6. Which mollusk has the largest brain? [p.267]
  a. bivalve
  b. cephalopod
  c. chiton
  d. gastropod

_____ 7. Roundworms can cause _____ . [p.267]
  a. elephantiasis
  b. trichinosis
  c. both a. and b.
  d. neither a. nor b.

_____ 8. Which of the following is not an arthropod feature? [p.268]
  a. exoskeleton
  b. jointed legs
  c. specialized sensory structures
  d. no fusion of body segments

_____ 9. Insects usually have _____ . [p.270]
  a. three pairs of walking legs
  b. two pairs of wings
  c. both a. and b.
  d. neither a. nor b.

_____ 10. Which of the following has a brain? [pp.262, 272]
  a. cnidarian
  b. echinoderm
  c. both a. and b.
  d. neither a. nor b.

_____ 11. Which of the following is not a chordate characteristic? [p.273]
  a. gill slits
  b. notochord
  c. post-anal tail
  d. solid ventral nerve cord

___ 12. Amphibians exchange gasses through their _____ . [p.277]
   a. lungs
   b. skin
   c. both a. and b.
   d. neither a. nor b.

___ 13. Which of the following is not classified as an amniote? [p.278]
   a. amphibian
   b. bird
   c. mammal
   d. reptile

___ 14. Which of the following has/have a large four-chambered heart? [p.279]
   a. birds
   b. mammals
   c. both a. and b.
   d. neither a. nor b.

___ 15. The oldest hominid fossils are found in _____ . [p.283]
   a. Africa
   b. Asia
   c. Australia
   d. Europe

## Chapter Objectives/Review Questions

In questions 1–16, identify each of the animals pictured.

1 _____ [p.263]

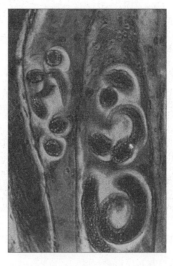

2 _____ [p.267]

3 _____ [p.269]

4 _____ [p.265]

**5** _____ [p.263]    **6** _____ [p.270]

**7** _____ [p.266]    **8** _____ [p.275]

9 _____ [p.271]   10 _____ [p.272]

11 _____ [p.281]   12 _____ [p.264]

164  Chapter Sixteen

13 _____ [p.266]   14 _____ [p.265]

15 _____ [p.275]   16 _____ [p.277]

17. What are the general characteristics of animals? [p.260]
18. Draw an evolutionary tree of animals showing the major changes at branch points. [p.260]
19. Describe the various types of body cavities or lack thereof seen in animals. [p.261]
20. Compare and contrast feeding in sponges and feeding in cnidarians. [p.262]
21. In what two forms can cnidarians exist? How are these forms similar? How are they different? [pp.262–263]
22. What are the three groups of flatworms? [pp.263–264]
23. Why is a closed circulatory system so important to cephalopods? [p.267]
24. How does a roundworm cause elephantiasis? [p.267]
25. What are the six adaptations of arthropods that made them so successful? [p.268]
26. What is metamorphosis? Describe metamorphosis in an insect. [pp.268–271]
27. What is odd about the symmetry of echinoderms? [p.272]
28. What are the general characteristics of chordates? [p.273]
29. Describe the evolution of amphibians from fish ancestors. [pp.276–277]
30. Why are amniote eggs so important to life on land? [pp.278–279]
31. Why is it thought that dinosaurs and many other terrestrial and marine organisms went extinct 65 million years ago? [p.279]
32. Why are the power and precision grips important to human evolution? [pp.281–282]
33. Compare and contrast the "out of Africa" and the "multiregional" models of the origin of modern humans. [p.286]

Animal Evolution

## Integrating and Applying Key Concepts

1. If you were a paleontologist and found only the fossilized teeth of an organism, what could you tell about the entire organism from just the teeth?
2. It has often been said that the best parasites do not kill their host. Why do think that statement is true?

# 17

# PLANTS AND ANIMALS: COMMON CHALLENGES

## INTRODUCTION

Chapter 17 looks at the processes that plants and animals use to maintain homeostasis. It also looks at some of the mechanisms used to achieve this state.

## FOCAL POINTS

- Figures 17.6 and 17.7 [p.294] look at homeostatic control mechanisms.
- Section 17.5 [pp.297, 298] describes the mechanisms of signal reception, transduction, and response.

---

# Interactive Exercises

*Too Hot to Handle* [p.289]

## 17.1. LEVELS OF STRUCTURAL ORGANIZATION [pp.290–291]

**Selected Words:** *heat stroke* [p.289], morphology [p.289], *division of labor* [p.290], structural organization [p.290], extracellular fluid [p.291]

**Boldfaced Terms**

[p.289] anatomy _____

[p.289] physiology _____

[p.290] tissue _____

[p.290] organ _____

[p.290] organ system _____

[p.290] growth _____

_____

[p.290] development _____

_____

[p.291] internal environment _____

_____

[p.291] homeostasis _____

_____

*Fill in the Blanks*

A body temperature of (1) _____ to _____ [p.289] is considered best for humans. As temperatures climb above that, (2) _____ [p.289] transports heat away from internal organs to the (3) _____ [p.289] from which heat is dissipated into the environment. Profuse (4) _____ [p.289] can dissipate even more heat. Above (5) _____ [p.289], the normal cooling mechanisms fail and temperature climbs, often leading to (6) _____ _____ [p.289] which, if not countered fast enough, can lead to (7) _____ _____ [p.289] or even (8) _____ [p.289]. (9) _____ [p.289], the study of a body's morphology, and (10) _____ [p.289], the study of a body's processes, are usually linked. In fact, it is often stated that the (11) _____ [p.289] of a body part almost always has something to do with its past or present (12) _____ [p.289].

*True–False*

If the statement is true, write a T in the blank. If the question is false, correct it by changing the underlined word(s) and writing the correct word(s) in the answer blank.

_____ 13. Both leaves and eyes fit the definition of organs. [p.290]

_____ 14. Development refers to an increase in the number, size, and volume of the body's cells. [p.290]

_____ 15. The major challenge faced by plants when they moved to land was too much oxygen. [p.290]

_____ 16. The body's internal environment consists of all of the fluid inside the body's cells. [p.291]

_____ 17. Keeping operating conditions within certain favorable limits is a state called homeostasis. [p.291]

## 17.2. RECURRING CHALLENGES TO SURVIVAL [pp.292–293]
## 17.3. HOMEOSTASIS IN ANIMALS [pp.293–295]
## 17.4. DOES HOMEOSTASIS OCCUR IN PLANTS? [pp.295–296]

*Selected Words:* vascular tissues [p.292], solute–water balance [p.293], osmotic gradients [p.293], intercellular signaling mechanisms [p.293], *altitude sickness* [p.294], resins [p.295]

## Boldfaced Terms

[p.292] diffusion _____

[p.292] surface-to-volume ratio _____

[p.293] active transport _____

[p.293] habitat _____

[p.293] interstitial fluid _____

[p.293] plasma _____

[p.293] sensory receptors _____

[p.293] integrator _____

[p.294] effectors _____

[p.294] negative feedback mechanism _____

[p.295] positive feedback mechanism _____

[p.295] system acquired resistance _____

[p.295] compartmentalization _____

[p.296] circadian rhythm _____

## Fill in the Blanks

Gas exchange is an important challenge in both plants and animals. In animals, needed (1) _____ [p.292] must move into cells while the waste (2) _____ _____ [p.292] is removed. Both of these gasses are able to move across cell membranes by (3) _____ [p.292]. Plants also use (3) to move (4) _____ _____ [p.292] into cells for photosynthesis and remove the excess (5) _____ [p.292] that is not needed for respiration. In order for this to happen, there must be a favorable (6) _____-_____-_____ ratio [p.292]. Any cell or organism

Plants and Animals: Common Challenges   169

without sufficient surface area could not survive. The favorable ratio is easy to see in (7) _____ [p.292] organs and organisms like leaves and flatworms but is more complex in larger, more solid organisms. Plants use (8) _____ _____ [p.292] to transport gasses and other substances from cell to cell. A similar system in humans picks up oxygen at the (9) _____ [p.292] and releases it from (10) _____ [p.292] into cells and interstitial fluids. In animals, other systems use this highway for transport like the (11) _____ [p.292] system that uses it to move white cells and chemical weapons to tissues under attack. To maintain a/an (12) _____-_____ [p.293] balance, cells use both diffusion and (13) _____ _____ [p.293] of solutes. Both are very important to the proper functioning of human organs like the (14) _____ [p.293] which produce urine. To coordinate and control these activities, cells often release (15) _____ _____ [p.293] which guide many activities within the body.

## Matching

Match each component of negative feedback systems of multicellular organisms with its letter on the diagram; then select an example (a–e) of each component and enter it in the parentheses. [p.294]

16. _____ effector ( )
17. _____ integrator ( )
18. _____ receptor ( )
19. _____ response ( )
20. _____ stimulus ( )

a. Brain
b. Overexertion in the heat
c. Pituitary gland
d. Slowing down of body activities
e. Temperature sensors in the skin

## Short Answer

21. What mechanisms do plants use to fight infection? [p.295] _____

## Fill in the Blanks

The (22) _____ _____ _____ [p.296] has become established along sandy shores in the northwestern Pacific coast of the United States. It manages to meet the challenges of water conservation by producing fine (23) _____ _____ [p.296] which trap moisture and by having leaves that can (24) _____ [p.296] along their length to reduce (25) _____

_____ [p.296] caused by the wind. This strategy also is employed in response to
(26) _____ [p.296] when the folded leaves reflect the (27) _____ _____
[p.296] from their surface. Many other plants also show leaf folding, especially at (28) _____
[p.296], an example of (29) _____ [p.296] rhythm.

## 17.5. HOW CELLS RECEIVE AND RESPOND TO SIGNALS [pp.297–298]

*Selected Words:* plasmodesmata [p.297], gap junctions [p.297], signal reception [p.297], functional response [p.297], signal transduction [p.297], *bcl-2* [p.298]

*Boldfaced Term*

[p.297] apoptosis _____

_____

*Fill in the Blanks*

Many cells receive and respond to signals. The main pathway begins with the activation of a(n)
(1) _____ _____ [p.297], often by the reversible binding of a signaling molecule. This
is followed by (2) _____ [p.297] of the signal into a molecular form that can operate in the cell.
Finally, a(n) (3) _____ [p.297] response is made. (4) _____ [p.297] or cell suicide
follows this pathway. This process was seen in the development of the fingers of your hand. Cells that will
undergo this process stockpile enzymes such as (5) _____ [p.297] which can break apart the
cytoskeletal elements of a cell. (6) _____ [p.297] cells then engulf the remnants. In some
(7) _____ [p.298], cells that were programmed to die do not, perhaps because they lose normal
(8) _____ [p.298] or perhaps because they do not receive the proper (9) _____ [p.298].

## Self-Quiz

___ 1. Over this temperature (in degrees Fahrenheit), cooling mechanisms fail. [p.289]
   a. 97
   b. 100
   c. 102
   d. 105

___ 2. A community of cells and intercellular substances that interact in one or more tasks is a/an _____. [p.290]
   a. organ
   b. organ system
   c. tissue
   d. none of the preceding

___ 3. The major challenge facing land plants is conservation of _____. [p.290]
   a. carbon dioxide
   b. energy
   c. oxygen
   d. water

___ 4. Cells and organs work best if the surface is _____ the volume. [p.292]
   a. larger than
   b. smaller than
   c. the same size as

Plants and Animals: Common Challenges   171

_____ 5. To maintain solute–water balance plants and animals use _____. [p.293]
   a. active transport
   b. diffusion
   c. both a. and b.
   d. neither a. nor b.

_____ 6. A habitat includes _____ components. [p.293]
   a. biotic
   b. physical and chemical
   c. both a. and b.
   d. neither a. nor b.

_____ 7. The extracellular fluid of an average human is about _____. [p.293]
   a. 500mL
   b. 5L
   c. 15L
   d. 50L

_____ 8. In the feedback system, the brain usually acts as a/an _____. [pp.294–295]
   a. effector
   b. integrator
   c. receptor
   d. stimulator

_____ 9. Plants can protect themselves by _____. [p.295]
   a. compartmentalization
   b. immune response
   c. both a. and b.
   d. neither a. nor b.

_____ 10. Circadian rhythms may be controlled by light-sensitive pigments called _____. [p.296]
   a. anthocyanins
   b. carotenoids
   c. chlorophylls
   d. phytochromes

## Chapter Objectives/Review Questions

1. What are some of the consequences of heat stroke? [p.289]
2. What is meant by division of labor? [p.290]
3. What is the difference between growth and development? [p.290]
4. What is meant by the internal environment of an organism? [p.291]
5. What is homeostasis? Why is it important? Give examples. [p.291]
6. Compare and contrast vascular tissues of plants and animals. [p.292]
7. Compare and contrast negative and positive feedback mechanisms. [pp.294–295]
8. What is meant by compartmentalization in plants? [p.295]
9. What is apoptosis? Why is it important to organisms? [pp.297–298]

## Integrating and Applying Key Concepts

The text states "The structure of a given body part almost always has something to do with a present or past function." What are some examples of parts that show present function and parts that show past function?

# 18

# PLANT FORM AND FUNCTION

## INTRODUCTION

This chapter looks at the anatomy of plants, covering both plant tissues and organs. The chapter also describes the functions of the various plant parts.

## FOCAL POINTS

- Figure 18.1 [p.301] diagrams plant structure.
- Figures 18.2 and 18.3 [p.302] show plant tissues.
- Figure 18.4 [p.303] compares/contrasts monocots and eudicots.
- Figure 18.5 [p.304] looks at stems.
- Figure 18.6 [p.305] diagrams the internal structure of leaves.
- Figure 18.10 [p.307] shows the internal structure of roots.
- Figure 18.11 [p.308] describes secondary growth.
- Table 18.2 [p.310] looks at plant nutrients.
- Figures 18.19 and 18.20 [pp.314, 315] look at water transport through xylem.
- Figure 18.26 [p.318] diagrams translocation in phloem.

---

## Interactive Exercises

*Drought Versus Civilization* [p.300]

### 18.1. OVERVIEW OF THE PLANT BODY [pp.301–303]

**Selected Words:** *droughts* [p.300], *ground tissue* system [p.301], *vascular tissue* system [p.301], *dermal tissue* system [p.301], *apical meristems* [p.301], *primary growth* [p.301], *secondary growth* [p.301], *fibers* [p.302], *sclereids* [p.302], *vessel members* [p.302], *tracheids* [p.302], *sieve-tube members* [p.303], *companion cells* [p.303], *guard cells* [p.303], *periderm* [p.303], *eudicots* [p.303], *monocots* [p.303]

## Boldfaced Terms

[p.301] shoots _____

[p.301] roots _____

[p.301] lateral meristems _____

[p.301] parenchyma _____

[p.302] collenchyma _____

[p.302] sclerenchyma _____

[p.302] xylem _____

[p.303] phloem _____

[p.303] epidermis _____

[p.303] cuticle _____

[p.303] stoma (plural, stomata) _____

[p.303] cotyledons _____

## Short Answer

1. List the ways that drought can affect plants. [p.300]

## Identification

Match the terms in exercises 2–6 with a letter on the diagram. Then complete the exercise by noting the letter of each term's description in the parentheses. [p.301]

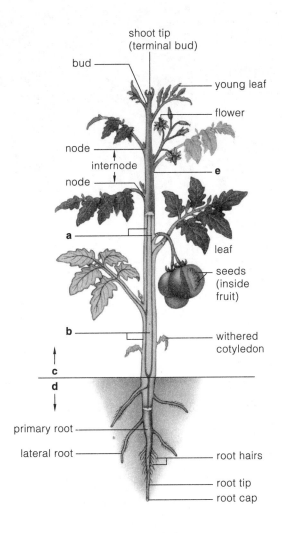

2. _____ ground tissue ( )
3. _____ vascular tissue ( )
4. _____ dermal tissue (epidermis) ( )
5. _____ shoot system ( )
6. _____ root system ( )

a. Aboveground, includes stems, leaves, and flowers
b. Typically grows below ground; anchors aboveground parts; absorbs soil, water, and minerals; stores and releases food; and anchors the aboveground parts
c. Serves basic functions, such as food and water storage
d. Covers and protects plant surfaces
e. Has pipelines that distribute water and solutes through the plant

On these micrographs, identify the tissue type in questions 7–12 and the cell type in questions 13–18. [p.302]

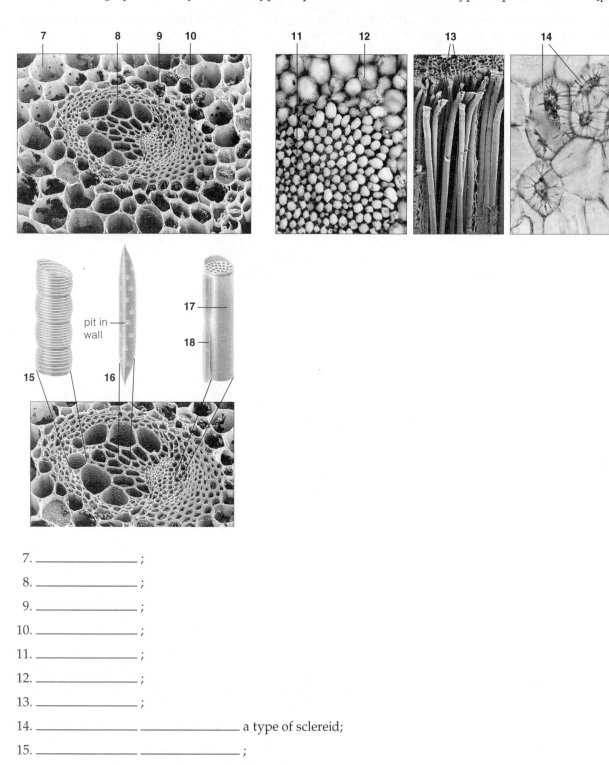

7. _____ ;
8. _____ ;
9. _____ ;
10. _____ ;
11. _____ ;
12. _____ ;
13. _____ ;
14. _____ _____ a type of sclereid;
15. _____ _____ ;
16. _____ ;
17. _____ _____ ;
18. _____ _____ .

176   Chapter Eighteen

*Complete the Table*

19. There are two classes of flowering plants, monocots and dicots. Complete the table below to summarize information about the two groups of flowering plants. [p.303]

| Class | Number of Cotyledons | Floral Parts in Multiples of | Leaf Venation | Pollen Grains | Vascular Bundles |
|---|---|---|---|---|---|
| a. Monocots | | | | | |
| b. Dicots | | | | | |

## 18.2. PRIMARY STRUCTURE OF SHOOTS [pp.304–306]

**Selected Words:** *cortex* [p.304], *pith* [p.304], *palisade* mesophyll [p.305], *spongy* mesophyll [p.305], meristem tissue [p.306], *terminal buds* [p.306], *lateral buds* [p.306], *apical meristem* [p.306]

*Boldfaced Terms*

[p.304] vascular bundles _____

_____

[p.305] leaf _____

_____

[p.305] veins _____

_____

[p.306] buds _____

_____

*Labeling*

Label each of the stem cross sections as either *eudicot* or *monocot*. [p.304]

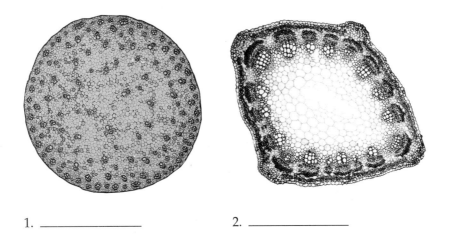

1. _____    2. _____

Plant Form and Function 177

Label each numbered structure in this illustration of leaf organization. [p.305]

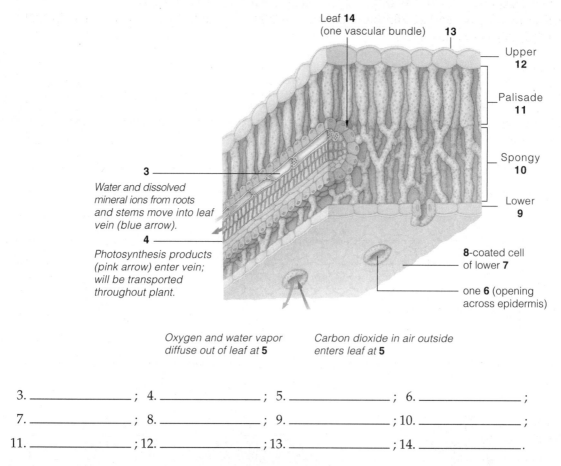

3. _____ ; 4. _____ ; 5. _____ ; 6. _____ ;
7. _____ ; 8. _____ ; 9. _____ ; 10. _____ ;
11. _____ ; 12. _____ ; 13. _____ ; 14. _____ .

Label each numbered structure in this illustration of a eudicot stem at successive stages of primary growth. [p.306]

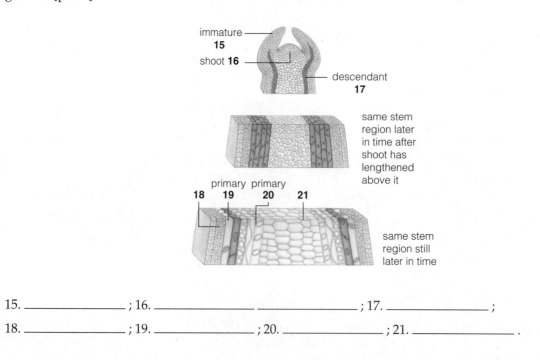

15. _____ ; 16. _____ _____ ; 17. _____ ;
18. _____ ; 19. _____ ; 20. _____ ; 21. _____ .

178   Chapter Eighteen

### 18.3. PRIMARY STRUCTURE OF ROOTS [pp.306–307]

**Selected Words:** *adventitious roots* [p.306], *apical meristem* [p.307], *root cap* [p.307], *primary xylem* [p.307], *primary phloem* [p.307], *pericycle* [p.307]

*Boldfaced Terms*

[p.306] taproot system _____

_____

[p.306] fibrous root system _____

_____

[p.307] vascular cylinder _____

_____

[p.307] root hairs _____

_____

[p.307] endodermis _____

_____

*Labeling*

Label the numbered parts of these illustrations. (*Note*: Some numbers are used more than once; they refer to the same structure.)

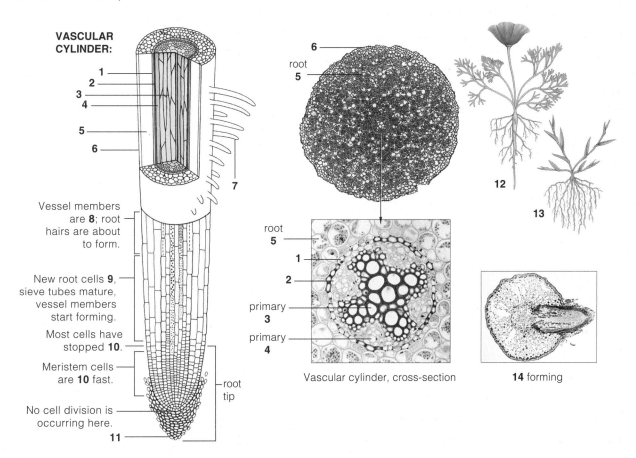

1. _____ [p.307]; 2. _____ [p.307]; 3. _____ [p.307]; 4. _____ [p.307]; 5. _____ [p.307]; 6. _____ [p.307]; 7. _____ [p.307]; 8. _____ [p.307]; 9. _____ [p.307]; 10. _____ [p.307]; 11. _____ _____ [p.307]; 12. _____ system [p.306]; 13. _____ root system [p.306]; 14. _____ _____ [p.307].

## 18.4. SECONDARY GROWTH—THE WOODY PLANTS [pp.308–309]

***Selected Words:*** "nonwoody," or herbaceous [p.308], biennials [p.308], perennials [p.308], woody plants [p.308], *inner* surface of the vascular cambium [p.308], *outer* surface [p.308], *periderm* [p.309], *early* wood [p.309], *late* wood [p.309], "tree rings" [p.309]

### *Boldfaced Terms*

[p.308] vascular cambium _____
_____

[p.308] cork cambium _____
_____

[p.308] wood _____
_____

[p.309] bark _____
_____

[p.309] cork _____
_____

[p.309] heartwood _____
_____

[p.309] sapwood _____
_____

[p.309] sap _____
_____

[p.309] hardwoods _____
_____

[p.309] softwoods _____
_____

## Labeling

Label each part of these illustrations that show woody stems as well as primary and secondary growth. [p.309]

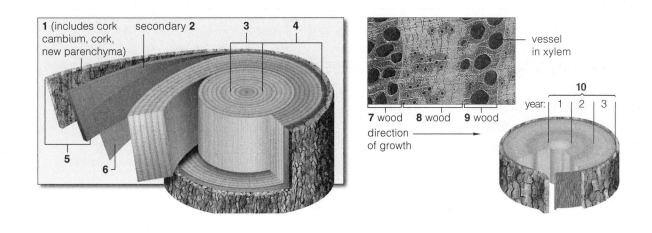

1. _____ ; 2. _____ ; 3. _____ ; 4. _____ ; 5. _____ ;
6. _____ ; 7. _____ ; 8. _____ ; 9. _____ ;
10. _____ .

## 18.5. PLANT NUTRIENTS AND AVAILABILITY IN SOIL [pp.310–311]

**Selected Words:** *macro*nutrients [p.310], *micro*nutrients [p.310], horizons [p.311]

### Boldfaced Terms

[p.310] nutrient _____

[p.310] soil _____

[p.310] humus _____

[p.310] loams _____

[p.311] topsoil _____

[p.311] leaching _____

[p.311] soil erosion _____

*Matching*

Match each term with its definition.

1. _____ soil [p.310]
2. _____ humus [p.310]
3. _____ loams [p.310]
4. _____ horizon [p.311]
5. _____ topsoil [p.311]
6. _____ nutrients [p.310]
7. _____ macronutrients [p.310]
8. _____ micronutrients [p.310]
9. _____ leaching [p.311]
10. _____ erosion [p.311]

a. Elements essential in an organism's life, because no other element can indirectly or directly fulfill its metabolic role
b. Layers of soils, which are in different stages of development in different places—for example, topsoil
c. The decomposing organic material in soil
d. The movement of land under the force of wind, running water, and ice
e. Elements other than the macronutrients that are essential for plant growth; present in traces of the plant's dry weight
f. Consists of particles of minerals mixed with variable amounts of decomposing organic material
g. Refers to the removal of some of the nutrients in soil as water percolates through it
h. Uppermost part of the soil that is highly variable in depth in different habitats (the A horizon); the best soils
i. Soils having more or less equal proportions of sand, silt, and clay; best for plant growth
j. Nine of the essential elements required for plant growth; required in amounts above 0.5 percent of the plant's dry weight

## 18.6. HOW DO ROOTS ABSORB WATER AND MINERAL IONS? [pp.312–313]

**Selected Words:** "fungus roots," or mycorrhizae (singular, mycorrhiza) [p.312], hyphae [p.312], nitrogen fixation [p.312]

### Boldfaced Terms

[p.312] root hairs _____

[p.312] root nodules _____

[p.313] Casparian strip _____

[p.313] exodermis _____

### Dichotomous Choice

Circle the words in parentheses that make each statement correct. [p.312]

1. Mycorrhizae and root nodules are two cases of (mutualism/parasitism).
2. (Gaseous nitrogen/Nitrogen "fixed" by bacteria) represents the chemical form of nitrogen that plants can use in their metabolism.
3. Nitrogen-fixing bacteria reside in localized swellings on legume plants known as (root hairs/root nodules).

4. Mycorrhizae represent symbiotic relationships in which fungi and the roots they cover both benefit; the root cells receive (sugars and nitrogen-containing compounds/scarce minerals that the fungus is better able to absorb).

5. In a mycorrhizal interaction between a young root and a fungus, the fungus benefits by obtaining (sugars and nitrogen-containing compounds/scarce minerals that the fungus is better able to absorb from root cells).

*Sequence*

Arrange these structures in chronological sequence to trace the path that water and nutrients take from the soil to cells in plant tissues. [p.313]

6. _____ a. Vascular cylinder (xylem and phloem)
7. _____ b. Cortex cells lacking Casparian strips
8. _____ c. Endodermis cells with Casparian strips
9. _____ d. Root epidermis

## 18.7. WATER TRANSPORT THROUGH PLANTS [pp.314–315]

*Selected Words:* tension [p.314], *cohesion* [p.314]

*Boldfaced Terms*

[p.314] transpiration _____

_____

[p.314] tracheids _____

_____

[p.314] vessel members _____

_____

[p.314] cohesion–tension theory _____

_____

*Fill in the Blanks*

Plants lose much of their water through the (1) _____ [p.314] of leaves. Evaporation of water molecules from leaves, stems, and other plant parts is a process called (2) _____ [p.314]. Inside vascular tissues, water moves through a complex tissue called (3) _____ [p.314]; its water-conducting cells are known as (4) _____ [p.314] and (5) _____ [p.314] members.

Botanist Henry Dixon came up with his (6) _____–_____ [p.314] theory to explain water transport in plants. The main points to the theory are, first, air's drying power causes (7) _____ [p.314]. This places water molecules that are confined in xylem's waterproofed conducting tubes in a state of (8) _____ [p.314]; that (8) extends from veins inside leaves, down through stems, and on into young roots where the water is being [9] _____ [p.314]. Second, the unbroken, fluid columns of water show (10) _____ [p.314]; they resist rupturing while they are being pulled up under (11) _____ [p.314]. Third, for as long as molecules of water continue to escape from a plant, the continuous (11) inside the (12) _____ [p.315] permits more molecules to be pulled upward from the roots, and therefore to replace them. (13) _____ [p.315] bonds are strong enough to hold water molecules together inside the water-conducting tubes of xylem. However, they are not strong enough to prevent the (14) _____ [p.315] molecules from breaking away from one another during transpiration and then escaping from leaves, through stomata.

### 18.8. HOW DO STEMS AND LEAVES CONSERVE WATER? [pp.316–317]

*Selected Words:* turgor [p.316], cuticle [p.316], *guard cells* [p.316], C4 pathway [p.317]

*Boldfaced Term*

[p.317] CAM plants _____

_____

*Dichotomous Choice*

Circle the words in parentheses that make each statement correct.

1. When plant cells lose water, the plant (wilts/becomes erect). [p.316]
2. Plant epidermal cells secrete a translucent, water-impermeable layer, the (cutin/cuticle), which coats cells exposed to the air. [p.316]
3. Stomata (open/close) mainly in response to water loss. [p.316]
4. When a pair of guard cells swells with turgor pressure, the opening between them (opens/closes). [p.316]
5. In most plants, stomata remain (closed/open) during the daylight photosynthetic period; water is lost from plants, but leaves accumulate carbon dioxide for aerobic respiration. [p.316]
6. Stomata stay (open/closed) at night in most plants. [p.316]
7. Guard cells are the only epidermal cells that (do/do not) have chloroplasts. [p.316]
8. Photosynthesis starts after the sun comes up. As the morning progresses, carbon dioxide levels (increase/decline) in photosynthetic cells, including the guard cells. [p.317]
9. CAM plants such as cacti and other succulents open stomata during the (day/night) when they fix carbon dioxide by way of a special C4 metabolic pathway. [pp.316–317]

### 18.9. HOW ORGANIC COMPOUNDS MOVE THROUGH PLANTS [pp.317–318]

*Selected Words:* source [p.318], sink [p.318]

*Boldfaced Terms*

[p.317] sieve tubes _____

_____

[p.317] companion cells _____

[p.317] translocation _____

[p.318] pressure flow theory _____

## Matching

Match each term with its description.

1. ____ translocation [p.317]
2. ____ sieve tubes [p.317]
3. ____ companion cells [p.317]
4. ____ source [p.318]
5. ____ sink [p.318]
6. ____ pressure flow theory [p.318]

a. Any plant region where organic compounds are being loaded into the sieve tubes
b. Help load organic compounds into sieve tubes
c. Any plant region where organic compounds are being unloaded, used, or stored
d. Transport of sucrose and all other organic compounds through the phloem of a vascular plant (under high pressure)
e. Internal pressure builds up at the source end of the sieve-tube system and pushes the solute-rich solution on toward any sink, where solutes are being removed.
f. Living cells in phloem through which organic compounds rapidly flow

# Self-Quiz

____ 1. Which of these is *not* considered a type of simple tissue? [p.302]
   a. xylem
   b. parenchyma
   c. collenchyma
   d. sclerenchyma
   e. a. and d.

____ 2. Which of these cell types does *not* appear in vascular tissues? [p.302]
   a. vessel members
   b. cork cells
   c. tracheids
   d. sieve-tube members
   e. companion cells

____ 3. Which of these is *not* characteristic of monocots? [p.303]
   a. scattered vascular bundles in the stem
   b. one cotyledon on the embryo
   c. vascular bundles in a ring in the stem
   d. parallel leaf venation
   e. floral parts in threes, or multiples thereof

____ 4. New stems and leaves form on plants by the activity of _____ . [p.306]
   a. internodes
   b. apical meristems
   c. lateral meristems
   d. epidermal cells
   e. cotyledons of the embryo

____ 5. A primary root and its lateral branchings represent a(n) _____ system. [p.306]
   a. lateral root
   b. adventitious root
   c. taproot
   d. branch root
   e. fibrous root

____ 6. Plant growth depends on _____ essential elements. [p.310]
   a. nine
   b. ten
   c. twenty-one
   d. nineteen
   e. sixteen

_____ 7. Gaseous nitrogen is converted to a plant-usable form by _____ . [p.312]
   a. root nodules
   b. mycorrhizae
   c. nitrogen-fixing bacteria
   d. Venus flytraps
   e. fungus-roots

_____ 8. _____ forces inward-moving water through the cytoplasm of endodermal cells. [p.313]
   a. Cytoplasm
   b. Plasma membranes
   c. Osmosis
   d. Casparian strips
   e. Diffusion

_____ 9. Water movement from the soil to the top of a tree is explained by _____ . [p.314]
   a. osmotic gradients being established
   b. transpiration alone
   c. pressure flow forces
   d. cohesion-tension theory
   e. cuticle layers impermeable to water

_____ 10. Stomata remain _____ during daylight, when photosynthesis occurs, but remain _____ during the night when carbon dioxide accumulates through aerobic respiration. [p.316]
   a. open; open
   b. closed; open
   c. closed; closed
   d. open; closed

_____ 11. Mature leaves represent _____ regions; growing leaves, stems, fruits, seeds, and roots usually represent _____ regions. [p.318]
   a. source; source
   b. sink; source
   c. source; sink
   d. sink; sink

---

## Chapter Objectives/Review Questions

1. Be able to list ways that a drought can affect civilization. [p.300]
2. Distinguish between the ground tissue system, the vascular tissue system, and the dermal tissue system. [p.301]
3. Lengthening of stems and roots originates at _____ meristems and all dividing tissues derived from them; this is called _____ growth. [p.301]
4. Increases in the girth of a plant start at _____ meristems. [p.301]
5. Be able to visually identify and generally describe the simple tissues and cells called parenchyma, collenchyma, and sclerenchyma. [pp.301–302]
6. _____ tissue conducts soil water and dissolved minerals, and it structurally supports the plant. [p.302]
7. _____ tissue transports sugars and other solutes. [p.303]
8. Name and describe the functions of the conducting cells in xylem and phloem. [p.302]
9. What is the general function of guard cells and stomata found within the epidermis of young stems and leaves? [p.303]
10. Distinguish between monocots and eudicots by listing their characteristics and citing examples of each group. [p.303]
11. Be able to visually distinguish between monocot stems and eudicot stems, as seen in cross section; identify the cells present by name and function. [p.304]
12. Describe the structure (cells and layers) and major functions of leaf epidermis, mesophyll, and vein tissue. [pp.305–306]
13. Distinguish a taproot system from a fibrous root system; define *adventitious root*. [p.306]
14. Describe the origin and function of root hairs. [p.307]
15. Tell how the formation of cork and bark occurs by cork cambium activity. [pp.308–309]

16. Distinguish early wood from late wood; heartwood from sapwood. [p.309]
17. Explain the origin of the annual growth rings (tree rings) seen in a cross section of a tree trunk. [p.309]
18. Define the terms *nutrient, soil,* and *humus*. [p.310]
19. Distinguish between macronutrients and micronutrients in relation to their role in plant nutrition. [p.310]
20. Define *leaching* and *erosion*. [p.311]
21. Explain why root hairs are so valuable in root absorption. [p.312]
22. Describe the mutualistic roles of root nodules and mycorrhizae in plant nutrition. [pp.312–313]
23. Trace (describe) the path of water and mineral ions into roots; name the structures and cells involved. [p.313]
24. The evaporation of water molecules from leaves as well as from stems and other plant parts is known as _____ . [p.314]
25. List and describe the main points of Henry Dixon's cohesion–tension theory. [pp.314–315]
26. Describe the means that plant leaves and stems have to conserve water; include turgor pressure, cuticle, guard cells, and CAM plants. [pp.316–317]
27. Tell how organic compounds are distributed in plants; include sieve tubes, companion cells, translocation, source, sink, and pressure flow theory. [pp.317–318]

## Integrating and Applying Key Concepts

How do you think maple syrup is made from maple trees? Which specific systems of the plant are involved, and why are maple trees tapped only at certain times of the year?

# 19

# PLANT REPRODUCTION AND DEVELOPMENT

## INTRODUCTION

This chapter looks at pollination, fertilization, and seed and fruit formation. It also covers growth and development of flowering plants.

## FOCAL POINTS

- Figure 19.1 [p.322] diagrams both flower structure and the generalized life cycle of flowering plants.
- Figure 19.4 [p.324] details the life cycle of flowering plants.
- Figure 19.5 [p.325] illustrates the development of seeds.
- Figure 19.8 [p.328] looks at seed structure.
- Figures 19.9 and 19.10 [p.329] diagram the growth and development of monocots and eudicots, respectively.
- Figure 19.12 [p.331] shows the effect of auxins on plant growth.

## Interactive Exercises

*Imperiled Sexual Partners* [p.321]

### 19.1. SEXUAL REPRODUCTION IN FLOWERING PLANTS [pp.322–325]

*Selected Words:* receptacle [p.322], sepals [p.322], petals [p.322], pollen [p.322], stigma [p.322], embryo sac [p.324], *double* fertilization [p.325]

## Boldfaced Terms

[p.322] flowers _____

[p.322] sporophyte _____

[p.322] gametophytes _____

[p.322] stamens _____

[p.322] anther _____

[p.322] carpels _____

[p.322] germinate _____

[p.322] ovary _____

[p.322] ovules _____

[p.323] pollinators _____

[p.324] microspores _____

[p.324] megaspores _____

[p.325] endosperm _____

[p.325] pollination _____

## Identification

Identify each numbered part on this diagram of the flowering plant life cycle. Then enter the letter of each term's description in the parentheses. [p.322]

1. _____ ( )
2. _____ ( )
3. _____ ( )
4. _____ ( )
5. _____ ( )
6. _____ ( )

a. An event that produces a young sporophyte
b. A floral shoot produced by the sporophyte
c. The "plant"; a vegetative body that develops from a fertilized egg
d. Cellular division event occurring within flowers to produce spores
e. Produces haploid eggs by mitosis
f. Produces haploid sperm by mitosis

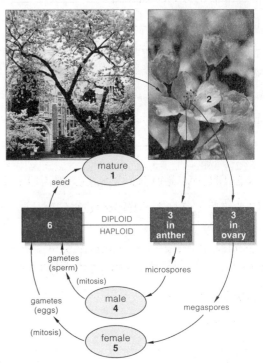

Identify each numbered part of this diagram of one type of a flower's structure. [p.322]

7. _____
8. _____
9. _____
10. _____
11. _____
12. _____
13. _____
14. _____
15. _____
16. _____
17. _____

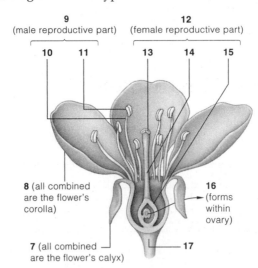

## Fill in the Blanks

Animals that transport pollen to the female part of a flower are called (18) _____ [p.323]. Flowers use many tactics to attract (18). Flowers pollinated by (19) _____ [p.323] are often red and offer

190 Chapter Nineteen

(20) _____ [p.323] as a reward. Flowers that produce the odor of decay attract (21) _____ [p.323]. Bats are attracted by intensely (22) _____ _____ [p.323] and pale-colored (23) _____ [p.323]. Bees are also attracted by (22) and by floral pigments that reflect (24) _____ _____ [p.323].

When an animal deposits pollen on the (25) _____ [p.325], the pollen germinates to form the (26) _____ _____ [p.325]. Inside the (26) are (27) _____ [p.323] sperm nuclei. One fertilizes the egg to form the (28) _____ _____ [p.325] while the other fuses with two haploid nuclei to form the (29) _____ _____ [p.325]. This process is called (30) _____ [p.325] fertilization.

## 19.2. FROM ZYGOTES TO SEEDS PACKAGED IN FRUIT [pp.325–326]
## 19.3. ASEXUAL REPRODUCTION OF FLOWERING PLANTS [p.327]

***Selected Words:*** *simple* fruits [p.325], *aggregate* fruits [p.325], *multiple* fruits [p.325], *accessory* fruits [p.325], *seed dispersal* [p.326], runners [p.327], clone [p.327]

### Boldfaced Terms

[p.325] cotyledons _____

_____

[p.325] seed _____

_____

[p.327] tissue culture propagation _____

_____

### Complete the Table

1. Complete the table by summarizing concepts associated with seeds and fruits. [pp.325–326]

*Structure* | *Origin*
--- | ---
a. Cotyledon(s) | 
b. Seeds | 
c. Seed coat | 
d. Fruits |

*Choice*

For questions 2–6, choose from these fruit types.

    a. simple dry fruits    b. simple fleshy fruits    c. aggregate fruits
    d. multiple fruits    e. accessory fruits

2. \_\_\_\_ pineapple [p.325]
3. \_\_\_\_ *Capsella*, acorns, pea pods [p.325]
4. \_\_\_\_ mountain ash, cacao [p.326]
5. \_\_\_\_ raspberries [p.325]
6. \_\_\_\_ apple, strawberry [pp.325–326]

For questions 7–15, choose from the following: [p.326]

    a. wind-dispersed fruits    b. fruits dispersed by animals    c. water-dispersed fruits

7. \_\_\_\_ heavy wax coats and air sacs
8. \_\_\_\_ coconut palms
9. \_\_\_\_ seed coats assaulted by digestive enzymes to assist in releasing embryos
10. \_\_\_\_ maples
11. \_\_\_\_ oranges, corn
12. \_\_\_\_ hooks, spines, hairs, and sticky surfaces
13. \_\_\_\_ cacao
14. \_\_\_\_ dandelion
15. \_\_\_\_ cockleburs, bur clover, and bedstraw

*Fill in the Blanks*

Quaking aspen produces shoots that arise from the parent plant's (16) _____ [p.327] system, and each becomes a shoot system. Barring rare mutations, all aspens in the area are (17) _____ [p.327] identical. A forest of aspen represents an example of vegetative reproduction called a (18) _____ [p.327]. Strawberry plants send out horizontal, aboveground stems called (19) _____ [p.327] with (20) _____ [p.327] from which new roots and shoots develop. African violets are propagated via leaf (21) _____ [p.327] from which adventitious roots develop. Culturing plants from bits of tissue or even from single cells is known as (22) _____ [p.327] propagation. This technique often yields thousands to millions of plants from (23) _____ [p.327] specimen. The techniques are being used to improve major (24) _____ [p.327] crops and also to increase the (25) _____ [p.327] of hybrid orchids, lilies, and other prized ornamental plants.

# 19.4. PATTERNS OF EARLY GROWTH AND DEVELOPMENT [pp.328–329]

*Selected Words:* germination [p.328], coleoptile [p.329], hypocotyl [p.329]

## Fill in the Blanks

Germination depends on (1) _____ [p.328] factors, such as the number of daylight hours and the temperature, (2) _____ [p.328], and oxygen level in soil. Water molecules move into a (3) _____ [p.328]. After the seed swells with enough water, its coat (4) _____ [p.328]. Once the seed coat splits, more (5) _____ [p.328] reaches the embryo, and (6) _____ [p.328] respiration moves into high gear. The embryo's (7) _____ [p.328] cells now divide rapidly. (8) _____ [p.328] is over when the (9) _____ [p.328] root breaks out of the seed coat.

## Labeling

Label the numbered parts of these drawings of early development of corn and beans. [p.329]

10. _____
11. _____
12. _____
13. _____
14. _____
15. _____
16. _____
17. _____
18. _____
19. _____
20. _____
21. _____
22. _____
23. _____
24. _____
25. _____
26. _____
27. _____
28. _____

## 19.5. CELL COMMUNICATION IN PLANT DEVELOPMENT [pp.330–331]
## 19.6. ADJUSTING RATES AND DIRECTIONS OF GROWTH [pp.332–333]

*Selected Words:* IAA (indoleacetic acid) [p.331], 2,4-D [p.331], 2,4,5-T [p.331], *Agent Orange* [p.331], dioxin [p.331], *statoliths* [p.332], *flavoproteins* [p.332]

*Boldfaced Terms*

[p.330] gibberellins _____

_____

[p.330] auxins _____

_____

[p.331] abscission _____

_____

[p.331] cytokinins _____

_____

[p.331] ethylene _____

_____

[p.331] abscisic acid (ABA) _____

_____

[p.332] gravitropism _____

_____

[p.332] phototropism _____

_____

*Choice*

For questions 1–7, choose from the following:

    a. auxins    b. gibberellins    c. cytokinins    d. abscisic acid    e. ethylene

1. _____ Make stems lengthen and help seeds germinate by stimulating cell division and elongation; also stimulate flowering of biennials and long-day plants [p.330]

2. _____ Stimulate rapid cell division in root and shoot meristems and in maturing fruits; keep leaves from aging too soon [p.331]

3. _____ Makes stomata close in water-stressed plants; induces and maintains dormancy in certain buds and seeds; stimulates the transport of photosynthetic products to seeds, stimulates protein synthesis and embryo formation in seeds [p.331]

4. _____ IAA, the most pervasive of its type in nature; used to thin flowers in spring so trees yield fewer but larger fruits; keeps fruit from dropping prematurely [p.331]

5. _____ Form in the apical meristems of shoots and coleoptiles and diffuse downward; stimulate shoots and coleoptiles into lengthening; induces cell division and in seeds, stimulates fruit formation; maintains apical dominance [p.330]

6. _____ Synthetic 2,4-D, widely used herbicide; 2,4-D mixed with 2,4,5-T produces Agent Orange [p.331]

7. _____ The only gaseous plant hormone; controls growth of most tissues, fruit ripening, leaf drop, and other aging responses [p.331]

For questions 8–13, choose from the following: [p.332]

a. phototropism    b. gravitropism

8. _____ Roots tend to grow down and young stems tend to grow up.
9. _____ Tomato seedlings start curving in the direction of the most light.
10. _____ Stems and leaves adjust the rate and direction of growth in response to light.
11. _____ Blue wavelengths stimulate the strongest response.
12. _____ Involves proteins that transport auxin across membranes.
13. _____ Lets an upside-down seedling grow in a room with no light and its roots curve down.

## 19.7. MEANWHILE, BACK AT THE FLOWER . . . [p.334]
## 19.8. LIFE CYCLES END, AND TURN AGAIN [p.335]
## 19.9. REGARDING THE WORLD'S MOST NUTRITIOUS PLANT [p.336]

*Selected Words:* *long-day* plants [p.334], *short-day* plants [p.334], *day-neutral* plants [p.334], *night length* [p.334], phytochrome [p.334], circadian rhythms [p.334], abscission zone [p.335], quinoa [p.336], *kwashiorkor* [p.336]

### Boldfaced Terms

[p.334] photoperiodism _____

_____

[p.335] senescence _____

_____

[p.335] dormancy _____

_____

[p.335] vernalization _____

_____

### Matching

Match each item with its description. [p.334]

1. _____ photoperiodism
2. _____ day-neutral plants
3. _____ phytochrome activation
4. _____ "long-day" plants
5. _____ circadian rhythms
6. _____ far-red wavelengths
7. _____ biological clocks
8. _____ phytochrome

a. Biological activities that recur in cycles of twenty-four hours or so
b. Flower in spring when daylength exceeds some critical value
c. Internal mechanisms in plants that preset the time for recurring changes in some aspect of their biochemistry
d. A biological response to a change in the relative length of daylight and darkness in the 24-hour cycle; phytochrome is the alarm
e. A blue-green pigment that is a component of mechanisms that promote or inhibit growth
f. Turn off phytochrome
g. Flower when mature enough to do so
h. Red wavelengths

## Choice

Choose from the following: [p.335]

a. senescence   b. abscission   c. entering dormancy
d. breaking dormancy   e. vernalization

9. _____ The shedding of flowers, leaves, and fruits
10. _____ Cues include plentiful water and milder temperatures
11. _____ Strong cues are short days, long, cold nights, and dry, nitrogen-poor soil [p.468]
12. _____ A zone consisting of thin-walled parenchyma cells at the base of a leaf stalk; cells produce ethylene
13. _____ The sum total of processes leading to the death of a plant or some of its parts [p.468]
14. _____ Low-temperature stimulation of flowering

## Fill in the Blanks

Alejandro Bonifacio has been studying the Andean plant (15) _____ [p.336]. This plant has (16) _____ [p.336] percent protein, more than wheat or rice and, unlike those grains, is not deficient in the amino acid (17) _____ [p.336]. (15) also contains more (18) _____ [p.336] than cereal grains. In addition, this plant is resistant to (19) _____ [p.336], (20) _____ [p.336], and saline soils. Bonifacio hopes that developing better strains of (1) will help alleviate (21) _____ [p.336], a form of malnutrition stemming from (22) _____-_____ [p.336] diets.

---

## Self-Quiz

_____ 1. The portion of the carpel that contains an ovule is the _____. [p.322]
  a. stigma
  b. anther
  c. style
  d. ovary
  e. filament

_____ 2. The phase in the life cycle of plants that gives rise to spores is known as the _____. [p.322]
  a. female gametophyte
  b. embryo
  c. sporophyte
  d. seed
  e. male gametophyte

_____ 3. In the double fertilization of flowering plants, one sperm nucleus fuses with the nucleus of the egg, and a zygote forms that develops into an embryo. The other sperm nucleus enters _____. [p.325]
  a. a cell inside the female gametophyte to produce three cells, each with one nucleus
  b. a cell inside the female gametophyte to produce one cell with two nuclei
  c. a cell inside the female gametophyte to form a cell with a single nucleus containing three chromosome sets
  d. one of the smaller megaspores to produce what will eventually become the seed coat
  e. endosperm cells that have already formed

_____ 4. "Simple, aggregate, multiple, and accessory" refer to types of _____. [p.325]
   a. carpels
   b. seeds
   c. fruits
   d. ovaries
   e. nuclei

_____ 5. An example of asexual reproduction in flowering plants is _____. [p.327]
   a. a clone of quaking aspen plants
   b. an African violet leaf callus
   c. tissue culture propagation
   d. strawberry runners
   e. all of the above

_____ 6. Germination depends on _____. [p.328]
   a. the number of daylight hours
   b. temperature
   c. moisture
   d. soil oxygen level
   e. all of the above

_____ 7. _____ exert(s) control over the growth of most plant tissues, fruit ripening, leaf drop, and other aging responses. [pp.330–331]
   a. Auxin
   b. Cytokinins
   c. Gibberellins
   d. Ethylene
   e. Abscisic acid

_____ 8. Which of the following is *not* associated with gravitropism? [p.332]
   a. Roots curve down on an upside-down seedling in a room without light
   b. Involves proteins that transport auxin across cell membranes
   c. Stems curve up on an upside-down seedling in a room without light
   d. Blue wavelengths of light
   e. Different rates of elongation on the opposing sides of stems and roots are enough to make the part bend

_____ 9. The sum of all the processes that lead to the death of a plant or some of its parts is called _____. [p.335]
   a. dormancy
   b. vernalization
   c. abscission
   d. senescence
   e. circadian rhythms

_____ 10. The Andean plant with up to 16 percent protein is _____. [p.336]
   a. corn
   b. potato
   c. quinoa
   d. rice
   e. wheat

## Chapter Objectives/Review Questions

1. Describe a possible scenario that could result from the destruction of the pollinators of flowering plants. [p.321]
2. Be able to name and describe the function of all the parts of a typical flower and know the general aspects of the flowering plant life cycle very well. [p.322]
3. List the pollinators of flowering plants and describe two examples of their coevolution with flowers. [p.323]
4. Review the life cycle of the eudicot as depicted in Figure 19.4 until you can "tell the story" of these events. [pp.324–325]
5. What are the four main categories of fruit? Be able to cite examples of each category. [pp.325–326]
6. Describe the various mechanisms of fruit and seed dispersal. [p.326]
7. Define *plant clone, runners,* and *tissue culture propagation;* describe an example of each. [p.327]
8. Be able to relate the sequence of events in monocot and eudicot development, as well as describing the structures involved. [pp.328–329]
9. Know the general functions of the major classes of plant hormones. [pp.330–331]

10. Define and cite examples of *gravitropism* and *phototropism*. [p.332]
11. Using your knowledge of photoperiodism, long-day plants, short-day plants, and day-neutral plants, explain why plants flower when they do. [p.334]
12. Define *abscission, senescence, dormancy,* and *vernalization*, and cite physical and chemical changes leading to these conditions. [p.335]

## Integrating and Applying Key Concepts

An oak tree has grown up in the middle of a forest. A lumber company has just cut down all of the surrounding trees except for a narrow strip of woods that includes the oak. How will the oak be likely to respond as it adjusts to its changed environment? To what new stresses will it be exposed? Which hormones will most probably be involved in the adjustment?

# 20

# ANIMAL TISSUES AND ORGAN SYSTEMS

## INTRODUCTION

Chapter 20 describes the various types of animal tissues. It also looks at how tissues interact to form organs and organ systems.

## FOCAL POINTS

- Figure 20.2 [p.341] illustrates junctions in epithelium.
- Figure 20.3 [p.342] shows various connective tissues.
- Figure 20.4 [p.343] illustrates the three types of muscle tissue.
- Figure 20.6a [p.344] diagrams the eleven major human organ systems.
- Figure 20.6b, c, d [p.345] describes anatomic terms.
- Figure 20.7 [p.346] diagrams skin.

## Interactive Exercises

*It's All About Potential* [p.339]

### 20.1. ORGANIZATION AND CONTROL IN ANIMAL BODIES [p.340]

*Selected Words:* undifferentiated cells [p.339], *embryonic stem cells* [p.339], adult stem cells [p.339]

*Boldfaced Terms*

[p.339] stem cells _____

_____

[p.340] tissue _____

_____

[p.340] organ _____

_____

[p.340] organ system _____

_____

[p.340] homeostasis _____

_____

*Short Answer*

1. List the advantages of the use of human embryonic cells for stem cell research as compared with the use of adult stem cells. [p.339]

_____

_____

_____

*Matching*

Match each term with its description. [p.340]

2. _____ internal environment
3. _____ homeostasis
4. _____ tissue
5. _____ organ
6. _____ organ system

a. Consists of two or more organs that are interacting physically, chemically, or both in a common task
b. Consists of blood and interstitial fluid (tissue fluids)
c. Consists of different tissues that are organized in specific proportions and patterns necessary to carry out particular tasks
d. A community of cells and intercellular substances that perform one or more tasks, including contraction by muscle tissue
e. A state in which the composition and volume of the internal environment are within ranges suitable for survival

## 20.2. FOUR BASIC TYPES OF TISSUES [pp.340–345]

**Selected Words:** *simple* epithelium [p.340], *stratified* epithelium [p.340], secretory cells [p.340], *peptic ulcer* [p.341], soft connective tissues [p.341], fibroblasts [p.341], collagen [p.341], tendons [p.342], ligaments [p.342], specialized connective tissues [p.342], striated [p.343], "voluntary" [p.343], "involuntary" [p.343], neuron [p.343], neuroglial cell [p.345]

### Boldfaced Terms

[p.340] epithelial tissue _____

_____

[p.341] glands _____

_____

[p.341] exocrine glands _____

_____

[p.341] endocrine glands _____

_____

[p.341] tight junctions _____

_____

[p.341] adhering junctions _____

_____

[p.341] gap junctions _____

_____

[p.341] loose connective tissue _____

[p.341] dense, irregular connective tissue _____

[p.341] dense, regular connective tissue _____

[p.342] cartilage _____

[p.342] bone tissue _____

[p.342] adipose tissue _____

[p.342] blood _____

[p.343] skeletal muscle tissue _____

[p.343] cardiac muscle tissue _____

[p.343] smooth muscle tissue _____

[p.343] nervous tissue _____

## Fill in the Blanks

(1) _____ [p.340] is a sheetlike tissue with a free surface facing the outside environment or a body fluid. (2) _____ [p.340] epithelium, with just one layer of cells, lines body cavities, ducts, and tubes. (3) _____ [p.340] epithelium has two or more layers and often functions in protection. (4) _____ [p.341] glands secrete mucus, saliva, earwax, milk, oil, digestive enzymes, and other cell products. These products are usually released onto a free surface of (5) _____ [p.341] through ducts or tubes. (6) _____ [p.341] glands have no ducts; their products are (7) _____ [p.341], which are secreted directly into some body fluid. The (8) _____ [p.341] molecules usually enter the bloodstream, which distributes them to specific (9) _____ [p.341] cells elsewhere in the body. (10) _____ [p.341] junctions stop leaks across tissues. (11) _____ [p.341] junctions are like spot welds that lock cells together. (12) _____ [p.341] junctions allow small ions and small molecules to flow between the cytoplasm of abutting cells. (13) _____ [p.341] junctions in the epithelium of your stomach help prevent a condition called (14) _____ [p.341] ulcer.

## Labeling

Identify each sketch of these cell junctions commonly seen in animal tissues. [p.341]

15. _____ junction
16. _____ junction
17. _____ junction

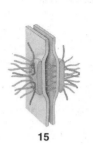

## Labeling

Label the accompanying illustrations as [pp.342–343]

18. ____
19. ____
20. ____
21. ____
22. ____
23. ____
24. ____
25. ____
26. ____
27. ____
28. ____

a. loose connective tissue
b. dense, irregular connective tissue
c. dense, regular connective tissue
d. cartilage
e. bone tissue
f. adipose tissue
g. blood
h. skeletal muscle tissue
i. cardiac muscle tissue
j. smooth muscle tissue
k. motor neuron of the nervous system

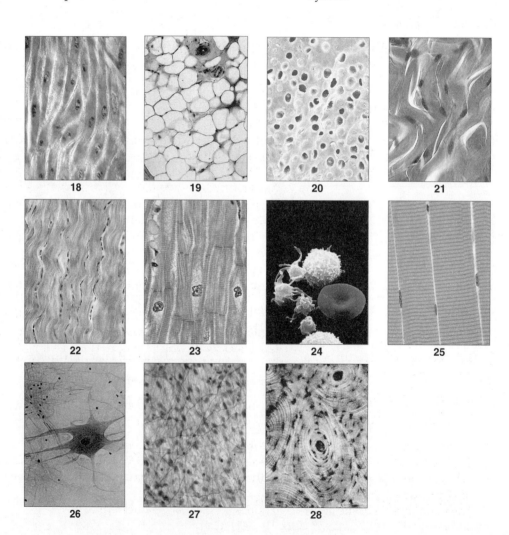

202  Chapter Twenty

*Choice*

Choose from these types of connective tissue:

          a. loose     b. dense, irregular     c. dense, regular

29. \_\_\_\_ Has fibroblasts and many collagen fibers oriented every which way [p.341]
30. \_\_\_\_ Rows of fibroblasts lie between many parallel bundles of fibers. [p.341]
31. \_\_\_\_ Contains fibers and cells loosely arranged in a semifluid ground substance [p.341]
32. \_\_\_\_ Has bundles of collagen fibers that help tendons resist being torn [p.341]
33. \_\_\_\_ Forms tough capsules around organs that don't stretch much [p.342]
34. \_\_\_\_ Often acts as a support framework for epithelium [p.341]
35. \_\_\_\_ Found in tendons, which attach skeletal muscle to bone [p.342]
36. \_\_\_\_ Besides fibroblasts, it contains white blood cells that defend against pathogens. [p.341]
37. \_\_\_\_ Occurs in elastic ligaments, which attach one bone to another [p.342]
38. \_\_\_\_ This tissue occurs in skin. [p.341]

*Complete the Table*

39. Complete the table by supplying the name of the specialized connective tissue described. [pp.341–342]

| Specialized Connective Tissue | Description |
| --- | --- |
| a. | The body's main reservoir of energy; fats move rapidly to and from the cells by way of blood vessels; insulates and cushions certain body parts |
| b. | In all vertebrate embryos, chondrocytes secrete and then become embedded in a solid, pliable intercellular material; resists compression; maintains ear, nose, and other body parts; protects and cushions joints |
| c. | Derived mainly from connective tissue, serves transport functions; circulating within a primarily water plasma are great numbers of proteins, ions, red blood cells, white blood cells, and platelets |
| d. | Mineral hardened; its collagen fibers and ground substance are strengthened with calcium salts; major tissue of bones; the weight-bearing tissue of vertebrate skeletons; interact with skeletal muscles to move the body |

*Dichotomous Choice*

Circle one of two possible answers given in parentheses in each statement.

40. Contractile cells of (skeletal/smooth) muscle tissue are tapered. [p.343]
41. The contractile walls of the heart are composed of (striated/cardiac) muscle tissues. [p.343]
42. Walls of soft internal organs such as the stomach and lungs of vertebrates contain (smooth/skeletal) muscle tissue. [p.343]
43. The main muscle tissue attached to bones is (skeletal/smooth). [p.343]

Animal Tissues and Organ Systems

44. (Smooth/Skeletal) muscle has many transverse stripes that define its repeating contractile units. [p.343]
45. "Involuntary" muscle action is associated with (smooth/skeletal) muscle tissue. [p.343]
46. The term *striated* means (bundled/striped). [p.343]
47. (Smooth/Skeletal) muscle units have orderly arrays of contractile units that interact and bring about contraction. [p.343]
48. The function of smooth muscle tissue is to (pump blood/move internal organs). [p.343]
49. (Smooth/Cardiac) muscle cells contract as a unit in response to signals that flow through gap junctions between plasma membranes. [p.343]
50. (Muscle/Nervous) tissue exerts the greatest control over how the body responds to changing conditions. [p.343]
51. Excitable cells are the (neuroglia/neurons). [p.343]
52. (Neuroglial/Muscle) cells metabolically support and protect neurons. [p.345]
53. When a (neuron/muscle cell) is suitably stimulated, it can propagate an electric disturbance along its plasma membrane to its endings, where the disturbance may trigger the stimulation or inhibition of neighboring neurons and other cells. [p.343]
54. Different types of (neuroglia/neurons) detect, assess, coordinate, and then deliver responses to specific kinds of change in the internal and external environments. [p.343]

## 20.3. ORGAN SYSTEMS MADE FROM TISSUES [p.345]
## 20.4. SKIN—EXAMPLE OF AN ORGAN SYSTEM [pp.346–347]

***Selected Words:*** *germ* cells [p.345], *somatic* [p.345], midsagittal plane [p.345], dorsal [p.345], ventral [p.345], frontal plane [p.345], transverse plane [p.345], anterior [p.345], posterior [p.345], superior [p.345], inferior [p.345], distal [p.345], proximal [p.345], keratinocytes [p.346], melanocytes [p.346], vitamin D [p.346], folate [p.346], *cold sores* [p.347], *Herpes simplex* [p.347]

### Boldfaced Terms

[p.345] ectoderm _____

_____

[p.345] mesoderm _____

_____

[p.345] endoderm _____

_____

[p.346] skin _____

_____

[p.346] melanin _____

_____

[p.346] epidermis _____

_____

[p.346] dermis _____

_____

[p.347] Langerhans cells _____

_____

*Fill in the Blanks*

Refer to text Figure 20.6d. [p.345]

The brain is housed in the (1) _____ cavity. The (2) _____ cavity contains the spinal cord and the beginnings of spinal nerves. The heart and lungs are found within the (3) _____ cavity. The stomach, spleen, liver, gallbladder, pancreas, small intestine, most of the large intestine, the kidneys, and ureters lie inside the (4) _____ cavity. The (5) _____ cavity contains the urinary bladder, sigmoid colon, rectum, and reproductive organs.

*Complete the Table*

6. Enter the name of the primary embryonic tissue that performs the function indicated by becoming specialized in particular ways. [p.345]

| Primary Tissue | Functions |
| --- | --- |
| a. | Gives rise to the body's muscles, bones, and most of the circulatory, reproductive, and urinary systems |
| b. | Gives rise to the lining of the digestive tract and to organs derived from it |
| c. | Gives rise to the skin's outer layer and to tissues of the nervous system |

*Identification*

For questions 7–13, identify each numbered part of this illustration, which reviews the directional terms and planes of symmetry for the human body. For all items, refer to Figure 20.6c. [p.345]

7. _____
8. _____
9. _____
10. _____
11. _____
12. _____
13. _____

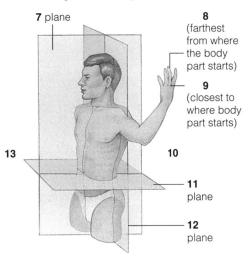

7 plane
8 (farthest from where the body part starts)
9 (closest to where body part starts)
10
11 plane
12 plane
13

For questions 14–24, identify the organ system described. For all items refer to Figure 20.6a. [p.344]

14. _____ Rapidly transports many materials to and from cells; helps stabilize internal pH and temperature

15. _____ Rapidly delivers oxygen to the tissue fluid that bathes all living cells; removes carbon dioxide wastes of cells; helps regulate pH

16. _____ Maintains the volume and composition of internal environment; excretes excess fluid and blood-borne wastes

17. _____ Supports and protects body parts; provides muscle attachment sites; produces red blood cells; stores calcium, phosphorus

18. _____ Hormonally controls body functions; works with nervous system to integrate short-term and long-term activities

19. _____ *Female:* produces eggs; after fertilization, affords a protected, nutritive environment for the development of new individual. *Male:* produces and transfers sperm to the female. Hormones of both systems also influence other organ systems

20. _____ Ingests food and water; mechanically, chemically breaks down food, and absorbs small molecules into internal environment; eliminates food residues

21. _____ Moves body and its internal parts; maintains posture; generates heat (by increases in metabolic activity)

22. _____ Detects both external and internal stimuli; controls and coordinates responses to stimuli; integrates all organ system activities

23. _____ Protects body from injury, dehydration, and some pathogens; controls its temperature; excretes some wastes; receives some external stimuli

24. _____ Collects and returns some tissue fluid to the bloodstream; defends the body against infection and tissue damage

## Matching

Match each term with its description.

25. _____ skin [p.346]
26. _____ melanocytes [p.346]
27. _____ keratinocytes [p.346]
28. _____ melanin [p.346]
29. _____ epidermis [p.346]
30. _____ dermis [p.346]
31. _____ vitamin D [p.346]
32. _____ folate [p.346]
33. _____ suntans [p.347]
34. _____ shoe-leather skin [p.347]
35. _____ skin's front line of defense [p.347]
36. _____ cold sores [p.347]

a. Steroid-like compounds that help the body absorb calcium from food
b. A stratified epithelium with many cell junctions and no extracellular matrix; ongoing mitotic divisions push cells from deeper layers to the free surface
c. Often triggered by sunburns; small, painful blisters announce this recurring *Herpes simplex* infection
d. Phagocytic Langerhans cells migrate through epidermis and engulf virus particles or bacteria; mobilize the immune system
e. Cells that make and give up melanin to keratinocytes
f. A B vitamin whose breakdown is blocked by dark skin that blocks UV radiation
g. A brownish-black pigment; one of the body's barriers to harmful UV radiation from the sun
h. Consists of an outer epidermis and an underlying dermis of dense connective tissue; the body's largest organ; performs many beneficial functions for the body and has exocrine glands, including sweat and mammary glands
i. Mainly a dense connective tissue; has stretch-resisting elastin and supportive collagen fibers; anchored by hypodermis consisting of loose connective and adipose tissues
j. Skin cells that make the water-resistant protein, keratin
k. Produced by sunlight exposure; coveted by many light-skinned people
l. Caused by ongoing UV exposure; elastin fibers in dermis clump; skin loses resiliency

# Self-Quiz

1. Which of the following is *not* included in connective tissues? [pp.341–342]
   a. bone
   b. blood
   c. cartilage
   d. skeletal muscle
   e. adipose tissue

2. Gland cells are contained in _____ tissues. [p.341]
   a. muscular
   b. epithelial
   c. connective
   d. nervous
   e. skeletal

3. Blood is considered to be a(n) _____ tissue. [p.342]
   a. epithelial
   b. muscular
   c. connective
   d. nervous
   e. skeletal

4. _____ allow ions and small molecules to flow between the cytoplasm of abutting cells. [p.341]
   a. Adhering junctions
   b. Filter junctions
   c. Gap junctions
   d. Tight junctions
   e. Loose junctions

5. Muscle that is not striped and is involuntary is _____ . [p.343]
   a. rough
   b. cardiac
   c. skeletal
   d. striated
   e. smooth

6. A(n) _____ is a community of cells and intercellular substances that perform one or more tasks. [p.340]
   a. organ
   b. organ system
   c. tissue
   d. cuticle
   e. adhering junction

7. A graduate student in developmental biology accidentally stabbed a fish embryo. Later the embryo developed into a creature that could not move and had no supportive or circulatory systems. Which embryonic tissue had suffered the damage? [p.345]
   a. ectoderm
   b. endoderm
   c. mesoderm
   d. protoderm

8. _____ is a generic name for steroid-like compounds that help the body absorb calcium from food. [p.346]
   a. Vitamin C
   b. Cholesterol
   c. Vitamin B
   d. Vitamin D
   e. Vitamin E

9. The secretion of tears, milk, sweat, and oil are functions of _____ tissues. [p.341]
   a. glandular epithelium
   b. loose connective
   c. smooth epithelium
   d. nervous
   e. soft connective

10. Of all types of animal tissues, _____ tissue exerts the most control over how the body responds to changing conditions. [p.343]
    a. connective
    b. epithelial
    c. muscle
    d. nervous
    e. bone

## Chapter Objectives/Review Questions

1. Be prepared to present arguments for and against stem cell research. [p.339]
2. Describe the internal environment of animals; define *homeostasis*. [p.340]
3. Define *tissue, organ,* and *organ system*. [p.340]
4. Be able to describe and give the functions of tight junctions, adhering junctions, and gap junctions. [p.341]
5. Define *glandular epithelium, glands, secretion,* and *excretion;* distinguish between exocrine and endocrine glands. [p.341]
6. Be familiar with the descriptions and appearance of loose connective tissue; dense, irregular connective tissue; dense, regular connective tissue; cartilage; bone tissue; adipose tissue; and blood. [pp.341–342]
7. _____ muscle tissue, attached to bones, helps the body move and maintain posture; _____ muscle tissue is present only in the heart wall; _____ muscle tissue is in the walls of soft internal organs, such as the stomach and lungs of vertebrates. [p.343]
8. _____ tissue exerts the most control over how the body responds to changing conditions. [p.343]
9. Be able to list and describe the eleven organ systems of a typical vertebrate, an adult human. [p.344]
10. Be familiar with the descriptive terms illustrated in Figure 20.6. [p.345]
11. Know the major tissue, organ, and system derivatives of ectoderm, mesoderm, and endoderm. [p.345]
12. Be able to define *melanocytes, keratinocytes, melanin, epidermis, dermis, vitamin D,* and *folate*. [p.346]
13. Describe the causes of suntans and shoe-leather skin; list characteristics of aging skin. [p.347]
14. Be able to give the role of Langerhans cells in the skin; relate cold sores to *Herpes simplex*. [p.347]

## Interpreting and Applying Key Concepts

Speculate as to why, of all places in the human body, marrow with its stem cells is located on the interior of long bones. Explain why your bones are remodeled after you reach maturity. Why does your body not keep the same mature skeleton throughout life?

# 21

# HOW ANIMALS MOVE

## INTRODUCTION
Chapter 21 describes the anatomy and physiology of the skeletal and muscular systems.

## FOCAL POINTS
- Figure 21.3 [p.351] shows the major bones of the human skeleton.
- Figures 21.4 and 21.5 [p.352] look at bone structure.
- Figures 21.7 and 21.8 [pp.354, 355] describe muscle function.

## Interactive Exercises

*Pumping Up Muscles* [p.349]

### 21.1. SO WHAT IS A SKELETON? [pp.350–353]
### 21.2. HOW DO BONES AND MUSCLES INTERACT? [p.353]

*Selected Words:* androstenedione [p.349], anabolic [p.349], creatine phosphate [p.349], *herniate* [p.350], *osteoblasts* [p.352], *osteocytes* [p.352], *osteoclasts* [p.352], *compact* bone [p.352], *spongy* bone [p.352], *osteoporosis* [p.352], *cartilaginous* joints [p.352], *synovial* joints [p.352], *strain* [p.352], *sprain* [p.352], *osteoarthritis* [p.353], *rheumatoid arthritis* [p.353]

### Boldfaced Terms

[p.350] hydrostatic skeleton _____

[p.350] exoskeleton _____

[p.350] endoskeleton _____

[p.350] vertebrae _____

[p.350] intervertebral disks _____

[p.352] red marrow _____

[p.352] yellow marrow _____

[p.352] joints _____

[p.352] ligaments _____

[p.353] skeletal muscles _____

[p.353] tendon _____

## Fill in the Blanks [p.349]

The dietary supplement (1) _____ is an intermediate in the production of the male sex hormone (2) _____ and the female sex hormone (3) _____. Although it has been touted as a way to increase (4) _____ mass and (5) _____, (6) _____ studies showed it had no greater effect than a (7) _____. Possible side effects in males include (8) _____ testicles and female-like (9) _____. In women, the (10) _____ cycle can be disrupted and a masculinized (11) _____ pattern can develop. Both sexes are at risk for (12) _____ damage, (13) _____, and a decline in (14) _____ cholesterol.

## Choice

Choose the correct skeletal type for each organism. [p.350]

        a. hydrostatic skeleton     b. exoskeleton     c. endoskeleton

15. _____ cockroach
16. _____ earthworm
17. _____ frog
18. _____ lobster
19. _____ sea star
20. _____ shark

## True–False

If the statement is true, write a T in the blank. If the statement is false, correct it by changing the underlined word(s) and writing the correct word(s) in the answer blank. [p.350]

21. _____ In humans, the most common bone broken is the femur.
22. _____ Human males have 11 pairs of ribs and human females have 12 pairs.
23. _____ The shock absorbers of the backbone are the vertebrae and they are made of bone.

210   Chapter Twenty-One

## Identification

Identify each numbered bone or bone groups in the figure. [p.351]

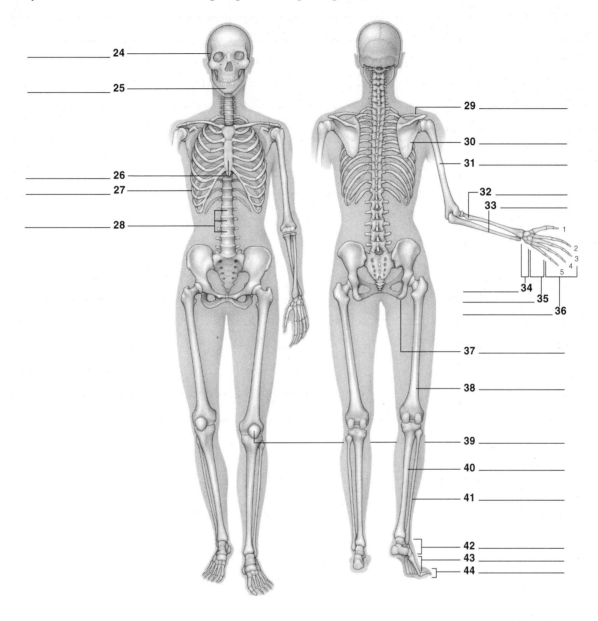

## Choice

45. _____ Bones do which of the following? [p.352]

   a. support and anchor muscles   b. enclose and protect soft tissue
   c. both a. and b.   d. neither a. nor b.

46. _____ The mature cells in adult bones are called _____. [p.352]

   a. osteoblasts   b. osteoclasts   c. osteocytes   d. osteophytes

47. _____ Blood cells are produced by _____ marrow. [p.352]

   a. red   b. yellow   c. both a. and b.   d. neither a. nor b.

How Animals Move   211

48. _____ Bones and teeth store approximately how much of the body's calcium? [p.352]
    a. 25%   b. 50%   c. 75%   d. almost all

49. _____ Ligaments usually connect _____. [p.352]
    a. bone to bone   b. bone to muscle   c. muscle to muscle

50. _____ Damage to cartilage occurs in _____. [p.353]
    a. osteoarthritis   b. rheumatoid arthritis   c. both a. and b.   d. neither a. nor b.

51. _____ Tendons usually connect _____. [p.353]
    a. bone to bone   b. bone to muscle   c. muscle to muscle

52. _____ The human body has approximately _____ skeletal muscles. [p.353]
    a. 100   b. 300   c. 500   d. 700

## 21.3. HOW DOES SKELETAL MUSCLE CONTRACT? [pp.354–356]
## 21.4. PROPERTIES OF WHOLE MUSCLES [pp.356–358]

*Selected Words:* Z band [p.354], *isotonically* contracting muscles [pp.356–357], *isometrically* contracting muscles [p.357], acetylcholine [p.357], *botulism* [p.357], *tetanus* [p.357], *muscle cramp* [p.357], *muscular dystrophies* [p.357], *aerobic exercise* [p.358], *strength training* [p.358]

### Boldfaced Terms

[p.354] sarcomeres _____
_____

[p.354] myofibrils _____
_____

[p.354] actin _____
_____

[p.354] myosin _____
_____

[p.355] sliding-filament model _____
_____

[p.355] sarcoplasmic reticulum _____
_____

[p.356] motor unit _____
_____

[p.356] muscle twitch _____
_____

[p.356] tetanus _____

[p.356] muscle tension _____

[p.357] muscle fatigue _____

## Fill in the Blanks

Both (1) _____ [p.354] and (2) _____ [p.354] muscle have alternating light and dark bands which make the muscle look striped or (3) _____ [p.354]. Each unit of contraction, called a (4) _____ [p.354], has thick filaments of the protein (5) _____ [p.354] and thinner filaments of the protein (6) _____ [p.354]. These filaments run (7) _____ [p.354] to the long axis of the myofibril. To contract, the muscles need the energy molecule (8) _____ [p.355]. The mineral (9) _____ [p.355] is also important to the continued contraction. To produce enough quickly available energy, the phosphate group from (10) _____ [p.355] phosphate can be donated to (11) _____ [p.355] to produce more ATP. Additional ATP is produced by (12) _____ _____ [p.356]. Initially (13) _____ [p.356] and (14) _____ [p.356] serve as initial substrates for (12) but later, (15) _____ _____ [p.356] become the primary substrates for this process.

## True–False

If the statement is true, write a T in the blank. If the question is false, correct it by changing the underlined word(s) and writing the correct word(s) in the answer blank.

_____ 16. When intensive exercise exceeds the oxygen supply, muscles <u>stop functioning</u>. [p.356]

_____ 17. In an <u>isometric</u> contraction, muscles shorten and move a load. [p.357]

_____ 18. The repeated stimulation of a muscle causes a sustained reaction called <u>a muscle twitch</u>. [p.356]

_____ 19. In a *Clostridium tetani* infection, the bacterium <u>spreads from the site of infection</u> causing disease throughout the organism. [p.357]

_____ 20. Tetanus is not a serious threat in this country because <u>cleanliness has all but eradicated the bacterium</u>. [p.357]

## Matching

Match each numbered phrase with the proper definition.

21. _____ muscle cramp [p.357]
22. _____ muscular dystrophy [pp.357–358]
23. _____ muscle fatigue [p.357]

a. decline in muscle capacity to generate force caused by ongoing stimulation
b. painful muscle contractions that can last 15 minutes or more
c. genetic disorder in which muscles progressively weaken and degenerate

*Short Answer*

24. What are the differences between aerobic and strength training exercises? What are the benefits of each? [p.358]

_____

_____

_____

_____

## Self-Quiz

_____ 1. A skeleton in which muscles work against fluids is called a/an _____. [p.350]
   a. bony endoskeleton
   b. cartilaginous endoskeleton
   c. exoskeleton
   d. hydrostatic skeleton

_____ 2. When our hominid ancestors began walking upright, our vertebral column took on the shape of a/an _____. [p.350]
   a. inverted C
   b. C
   c. I
   d. S

_____ 3. The bone that protects the knee joint is called the _____. [p.351]
   a. clavicle
   b. patella
   c. scapula
   d. sternum

_____ 4. Bone is removed by _____. [p.352]
   a. osteoblasts
   b. osteoclasts
   c. osteocytes
   d. osteoplasts

_____ 5. _____ joints allow free movement. [p.352]
   a. Cartilaginous
   b. Synovial
   c. Both a. and b.
   d. Neither a. nor b.

_____ 6. Bending the arm upward at the elbow joint requires contraction of the _____. [p.353]
   a. biceps
   b. triceps
   c. both a. and b.
   d. neither a. nor b.

_____ 7. The base units of muscle contraction are called _____. [p.354]
   a. actins
   b. myofibrils
   c. myosins
   d. sarcomeres

_____ 8. Muscles can produce ATP _____. [p.356]
   a. aerobically
   b. anaerobically
   c. both a. and b.
   d. neither a. nor b.

_____ 9. A motor unit consists of _____. [p.356]
   a. motor neuron
   b. muscle fiber
   c. both a. and b.
   d. neither a. nor b.

_____ 10. Regular exercise increases the _____ of muscle cells. [p.358]
   a. number
   b. size
   c. both a. and b.
   d. neither a. nor b.

# Chapter Objectives/Review Questions

1. What are the risks associated with ingesting androstenedione? [p.349]
2. Describe the three types of skeletons. Give examples of several animals with each type of skeleton. [p.350]
3. What are intervertebral disks? Why are they important? [p.350]
4. What are the differences between compact and spongy bone? [p.352]
5. Compare and contrast strains and sprains. [p.352]
6. Why do severed ligaments need to be surgically reconnected within ten days of being severed? [pp.352–353]
7. Compare and contrast osteo- and rheumatoid arthritis. [p.353]
8. Describe the process of muscle contraction. [pp.354–355]
9. What energy source(s) do muscles use for contraction? How is this energy supplied to muscle tissue? [pp.355–356]
10. How do *Clostridium* bacteria affect muscle contraction? [p.357]
11. Compare the benefits of exercise in older and younger individuals. [p.358]

# Integrating and Applying Key Concepts

1. What type of exercise plan would be of most benefit to young adults? What about for the elderly?
2. What are the similarities and differences between the skeleton of a lizard and the skeleton of a human? How do the differences support the lifestyles of the two organisms?

# 22

# CIRCULATION AND RESPIRATION

## INTRODUCTION

Chapter 22 describes both the circulatory and respiratory systems. It also looks at the integration of the two systems.

## FOCAL POINTS

- Figure 22.1 [p.361] compares open and closed circulatory systems.
- Figure 22.2 [p.362] contrasts blood flow in fish, amphibians, and mammals/birds.
- Figure 22.3 [p.363] lists the typical components of blood.
- Figures 22.4 and 22.5 [pp.364, 365] look at the human circulatory system.
- Figure 22.6 [p.365] diagrams the human heart.
- Figure 22.9 [p.367] contrasts various blood vessels.
- Figures 22.14 and 22.15 [p.370] look at blood clots.
- Figure 22.18 [p.372] describes gas exchange in fish gills.
- Figure 22.21 [p.374] diagrams the human respiratory system.
- Figure 22.24 [p.377] details the structure of the respiratory membrane.
- Figure 22.27 [p.379] looks at the risk of smoking and how quitting reduces these risks.

## Interactive Exercises

*Up in Smoke* [p.360]

### 22.1. THE NATURE OF BLOOD CIRCULATION [pp.361–362]
### 22.2. CHARACTERISTICS OF HUMAN BLOOD [pp.362–363]

**Selected Words:** secondhand smoke [p.360], *internal* environment [p.361], *open* circulatory system [p.361], *closed* circulatory system [p.361], platelets [p.363], stem cells [p.363], plasma proteins [p.363], erythrocytes [p.363], hemoglobin [p.363], leukocytes [p.363], neutrophils [p.363], basophils [p.363], macrophages [p.363], dendritic cells [p.363], lymphocytes [p.363], B and T cells [p.363], natural killer cells [p.363], megakaryocytes [p.363]

## Boldfaced Terms

[p.361] interstitial fluid _____

[p.361] blood _____

[p.361] heart _____

[p.361] capillaries _____

[p.362] pulmonary circuit _____

[p.362] systemic circuit _____

[p.362] lymphatic system _____

[p.363] plasma _____

[p.363] red blood cells _____

[p.363] cell count _____

[p.363] white blood cells _____

## Fill in the Blanks

Each day, (1) _____ [p.360] teenagers begin smoking. Damage from smoking starts (2) _____ [p.360]. (3) _____ [p.360] are immobilized for hours and (4) _____ _____ [p.360] cells are killed. Young smokers also build up (5) _____ [p.360] in their airways. This leads to increased numbers of (6) _____ [p.360], (7) _____ [p.360], and bronchitis. (8) _____ [p.360] in cigarette smoke can trigger (9) _____ [p.360] in many body organs. For example, females who start smoking as teenagers are (10) _____ [p.360] more likely to develop breast cancer than nonsmokers. Smoking also raises the level of (11) _____ _____ [p.360] which makes blood (12) _____ [p.360] promoting the formation of (13) _____ [p.360].

*Matching*

Match each organism with its circulatory system. [pp.361–362]

14. _____ amphibian
15. _____ arthropod
16. _____ fish
17. _____ mammal

a. closed, two-chambered heart
b. closed, three-chambered heart
c. closed, four-chambered heart
d. open

*Complete the Table* [p.363]

| Component | Number/amount in blood | Function |
| --- | --- | --- |
| Plasma | 18. | 19. |
| 20. | Approximately 5 million/cc | 21. |
| Neutrophils | 22. | 23. |
| Lymphocytes | 24. | 25. |
| 26. | 27. | Initiate blood clotting |

### 22.3. HUMAN CARDIOVASCULAR SYSTEM [pp.364–366]

**Selected Words:** pericardium [p.364], myocardium [p.365], endothelium [p.365], coronary arteries [p.365], atrium (atria) [p.365], ventricle [p.365], atrioventricular (AV) valve [p.365], semilunar valve [p.365], sinoatrial (SA) node [p.366], atrioventricular (AV) node [p.366]

**Boldfaced Terms**

[p.364] aorta _____

_____

[p.365] cardiac cycle _____

_____

[p.366] cardiac conduction system _____

_____

[p.366] cardiac pacemaker _____

_____

## Labeling

Label the numbered structures on these diagrams. [p.365]

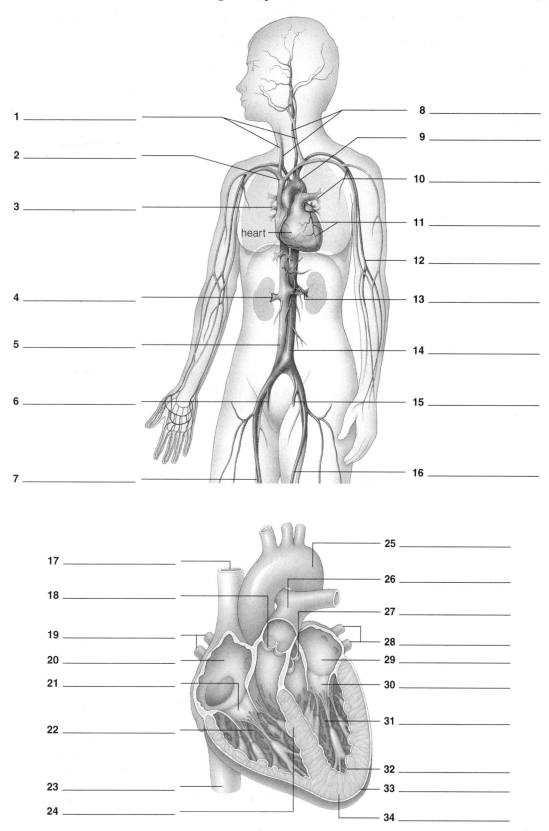

Circulation and Respiration

## Matching

Match each blood vessel with its function. [p.365]

35. _____ carotid artery
36. _____ inferior vena cava
37. _____ jugular vein
38. _____ coronary artery
39. _____ renal vein
40. _____ brachial artery

a. drains blood from kidneys
b. supplies blood to arms
c. supplies blood to head and brain
d. drains blood from the lower body
e. drains blood from the head
f. supplies blood to the heart

## Fill in the Blanks

Cardiac muscle depends on its abundant (41) _____ [p.366] to produce enough (42) _____ [p.366]. Like (43) _____ [p.366] muscle, cardiac muscle has sarcomeres, but unlike (43) muscle cells, cardiac muscle cells are (44) _____ [p.366] and connect to several other cells at their (45) _____ [p.366]. The (46) _____ _____ [p.366] serves as the heart's (47) _____ [p.366], since it is here that the first signals for muscle contraction are generated. The (48) _____ _____ [p.366] is the only electrical connection between the atria and ventricles which are separated everywhere else by connective tissue.

## 22.4. STRUCTURE AND FUNCTION OF BLOOD VESSELS [pp.367–369]
## 22.5. CARDIOVASCULAR DISORDERS [pp.370–371]

**Selected Words:** arteries [p.367], arterioles [p.367], capillaries [p.367], veins [p.367], *systolic* pressure [p.368], *diastolic* pressure [p.368], *ultrafiltration* [p.369], *reabsorption* [p.369], *edema* [p.369], *elephantiasis* [p.369], *venules* [p.369], *spasm* [p.370], *coagulate* [p.370], *hypertension* [p.370], *arteriosclerosis* [p.370], *atherosclerosis* [p.370], *low-density lipoprotein* [p.370], *high-density lipoprotein* [p.370], *atherosclerotic plaque* [p.370], *thrombus* [p.370], *embolus* [p.371], *stroke* [p.371], *angina pectoris* [p.371], *coronary bypass surgery* [p.371], *laser angioplasty* [p.371], *balloon angioplasty* [p.371]

## Boldfaced Terms

[p.367] blood pressure _____

[p.367] vasodilation _____

[p.368] vasoconstriction _____

[p.370] hemostasis _____

## Complete the Table

Fill in each box in the table with either Yes or No. [p.367]

| Vessel | Endothelium | Smooth Muscle | Outer Coat | Valves |
|---|---|---|---|---|
| Arteries | 1. | 5. | 9. | 13. |
| Arterioles | 2. | 6. | 10. | 14. |
| Capillaries | 3. | 7. | 11. | 15. |
| Veins | 4. | 8. | 12. | 16. |

## Matching

Match each term with its definition. [pp.368–369]

17. _____ vasoconstriction
18. _____ systolic
19. _____ diastolic
20. _____ edema
21. _____ ultrafiltration

a. swelling caused by fluid in interstitial spaces
b. smooth muscle contraction causes vessel diameter to decrease
c. pressure forces fluid out of capillaries
d. highest pressure in the arteries
e. lowest pressure in the arteries

## Fill in the Blanks

The three steps in preventing blood loss from vessels are blood vessel (22) _____ [p.370], clumping of (23) _____ [p.370], and blood (24) _____ [p.370]. This process is necessary but can cause a problem if the clot completely blocks (25) _____ _____ [p.370] through a vessel. (26) _____ [p.370] or high blood pressure is often known as a (27) _____ _____ [p.370] because affected individuals may not show any outward symptoms. Another problem is (28) _____ [p.370] in which lipids like (29) _____ [p.370] begin to deposit in the artery wall. Calcium deposits and a fibrous net can also form over the deposit forming an (30) _____ _____ [p.370]. Sometimes a (31) _____ [p.370] also forms in the same area. A stationary (31) is called a (32) _____ [p.370] while one that travels through the blood is called a/an (33) _____ [p.371]. When a/an (33) lodges in the brain, a/an (34) _____ [p.371] can occur. When a (30) narrows the coronary arteries, (35) _____ _____ [p.371] or a/an (36) _____ _____ [p.371] can occur. The plaque can be removed or reduced using laser or balloon (37) _____ [p.371].

## 22.6. THE NATURE OF RESPIRATION [pp.371–373]
## 22.7. HUMAN RESPIRATORY SYSTEM [pp.374–375]

**Selected Words:** *partial pressure* [p.371], surface-to-volume ratio [p.372], ventilation [p.372], *external* gills [p.372], *internal* gills [p.372], integumentary exchange [p.373], vocal cords [p.375], glottis [p.375], *laryngitis* [p.375], epiglottis [p.375], pleural membrane [p.375], *pleurisy* [p.375]

### Boldfaced Terms

[p.371] respiration _____

[p.371] respiratory surface _____

[p.372] gills _____

[p.372] hemoglobin _____

[p.372] myoglobin _____

[p.373] countercurrent exchange _____

[p.373] lungs _____

[p.374] alveoli _____

[p.375] pharynx _____

[p.375] larynx _____

[p.375] trachea _____

[p.375] bronchus _____

[p.375] diaphragm _____

[p.375] bronchioles _____

*Fill in the Blanks*

Unless they have a respiratory system, organisms need a high (1) _____-_____-_____ [p.372] ratio so gases can be exchanged by (2) _____ [p.372] across the skin. For large-bodied aquatic animals, respiration usually takes place via (3) _____ [p.372] and for large-bodied terrestrial animals, respiration is through (4) _____ [p.372]. In addition to (4), many amphibians use (5) _____ [p.372] as a respiratory surface. In humans, movement of air into and out of the lungs is caused by contractions of several muscles like the (6) _____ [p.375] located between the chest and abdominal cavities. Air moves into and out of the lungs by (7) _____ _____ [p.376]. Diffusion occurs through specialized air sacs called (8) _____ [p.374] where gas exchange occurs between them and (9) _____ [p.375] capillaries. Vocalization requires air to pass through a narrow gap called the (10) _____ [p.375]. The air causes the thin membranous (11) _____ _____ [p.375] on either side of the gap to vibrate. Another thin flap, the (12) _____ [p.375], blocks food from entering the (13) _____ [p.375] and the rest of the respiratory system.

*Labeling*

Label the numbered parts of the human respiratory system; then note the function of each one by placing the proper letter in the parentheses. [p.374]

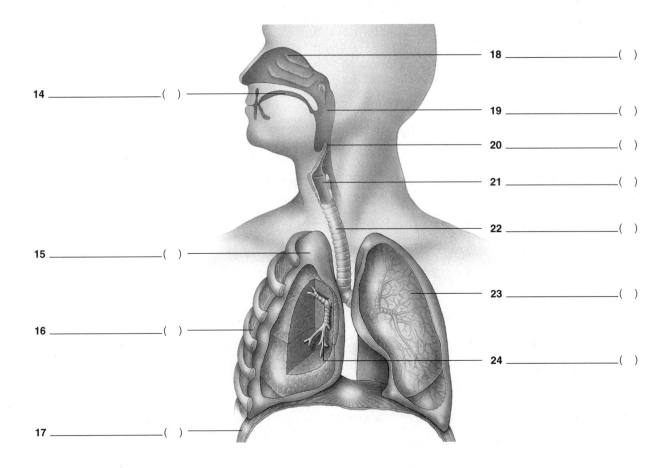

14. _____ ( )
15. _____ ( )
16. _____ ( )
17. _____ ( )
18. _____ ( )
19. _____ ( )
20. _____ ( )
21. _____ ( )
22. _____ ( )
23. _____ ( )
24. _____ ( )

a. Airway connecting larynx to bronchi
b. Airway where sounds are produced
c. Airway connecting nasal cavity/mouth to larynx
d. Chamber where air is moistened, warmed, and filtered; major intake point
e. Closes off larynx/trachea during swallowing
f. Double membrane that separates lungs from other organs
g. Increasingly branched airways that terminate in alveoli
h. Lobed, elastic organ of breathing
i. Muscle sheet between chest and abdominal cavities
j. Muscles at ribs with role in breathing
k. Supplemental air intake chamber

## 22.8. MOVING AIR AND TRANSPORTING GASES [pp.376–378]
## 22.9. WHEN THE LUNGS BREAK DOWN [pp.378–379]

*Selected Words:* inhalation [p.376], exhalation [p.376], respiratory membrane [p.376], oxyhemoglobin [p.377], carbaminohemoglobin [p.377], bicarbonate [p.377], bronchitis [p.378], emphysema [p.379], antitrypsin [p.379]

### Boldfaced Terms

[p.376] respiratory cycle _____

_____

[p.376] heme groups _____

_____

### Fill in the Blanks

A respiratory cycle consists of one (1) _____ [p.376] and one (2) _____ [p.376]. During (1), the movement of the diaphragm and rib cage enlarges the (3) _____ _____ [p.376] which causes the lungs to expand. The (4) _____ _____ [p.376] is quite thin and is made up of the alveolar and capillary (5) _____ [p.376] and their fused (6) _____ _____ [p.376]. Across this layer both (7) _____ [p.377] and (8) _____ _____ [p.376] are able to diffuse. Inside the circulatory system, most oxygen is bound to (9) _____ [p.376], each molecule of which can bind (10) _____ [p.376] oxygen molecules. The highest partial pressure of oxygen is found in the (11) _____ _____ [p.377]. The oxygen in oxyhemoglobin is bound weakly and oxyhemoglobin readily gives up its oxygen when the blood is (12) _____ [p.377], (13) _____ [p.377] is lower, and the partial pressure of (14) _____ _____ [p.377] is high. Although some carbon dioxide binds to (15) _____ [p.377], most is converted to (16) _____ [p.377] and transported in that form.

### Matching

Match each of the following respiratory problems with its best definition.

17. _____ bronchitis [p.378]
18. _____ emphysema [p.379]
19. _____ laryngitis [p.379]
20. _____ pleurisy [p.379]

a. inflammation of the membranes around the lungs
b. inflammation of the vocal cords
c. lungs inelastic; airflow is permanently compromised
d. lungs secrete excess mucus, trap bacteria, etc.

### True–False

If the statement is true, write a T in the blank. If the statement is false, correct it by writing the correct word(s) for the underlined word(s) in the answer blank. [p.379]

_____ 21. Tobacco use kills about <u>1 million</u> people yearly worldwide.

_____ 22. Cigarette smoking is a primary factor in <u>lung cancer, laryngeal cancer, bladder cancer, and heart disease</u>.

_____ 23. In <u>nonsmokers</u> it takes 30 percent longer for bones to heal.

_____ 24. It takes <u>20–25</u> years for a former smoker's lifespan to approach that of a nonsmoker's.

_____ 25. Marijuana cigarettes contain <u>less</u> tar than tobacco cigarettes.

---

## Self-Quiz

_____ 1. Nicotine in smoke _____ . [p.360]
   a. causes the heart to beat faster
   b. immobilizes respiratory cilia
   c. both a. and b.
   d. neither a. nor b.

_____ 2. An open circulatory system is found in _____ . [p.361]
   a. arthropods
   b. mammals
   c. both a. and b.
   d. neither a. nor b.

_____ 3. Unspecialized cells which give rise to all the types of blood cells are called _____ . [p.363]
   a. erythrocytes
   b. leukocytes
   c. megakaryocytes
   d. stem cells

_____ 4. The major artery that delivers oxygen-enriched blood to all body tissues is the _____ . [p.365]
   a. aorta
   b. carotid artery
   c. coronary artery
   d. pulmonary trunk

_____ 5. There is a one-way valve between the _____ in adult humans. [p.365]
   a. right atrium and right ventricle
   b. right atrium and left atrium
   c. both a. and b.
   d. neither a. nor b.

_____ 6. Both smooth muscle and elastic tissue are found in _____ . [p.367]
   a. arteries
   b. veins
   c. both a. and b.
   d. neither a. nor b.

_____ 7. The oxygen-carrying pigment that gives muscle cells most of their red color is _____ . [p.372]
   a. hemoglobin
   b. immunoglobin
   c. myoglobin
   d. xenoglobin

_____ 8. The tissue flap that prevents food from entering the respiratory track is the _____ . [p.375]
   a. alveolus
   b. epiglottis
   c. glottis
   d. vocal chord

_____ 9. Carbon dioxide is most commonly transported in the blood as _____ . [p.377]
   a. bicarbonate
   b. carbaminohemoglobin
   c. carbonate
   d. oxyhemoglobin

_____ 10. Individuals who lack the enzyme antitrypsin are more susceptible to _____ . [p.379]
   a. cancer
   b. emphysema
   c. both a. and b.
   d. neither a. nor b.

_____ 11. For which of the following does smoking increase your risk? [p.379]
   a. bladder cancer
   b. heart disease
   c. allergies
   d. all of the preceding

## Chapter Objectives/Review Questions

1. Why does smoking lead to increased respiratory infections? [p.360]
2. What are the differences in efficiency among fish, amphibian, and human respiratory systems? Relate this to structure. [pp.361–362]
3. What are the components of blood? What does each do? [pp.362–363]
4. Describe blood flow in humans. [p.364]
5. Why are valves important to the proper functioning of the heart? [p.365]
6. What is meant by blood pressure? How is it measured? [pp.368–369]
7. How can blood flow against gravity in veins? [p.369]
8. Why are clots good? Why are they bad? [pp.370–371]
9. What are the differences between HDL and LDL cholesterol? [p.370]
10. What are the symptoms of hypertension? [p.370]
11. Why is the surface-to-volume ratio so important to gas exchange? How do larger organisms increase this ratio for their respiratory surfaces? [p.372]
12. How do humans make vocal sounds? [p.375]
13. Describe the human respiratory membrane. [p.376]
14. How are oxygen and carbon dioxide transported in humans? [p.377]
15. How can bronchitis develop into emphysema? [p.378]
16. What are the risks associated with smoking? [p.379]
17. Why can smoking pot cause even more damage than smoking cigarettes? [p.379]

## Integrating and Applying Key Concepts

1. There are a few species of salamanders that have no lungs and use only integumentary respiration. Without seeing one, what do you guess they would look like?
2. You have been asked to give an antismoking talk to your high school. What are the messages/arguments that you think students of that age would listen to?

# 23

# IMMUNITY

## INTRODUCTION

Chapter 23 describes the function of the immune system. It details barrier, innate, antibody-mediated, and cell-mediated immune responses. It also discusses problems of the immune system.

## FOCAL POINTS

- Figure 23.2 [p.384] illustrates the various types of leukocytes.
- Table 23.2 [p.384] looks at several chemicals important to the immune system.
- Table 23.3 and Figure 23.3 [pp.384, 385] describe physical barriers to infection.
- Figure 23.7 [p.388] diagrams the inflammatory response.
- Figure 23.10 [p.390] outlines the antibody-mediated and cell-mediated immune responses.
- Figure 23.11 [p.391] illustrates and describes the immune organs.
- Figure 23.12 [p.392] diagrams antibody structure.
- Figure 23.14 [p.394] details the antibody-mediated immune response.
- Figure 23.15 [p.395] illustrates clonal selection.
- Figure 23.16 [p.396] details the cell-mediated immune response.
- Figure 23.20 [p.398] diagrams the replication cycle of HIV.

## Interactive Exercises

*The Face of AIDS* [p.382]

**23.1. INTEGRATED RESPONSES TO THREATS** [pp.383–385]

**23.2. SURFACE BARRIERS** [pp.385–387]

**23.3. THE INNATE IMMUNE RESPONSE** [pp.387–389]

**Selected Words:** "immune" [p.382], smallpox [p.382], *self* [p.383], *nonself* [p.383], interleukins [p.384], interferons [p.384], tumor necrosis factor [p.384], monocytes [p.384], B and T cells [p.385], *acne* [p.387], *periodontitis* [p.387], "activated" complement [p.387], cascading reactions [p.387], chemotactic [p.388], histamines [p.388], swelling [p.388], prostaglandins [p.389]

## Boldfaced Terms

[p.382] vaccination _____

[p.383] immunity _____

[p.383] antigen _____

[p.383] complement _____

[p.383] phagocytes _____

[p.383] innate immunity _____

[p.383] cytokines _____

[p.383] lymphocytes _____

[p.383] adaptive immunity _____

[p.384] neutrophils _____

[p.384] macrophages _____

[p.385] dendritic cells _____

[p.385] basophils _____

[p.385] mast cells _____

[p.385] eosinophils _____

[p.385] B and T lymphocytes _____

[p.385] natural killer cells (NK cells) _____

[p.386] lysozyme _____

[p.387] plaque _____

[p.388] acute inflammation _____

[p.389] fever _____

## Fill in the Blanks

Molecules that the body recognizes as nonself and elicit a/an (1) _____ [p.383] response are called (2) _____ [p.383]. An evolutionarily early chemical response to nonself is the release of (3) _____ [p.383] proteins. In addition to killing microbes, (3) can tag microbes for engulfment by (4) _____ [p.383]. This kind of immunity is called (5) _____ [p.383] immunity and is the only kind found in (6) _____ [p.383] animals. In jawed fish, chemicals called (7) _____ [p.383] and specialized white blood cells called (8) _____ [p.383] evolved. These could tailor a response to specific pathogens, a response called (9) _____ [p.383] immunity. Before any of these types of immunity evolved, physical barriers like (10) _____ [p.384] and (11) _____ _____ [p.384] helped to prevent pathogens from entering organisms.

## Matching

Match each type of blood cell with its function.

12. _____ basophils [p.385]
13. _____ dendritic cells [p.385]
14. _____ eosinophils [p.385]
15. _____ lymphocytes [p.385]
16. _____ macrophages [p.384]
17. _____ neutrophils [p.384]

a. most abundant, fast-acting phagocytes
b. can engulf up to 100 bacterial cells through phagocytosis
c. alert the adaptive system to the presence of antigens
d. release enzymes and cytokines in response to antigens
e. specialize in control of parasites
f. central to adaptive immunity; can shuffle genes to respond to many antigens

## Choice

Choose from the following:

a. body surface defenses  b. phagocytes and their kin
c. complement proteins    d. inflammatory response

18. _____ Develops when something damages or kills cells of any tissue region; fast-acting phagocytes leave the blood at capillary beds in an affected tissue [p.388]
19. _____ Vertebrate skin, lysozyme-rich mucus and cilia in airways sweep out trapped and digested bacteria [p.386]
20. _____ Abundant neutrophils ingest, kill, and digest bacterial cells before they themselves die [p.384]
21. _____ Tears and saliva contain lysozyme and other protective chemicals [p.386]

22. _____ Eosinophils secrete enzymes that put holes in parasitic worms [p.385]
23. _____ Urine's flushing action and low pH remove pathogens from the urinary tract; diarrhea can flush them out of the gut when gastric fluid doesn't kill them [p.386]
24. _____ Mast cells in connective tissues respond to damage by making and releasing histamine and other local signaling molecules into interstitial fluid [p.388]
25. _____ Histamine triggers vasodilation of arterioles as blood-engorged tissues redden and warm with blood-borne metabolic heat [p.388]
26. _____ Products of cascading reactions create concentration gradients that attract phagocytes; some bind en masse to different bacteria, parasites, and enveloped viruses [p.387]
27. _____ Following the first response to tissue damage, neutrophils squeeze across capillary walls and go to work; monocytes arrive and differentiate into macrophages; chemical mediators known as chemotaxins attract more phagocytes [p.388]

## 23.4. TAILORING RESPONSES TO SPECIFIC ANTIGENS [pp.389–391]

**Selected Words:** *self versus nonself recognition* [p.389], *specificity* [p.389], *diversity* [p.389], *memory* [p.389], *effector* cells [p.390], *memory* cells [p.390], *intracellular* pathogens [p.390], lymph node [p.390], tonsils [p.391], thymus gland [p.391], spleen [p.391]

### Boldfaced Terms

[p.389] MHC markers _____

_____

[p.389] T cell receptors _____

_____

[p.390] antibody-mediated immune response _____

_____

[p.390] cell-mediated immune response _____

_____

### Short Answer

1. What are the four defining features of adaptive immunity? [p.389]

_____

_____

### Fill in the Blanks

Recognition of a specific antigen causes repeated (2) _____ _____ _____ [p.389] in lymphocytes. T cells only recognize antigens that have been displayed by (3) _____-_____ [p.389] cells. (3) cells form when an antigen is taken in by (4) _____ [p.389]. (5) _____ [p.389] enzymes digest the antigen, and bits of it bind to (6) _____ [p.390] markers. This complex is then displayed on the surface of (3) cells. Activated lymphocytes that act immediately against the pathogen are called (7) _____ [p.390] cells whereas (8) _____ [p.390]

Immunity 231

cells are long-lived cells that act against future infections. (9) _____ [p.390] cells are the primary cells involved in the (10) _____-_____ [p.390] immune response which attacks circulating pathogens. T cells are very involved in the (11) _____-_____ [p.390] immune response which attacks (12) _____ [p.390] pathogens. During the immune response, many lymphocytes migrate to (13) _____ _____ [p.391] where they often remain, causing the (13) to become swollen.

## 23.5. ANTIBODIES AND OTHER ANTIGEN RECEPTORS [pp.392–393]
## 23.6. ANTIBODY-MEDIATED IMMUNE RESPONSE [pp.394–395]
## 23.7. THE CELL-MEDIATED IMMUNE RESPONSE [pp.395–396]

**Selected Words:** *variable* region [p.392], *constant* region [p.392], *IgG, IgA, IgE, IgM, IgD* [p.392], *clonal selection* [p.394], *helper T cells* [p.394], *cytotoxic T cells* [p.396], *perforin* [p.396], alters its MHC markers [p.396]

### Boldfaced Terms

[p.392] antibodies _____

_____

[p.392] immunoglobulins, or Igs _____

_____

### Matching

Match each term with its description. [p.392]

1. _____ immunoglobulins, or Igs
2. _____ IgM
3. _____ IgM and IgD
4. _____ IgG
5. _____ IgE
6. _____ IgA

a. Triggers inflammation after attacks by parasitic worms and other pathogens
b. The first to be secreted in a primary response and the first made by newborns; may help activate helper T cells
c. Long-lasting antibody that makes up 80 percent or so of the immunoglobulins in blood; best at activating complement proteins and neutralizing many toxins; crosses the placenta and helps protect a fetus
d. Refers to five classes of antibodies; protein products of gene shufflings that form when B cells mature and engage in immune responses
e. The most common membrane-bound antibodies on naïve B cells
f. The main immunoglobulin in exocrine gland secretions, including tears, saliva, and breast milk; also in mucous linings of the respiratory, digestive, and reproductive tracts

### Fill in the Blanks

Each Y-shaped antibody molecule has (7) _____ [p.392] regions which recognize and bind to specific (8) _____ [p.392]. The genes for these (8) receptors are formed from (9) _____ [p.393] assembled segments of DNA which is the source of antibody diversity. B cells form and mature in the (10) _____ _____ [p.393] while T cells form in the same place but mature in the

(11) _____ _____ [p.393]. In the (11), T cells that recognize (12) _____ [p.393] antigens are weeded out.

In the antibody-mediated response, foreign antigens and antigen complexes induce (13) _____-_____ [p.394] endocytosis by (14) _____ [p.394] cells. This activates the (14) cells. At the same time, (15) _____ [p.394] cells travel to the lymph nodes where they interact with (16) _____ [p.394] cells that have bound foreign antigen. The theory of (17) _____ _____ [p.394] states that the (15) cells were selected by the antigen. These cells divide forming a huge number of (18) _____ _____ [p.394] cells (clones) that develop into (19) _____ _____ [p.394] cells. (10) cells recognize activated B cells and cause the latter to divide into huge populations of both (20) _____ [p.395] B cells, which secrete (21) _____ [p.395], and (22) _____ [p.395] B cells. When (21) bind to invaders, the invaders are more easily recognized by (23) _____ [p.395] cells.

The cell-mediated response starts much the same way when T cells in the lymphocytes recognize (24) _____ [p.396] cells. This activates T cells which divide into populations of (25) _____ [p.396] T cells and immature (26) _____ [p.396] T cells. In response to (27) _____ [p.396] secreted by (25) T cells, (26) T cells mature and begin circulating throughout the body. They bind to infected body cells and induce them to undergo (28) _____ [p.396]. (27) also enhance the functions of both (29) _____ [p.396] and (30) _____ _____ [p.396] cells.

## 23.8. DEFENSES ENHANCED OR COMPROMISED [p.397]
## 23.9. AIDS: IMMUNITY LOST [pp.398–399]

**Selected Words:** *active* immunization [p.397], booster [p.397], *passive* immunization [p.397], *asthma* [p.397], *hay fever* [p.397], *anaphylactic shock* [p.397], *Graves' disease* [p.397], *multiple sclerosis* [p.397], *primary* immune deficiencies [p.397], *secondary* immune deficiencies [p.397], HIV (*Human Immunodeficiency Virus*) [p.398], *Kaposi's sarcoma* [p.398], reverse transcriptase [p.398], AZT [p.399]

### Boldfaced Terms

[p.397] immunization _____
_____

[p.397] vaccine _____
_____

[p.397] allergen _____
_____

[p.397] allergy _____
_____

[p.397] autoimmune response _____
_____

*Matching*

Match each term with its description. [p.397]

1. _____ active immunization
2. _____ passive immunization
3. _____ allergens
4. _____ allergy
5. _____ anaphylactic shock
6. _____ autoimmunity
7. _____ Graves' disease
8. _____ multiple sclerosis
9. _____ primary immune deficiencies
10. _____ secondary immune deficiencies

a. Substances that cause inflammation, excess mucus secretion, and sometimes lead to immune responses; examples are different drugs, foods, pollen, dust mites, fungal spores, cosmetics, perfumes, and various types of insect venom
b. A rare, life-threatening response to an allergen; allergy to bee venom is an example
c. Autoimmune disease; the body makes antibodies that can bind to receptors in the thyroid, thus increasing production of thyroid hormones, which increases metabolism
d. Losses of immune function following exposure to outside agents, such as viruses; AIDS is the most common one
e. An antigen-containing preparation known as vaccine is taken orally or is injected into the body; triggers an immune response
f. Immune deficiencies, present at birth; results from altered genes or abnormal developmental steps; *SCID* is like this
g. When self-recognition fails, the immune fighters start an inappropriate attack against the body; wrongly activated antibodies or T cells can severely damage tissues
h. Uses an injection of antibody purified from the blood of someone who already had the disease or from another source; antibodies fight the infection but don't activate the immune system
i. An autoimmune disease; autoreactive T cells trigger inflammation of myelin sheaths and disrupt nerves by slipping into the fluid that bathes the brain and spinal cord; symptoms range from mild numbness to paralysis and blindness
j. Condition caused by hypersensitivity to allergens

*Choice*

Choose from the following:

    a. HIV infection    b. transmitting HIV    c. drugs and vaccines for HIV

11. _____ HIV requires a medium that permits it to leave one host, survive in the outside environment, and then enter another host. [p.398]

12. _____ Current types can't cure infected people because we can't go after HIV genes already inserted in someone's DNA [p.399]

13. _____ AZT can block reverse transcription. [p.399]

14. _____ During some phases, the virus infects about 2 billion helper T cells and makes 100 million to 1 billion virus particles each day. [p.399]

15. _____ Having sex with an infected partner, in semen and vaginal secretions, enters through epithelial linings, vaginal birth, breast feeding, infected syringes [p.399]

16. _____ Gradually, the number of virus particles in blood rises and the battle tilts; the body produces fewer and fewer replacement helper T cells; the body loses its capacity to initiate effective immune responses [p.399]

# Self-Quiz

_____ 1. The ability of B and T cell populations to tailor their defensive responses so that they are specific to one of potentially limitless kinds of pathogens is known as _____ . [p.383]
   a. passive immunity
   b. autoimmunity
   c. active immunity
   d. immunization
   e. adaptive immunity

_____ 2. Which of the following is most involved with vertebrate adaptive defenses? [pp.383–389]
   a. skin
   b. complement proteins
   c. IgG
   d. the inflammatory response
   e. phagocytes

_____ 3. Antibodies are shaped like the letter _____ and the antigen binding site is located at the ends of the _____ . [p.392]
   a. Y; variable chains
   b. W; constant region of the heavy chains
   c. Z; variable chains
   d. H; flexible hinge region
   e. E; constant region of the light chains

_____ 4. _____ cells bear tremendously diverse receptors. [p.393]
   a. Macrophage
   b. Neutrophil
   c. B and T
   d. Eosinophil
   e. Basophil

_____ 5. After formation in bone marrow, T cells migrate to the thymus gland where they acquire unique antigen-binding receptors called _____ . [p.393]
   a. TCRs
   b. MHC complexes
   c. IgM
   d. perforins
   e. lysozymes

_____ 6. The clonal selection hypothesis explains _____ . [p.394]
   a. how self cells are distinguished from nonself cells
   b. how B cells differ from T cells
   c. how so many different kinds of antigen-specific receptors can be produced by lymphocytes
   d. how so many cells directed against a single antigen can be produced
   e. how antigens differ from antibodies

_____ 7. Perforins that induce apoptosis are produced by _____ . [p.396]
   a. helper T cells
   b. macrophage
   c. tumor
   d. cytotoxic T cells
   e. T and B cells

_____ 8. In T cell–mediated responses, a helper T cell binds to _____ . [p.396]
   a. TCR
   b. antigen–MHC complexes
   c. a cytotoxic T cell
   d. a B cell
   e. virus particles

_____ 9. Active immunization involves _____ . [p.397]
   a. oral and injected vaccines
   b. booster injections
   c. primary immune responses
   d. weakened or killed pathogens or inactivated natural toxins
   e. all of the above

_____ 10. AIDS is _____ . [pp.398–399]
   a. a disease for which there is a cure
   b. a group of disorders that follow infection by HIV
   c. another name for the HIV virus
   d. transmitted by food, air, water, casual contact, or insect bites
   e. also called CD4

## Chapter Objectives/Review Questions

1. List the important steps in the evolution of the body's immune system; include "self from nonself," lysozymes, cytokines, B and T lymphocytes, and adaptive immunity. [p.383]
2. List and discuss four defense responses not directed at a particular antigen that serve to exclude microbes from the body. [pp.384–389]
3. Be able to list and describe the four features that define the immune system of all jawed vertebrates. [p.389]
4. Distinguish the roles of B cells from T cells. [pp.394–396]
5. By a theory of _____ _____, the antigen "chooses" only the B or T cell that bears the receptor able to bind with it. [p.394]
6. Describe how B cells, T cells, and antigen receptors form. [pp.394–396]
7. Discuss the "battlegrounds" of the immunity system. [pp.394–396]
8. Distinguish between the antibody-mediated and the cell-mediated response patterns; be able to discuss details. [pp.394–396]
9. In _____ immunization, an antigen-containing preparation known as a vaccine is taken orally or is injected into the body. [p.397]
10. _____ immunization helps those who are battling hepatitis B, tetanus, rabies, and a few other diseases. It uses an injection of purified antibodies. [p.397]
11. Be able to define *allergens, allergy, asthma, hay fever,* and *anaphylactic shock.* [p.397]
12. When self-recognition fails, the immune fighters start an inappropriate attack against the body known as _____. [p.397]
13. Explain what is meant by primary immune response, in contrast to secondary immune response. [p.397]
14. Describe the symptoms of AIDS, methods of HIV infection, how HIV is transmitted, and the drugs and vaccines used in the battle. [pp.398–399]

## Integrating and Applying Key Concepts

Cancers of the brain, eye, and thyroid usually are not included in the category of head and neck cancers. Cancers of the scalp, skin, muscles, and bones of the head and neck are also not considered cancers of the head and neck. Tobacco (including smokeless tobacco) and alcohol use are the most important risk factors for head and neck cancers. List the anatomical portions of the head and neck that you think might be included in the category of head and neck cancers.

# 24

# DIGESTION, NUTRITION, AND EXCRETION

## INTRODUCTION

Chapter 24 describes the anatomy and physiology of the digestive and urinary systems.

## FOCAL POINTS

- Figures 24.1 and 24.2 [pp.403, 404] give examples of digestive systems.
- Figure 24.3 [p.405] diagrams the human digestive system.
- Figure 24.6 [p.407] looks at the structure of the small intestine.
- Figure 24.7 [p.408] illustrates digestion and absorption in the small intestine.
- Table 24.1 [p.411] is the USDA food guide.
- Table 24.2 [p.412] lists the most important human vitamins, their sources, uses, and effects of deficiencies/excesses.
- Table 24.3 [p.413] lists the most important human minerals, their sources, uses, and effects of deficiencies/excesses.
- Figure 24.10 [p.414] is a weight guideline for men and women.
- Figure 24.11 [p.415] diagrams the human urinary system.
- Figures 24.12 and 24.13 [pp.416, 417] look at the anatomy and function of the human kidney.
- Figure 24.14 [p.418] describes feedback control of ADH.

## Interactive Exercises

*Hips and Hunger* [p.402]

### 24.1. THE NATURE OF DIGESTIVE SYSTEMS [pp.403–404]
### 24.2. HUMAN DIGESTIVE SYSTEM [pp.404–409]

**Selected Words:** leptin [p.402], ghrelin [p.402], *incomplete* digestive system [p.403], *complete* digestive system [p.403], *mechanical processing and motility* [p.403], *secretion* [p.403], *digestion* [p.403], *absorption* [p.403], *elimination* [p.403], *crop* [p.403], *gizzard* [p.403], *ruminants* [p.404], *enamel* [p.404], *dentin* [p.404], *pulp* [p.404], *salivary glands* [p.404], *epiglottis* [p.406], *heartburn* [p.406], *acid reflux* [p.406], *chyme* [p.406], *peptic ulcers* [p.406], duodenum [p.406], "brush border" [p.407], emulsification [p.407], bile salts [p.408], "bulk" [p.409], colon [p.409], *appendicitis* [p.409], *constipation* [409], *colon polyps* [p.409], *colonoscopy* [p.409], *virtual colonoscopy* [p.409]

## Boldfaced Terms

[p.403] digestive system _____

_____

[p.406] esophagus _____

_____

[p.406] sphincter _____

_____

[p.406] stomach _____

_____

[p.406] gastric fluid _____

_____

[p.406] small intestine _____

_____

[p.406] liver _____

_____

[p.406] gall bladder _____

_____

[p.406] pancreas _____

_____

[p.406] villi _____

_____

[p.407] microvilli _____

_____

[p.407] bile _____

_____

[p.408] micelle formation _____

_____

[p.409] leptin _____

_____

[p.409] ghrelin _____

_____

[p.409] gastrin _____

[p.409] secretin _____

[p.409] cholecystokinin _____

[p.409] large intestine _____

[p.409] rectum _____

[p.409] appendix _____

*Fill in the Blanks*

With (1) _____ [p.402] percent of adults overweight or obese, (2) _____ [p.402] are among the fattest people in the world. Excess weight often leads to (3) _____ [p.402] disease, (4) _____ [p.402], and some types of (5) _____ [p.402]. When we take in more calories than we use, fat storage cells in (6) _____ [p.402] tissue increase in size.

To extract calories from food, most animals have a (7) _____ [p.403] system. A simple sac-like one with one opening is said to be (8) _____ [p.403] while a tubular one with both a (9) _____ [p.403] to take in food and an (10) _____ [p.403] to eliminate wastes is said to be (11) _____ [p.403]. In the latter, the tube is usually divided into (12) _____ [p.403] areas, each with a unique function. Examples of these areas in birds include the (13) _____ [p.403], which stores food, and the (14) _____ [p.403], which grinds it. Because meat and insects don't need much digestion, animals that eat them usually have short (15) _____ [p.403] while animals which depend on (16) _____ [p.403] have much longer (15). Certain hoofed mammals called (17) _____ [p.404] have a multi-chambered stomach in which cellulose slowly undergoes digestion.

*Labeling*

Label the numbered parts of this illustration of the human digestive system; then note, in the parentheses, the function of each part. [p.405]

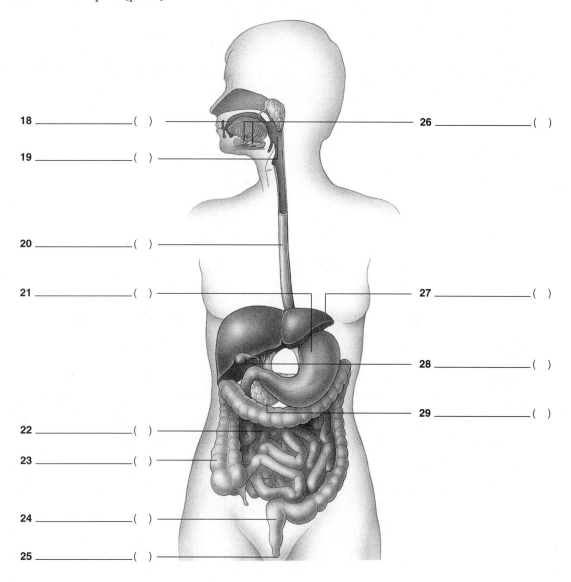

18 _____ ( )
19 _____ ( )
20 _____ ( )
21 _____ ( )
22 _____ ( )
23 _____ ( )
24 _____ ( )
25 _____ ( )
26 _____ ( )
27 _____ ( )
28 _____ ( )
29 _____ ( )

a. stores and concentrates bile
b. secretes a mixture of water, mucus, buffers, and a carbohydrate-digesting enzyme
c. opening through which feces are expelled
d. moves food from the pharynx to the stomach
e. major digestion and absorption of nutrients occurs here
f. entrance to the esophagus and the respiratory system
g. concentrates and stores undigested matter, absorbs water
h. secretes many digestive enzymes, buffers, and insulin
i. its distension stimulates expulsion of feces
j. entrance to system, food moistened, mechanical digestion begins
k. secretes bile to aid fat digestion, involved in carbohydrate and protein digestion
l. stores food, many pathogens killed by acidic fluid, protein digestion begins

*Choice*

30. \_\_\_\_ The complete digestive system of an adult human is about _____ meters long. [p.404]

    a. 2   b. 4   c. 7   d. 10

31. \_\_\_\_ Teeth specialized for grinding are called _____ . [p.404]

    a. incisors   b. molars   c. both a. and b.   d. neither a. nor b.

32. \_\_\_\_ Saliva contains _____ . [p.405]

    a. amylase   b. buffers   c. both a. and b.   d. neither a. nor b.

33. \_\_\_\_ The ring of smooth muscle that opens and closes the openings of the stomach is called a _____ . [p.406]

    a. duodenum   b. epiglottis   c. esophagus   d. sphincter

34. \_\_\_\_ The epithelium of the stomach secretes _____ . [p.406]

    a. amylase   b. hydrochloric acid   c. both a. and b.   d. neither a. nor b.

35. \_\_\_\_ Enzymes secreted by the stomach digest _____ . [p.406]

    a. proteins   b. lipids   c. both a. and b.   d. neither a. nor b.

36. \_\_\_\_ The first part of the small intestine is called the _____ . [p.406]

    a. colon   b. duodenum   c. ileum   d. jejunum

37. \_\_\_\_ Villi are most important because they increase _____ in the small intestine. [p.406]

    a. absorption   b. contractions   c. both a. and b.   d. neither a. nor b.

38. \_\_\_\_ Bile salts accelerate the digestion of _____ . [p.407]

    a. carbohydrate   b. fat   c. both a. and b.   d. neither a. nor b.

39. \_\_\_\_ A diet high in bulk can lower the chances of _____ . [p.409]

    a. colon cancer   b. appendicitis   c. both a. and b.   d. neither a. nor b.

## 24.3. HUMAN NUTRITIONAL REQUIREMENTS [pp.410–413]
## 24.4. WEIGHTY QUESTIONS, TANTALIZING ANSWERS [pp.413–414]

**Selected Words:** "empty calories" [p.410], "trans fats" [p.411], *complete* proteins [p.411], *incomplete* proteins [p.411], "low-carb" diet [p.411], antioxidants [p.412], *body mass index* (BMI) [p.414]

### Boldfaced Terms

[p.411] essential fatty acids _____

[p.411] essential amino acids _____

[p.412] vitamins _____

[p.412] minerals _____

[p.414] kilocalorie _____

## Fill in the Blanks

Refined sugars are often considered (1) _____ _____ [p.410] in the diet since they supply nothing but energy. Foods rich in (2) _____ _____ [p.410] supply the same energy in addition to (3) _____ [p.410], vitamins, and minerals. Although your body can make most of the fatty acids it needs, some, like linoleic acid, are considered (4) _____ [p.411]. (5) _____ _____ [p.411] are a good source of these fatty acids. Overconsumption of saturated fats, (6) _____ [p.411], and (7) _____ [p.411] fats also increases the risk of (8) _____ _____ [p.411] and (9) _____ [p.411]. Incomplete proteins lack high enough levels of one or more (10) _____ _____ _____ [p.411] which are needed to build your own proteins. Most (11) _____ [p.411] proteins are incomplete.

The USDA has replaced its food pyramid with a set of nutritional guidelines. In these guidelines, it is suggested that, for a 2000-calorie diet, you take in (12) _____ [p.411] cups of fruits, (13) _____ [p.411] cups of vegetables, (14) _____ [p.411] cups of dairy products, (15) _____ [p.411] ounces of grains, and (16) _____ [p.411] ounces of meat/beans/eggs each day. That means that approximately (17) _____ [p.411] percent of the calorie intake comes from complex carbohydrates and none comes from refined sugars.

## Complete the Tables [pp.412–413]

18.

| Vitamin | Common Source | Main Function |
|---|---|---|
| a. | b. | Visual pigments, bones, teeth |
| c. | Fish liver oils, egg yolk | d. |
| Vitamin E | e. | f. |
| g. | Enterobacteria, leafy vegetables | h. |
| Vitamin B$_1$ | i. | j. |
| Folic acid | k. | l. |
| m. | n. | Collagen synthesis, bone, teeth |

19.

| Mineral | Common Source | Main Function |
|---|---|---|
| a. | b. | Bones, teeth, blood clotting |
| c. | Marine fish, shellfish | d. |
| e. | f. | Hemoglobin, cytochromes |
| Magnesium | g. | h. |

*Fill in the Blanks*

In the United States, it is expected that life expectancy could decrease by (20) _____ [p.414] years if current weight trends continue. Some weight-related problems include (21) _____ _____ [p.414], (22) _____ [p.414] disease, and (23) _____ [p.414] and (24) _____ [p.414] cancers. At present, children are (25) _____ [p.414] as fat than in 1980, which does not bode well for the future. The (26) _____ [p.414] is an indicator of obesity-related health risk. (27) _____ [p.414] alone cannot reduce this risk measure since the body reduces the (28) _____ [p.414] rate to conserve energy. To reduce the (26), you must reduce (29) _____ [p.414] intake while (30) _____ [p.414] regularly.

## 24.5. URINARY SYSTEM OF MAMMALS [p.415]
## 24.6. HOW THE KIDNEYS MAKE URINE [pp.416–418]
## 24.7. WHEN KIDNEYS BREAK DOWN [p.419]

**Selected Words:** *extracellular* fluid [p.415], ureter [p.415], urinary bladder [p.415], urethra [p.415], *glomerular* capillaries [p.416], Bowman's capsule [p.416], proximal tubule [p.416], loop of Henle [p.416], distal tubule [p.416], collecting duct [p.416], *peritubular* capillaries [p.416], *countercurrent exchange system* [p.418], *thirst center* [p.418], *diabetes insipidus* [p.418], *hyperaldosteronism* [p.418], *kidney stones* [p.419], *renal failure* [p.419], *kidney dialysis machine* [p.419], *hemodialysis* [p.419], *peritoneal dialysis* [p.419]

*Boldfaced Terms*

[p.415] urinary excretion _____

[p.415] urea _____

[p.415] urinary system _____

[p.415] kidneys _____

[p.415] urine _____

[p.416] nephrons _____

_____

[p.416] glomerular filtration _____

_____

[p.416] tubular reabsorption _____

_____

[p.417] tubular secretion _____

_____

[p.418] ADH _____

_____

[p.418] aldosterone _____

_____

## Labeling

Label the numbered parts of this illustration of the human urinary system; then note, in the parentheses, the function of each part. [p.415]

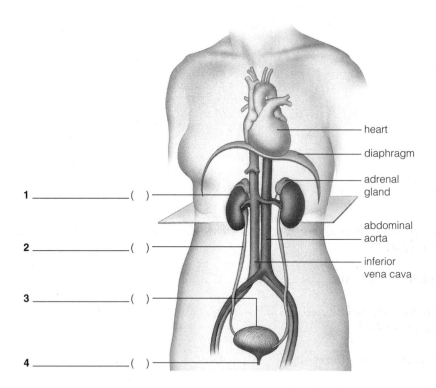

a. Temporarily stores urinary wastes
b. Channel for urine flow to the body's surface
c. Channel for urine flow from the kidney
d. Filters water, solutes. Reclaims some. Excretes wastes.

## Fill in the Blanks

All mammals gain water by absorption from the (5) _____ [p.415] and as a byproduct of (6) _____ [p.415]. The major control over water loss is exerted by the (7) _____ [p.415] system. Water is also lost from the (8) _____ [p.415] surface and from the skin through (9) _____ [p.415]. In addition to water, urinary wastes contain (10) _____ [p.415], produced as a waste product when amino acids are broken down and (11) _____ _____ [p.415], produced from the breakdown of nucleic acids.

## Labeling

Label the numbered parts of this illustration of the nephron. [p.416]

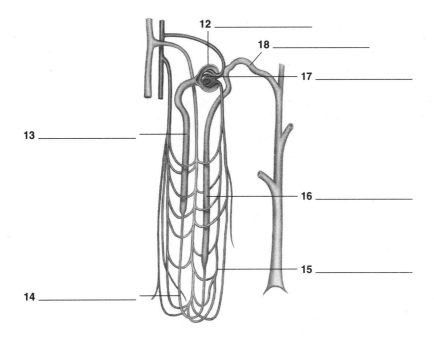

## Choice

19. _____ Blood pressure forces all but _____ from the glomerular capillaries into Bowman's capsule. [p.416]

   a. proteins   b. salts   c. sugars   d. water

20. _____ Most of the material filtered out at Bowman's capsule is reabsorbed by the _____ capillaries. [p.417]

   a. coronary   b. glomerular   c. peritubular   d. pulmonary

21. _____ About how much water that exits the glomerular capillaries is returned to the blood? [p.417]

   a. 10%   b. 33%   c. 66%   d. 99%

22. _____ Excess $H^+$ can be permanently eliminated by _____ . [p.417]

   a. buffers   b. the urinary system   c. both a. and b.   d. neither a. nor b.

23. _____ Which hormone decreases water loss from the human urinary system? [p.418]

   a. ADH   b. aldosterone   c. both a. and b.   d. neither a. nor b.

24. _____ Kidney problems can be caused by _____ . [p.419]

   a. arsenic poisoning   b. diabetes   c. both a. and b.   d. neither a. nor b.

25. _____ Kidney stones often form from _____ . [p.419]

   a. calcium salts   b. uric acid   c. both a. and b.   d. neither a. nor b.

## Self-Quiz

_____ 1. What percentage of Americans are overweight? [p.402]
   a. 20%
   b. 40%
   c. 60%
   d. 80%

_____ 2. A separate mouth and anus are found in _____ digestive systems. [p.403]
   a. complete
   b. incomplete
   c. both a. and b.
   d. neither a. nor b.

_____ 3. Which of the following are important in the mechanical processing of food? [pp.403–404]
   a. gizzards
   b. teeth
   c. both a. and b.
   d. neither a. nor b.

_____ 4. Bile is produced by the _____ and helps digest _____ . [p.407]
   a. liver; fats
   b. liver; proteins
   c. pancreas; fats
   d. pancreas; proteins

_____ 5. Which structure absorbs the majority of nutrients? [pp.406–407]
   a. liver
   b. pancreas
   c. small intestine
   d. stomach

_____ 6. The majority of digestion is completed in the _____ . [p.409]
   a. liver
   b. pancreas
   c. small intestine
   d. stomach

_____ 7. In the USDA recommended diet, the majority of calories come from _____ . [p.411]
   a. complex carbohydrates
   b. fats
   c. proteins
   d. sugars

_____ 8. Which vitamin is a precursor of visual pigments? [p.412]
   a. vitamin A
   b. vitamin C
   c. vitamin D
   d. vitamin K

_____ 9. The structure that transports urine from the kidney to the bladder is the _____ . [p.415]
   a. glomerulus
   b. loop of Henle
   c. urethra
   d. ureter

_____ 10. The glomerular capillaries are _____ other capillaries. [p.416]
   a. more leaky than
   b. less leaky than
   c. as leaky as

_____ 11. Sodium reabsorption by the kidneys is triggered by the hormone _____ . [p.418]
   a. ADH
   b. aldosterone
   c. ghrelin
   d. insulin

## Chapter Objectives/Review Questions

1. Compare and contrast complete and incomplete digestive systems. [p.403]
2. What five tasks are performed by a complete digestive system? [p.403]
3. What adaptations do ruminants have to enhance digestion of hard-to-digest plant materials? [p.404]
4. How do peptic ulcers form? [p.406]
5. Describe the adaptations found in the small intestine for increased absorption of nutrients. [pp.406–407]
6. How do bile salts aid the digestion of fats? [pp.407–408]
7. What are the primary functions of the large intestine? [p.409]
8. What problems can be caused by a low-fiber diet? [p.409]
9. How does eating simple sugars often lead to obesity? [p.410]
10. Describe the USDA recommended diet. [p.411]
11. What are vitamins? Why are they important? What can happen if you have too little or too much? [p.412]
12. Why are minerals needed in your body? [p.413]
13. Calculate your BMI. Does it indicate that you are at increased risk for weight-related health problems? [p.414]
14. Describe the human urinary system. [p.415]
15. Describe the functions of the nephron. [pp.416–417]
16. How do kidneys control acid–base balance? [p.417]
17. What are the main causes of kidney problems? [p.419]
18. Compare and contrast hemodialysis and peritoneal dialysis. [p.419]

## Integrating and Applying Key Concepts

1. Explain what happens to all the components of a fast-food burger, fries, and a shake as they make their way through your digestive track.
2. Giant pandas have a carnivore-style digestive track but eat mainly bamboo. What kinds of problems would this cause?
3. There are many different diets that claim to be good for weight loss. After reading this chapter, what do you think the best weight-loss regimen would be? Why?

# 25

# NEURAL CONTROL AND THE SENSES

## INTRODUCTION

This chapter focuses on the anatomy and physiology of the nervous system and includes discussion of the central and peripheral nervous systems and the special senses.

## FOCAL POINTS

- Figure 25.1 [p.423] diagrams a typical motor neuron.
- Figures 25.3 and 25.4 [pp.424, 425] show how ion flow and action potentials are related.
- Figure 25.5 [p.426] looks at the chemical synapse.
- Figure 25.7 [p.428] diagrams information flow in neurons.
- Figure 25.9 [p.429] illustrates a reflex arc.
- Figure 25.10 [p.430] shows various types of nervous systems.
- Figures 25.12 and 25.13 [pp.431, 432] diagram the human nervous system.
- Figures 25.15, 25.16, 25.17, and 25.18 [pp.434, 435] illustrate various parts of the human brain.
- Figure 25.22 [p.439] diagrams a sampling of sensory receptors found in human skin.
- Figures 25.23 and 25.24 [p.440] illustrate chemoreceptors.
- Figure 25.25 [p.441] shows the human eye.
- Figures 25.28 and 25.29 [p.443] look at rods and cones.
- Figure 25.32 [p.445] diagrams the human ear.

---

## Interactive Exercises

*In Pursuit of Ecstasy* [p.422]

### 25.1. NEURONS—THE GREAT COMMUNICATORS [pp.423–425]
### 25.2. HOW MESSAGES FLOW FROM CELL TO CELL [pp.426–427]
### 25.3. THE PATHS OF INFORMATION FLOW [pp.428–429]

**Selected Words:** MDMA [p.422], *stimulus* [p.423], *input* zones [p.423], *trigger* zone [p.423], *conducting* zone [p.423], *output* zones [p.423], *excitable* cells [p.424], *threshold potential* [p.424], "all-or-nothing" event [p.425], *presynaptic cell* [p.426], *postsynaptic cell* [p.426], acetylcholine (ACh) [p.426], norepinephrine [p.427], epinephrine (adrenaline) [p.427], dopamine [p.427], serotonin [p.427], GABA [p.427], *diverging* circuits [p.428], *converging* circuits [p.428], *reverberating* circuits [p.428], myelin sheath [p.428], *multiple sclerosis* [p.428], *stretch reflex* [p.429]

## Boldfaced Terms

[p.423] neurons _____

[p.423] sensory neurons _____

[p.423] interneurons _____

[p.423] motor neurons _____

[p.423] effectors _____

[p.423] dendrites _____

[p.423] axon _____

[p.423] neuroglia _____

[p.423] resting membrane potential _____

[p.424] action potential _____

[p.424] positive feedback _____

[p.426] chemical synapse _____

[p.426] neurotransmitter _____

[p.427] synaptic integration _____

[p.428] nerves _____

[p.429] reflex _____

[p.428] muscle spindles _____

*Short Answer*

1. Describe the lure and the serious dangers of experimenting with the drug Ecstasy. [p.422]
_____
_____
_____
_____

*Identification*

Identify the numbered parts of this illustration; then note, in the parentheses, the description or function of each part. [pp.423–424]

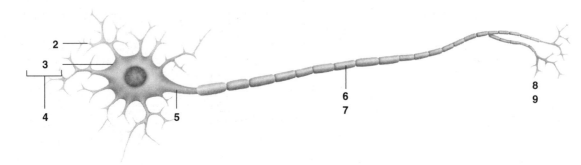

2. _____ ( ); 3. _____ _____ ( ); 4. _____ zone ( );

5. _____ zone ( ); 6. _____ ( ); 7. _____ zone ( );

8. _____ _____ ( ); 9. _____ zone ( ).

a. A slender, long extension for conduction located near the trigger zone
b. Output zones
c. The cell body and dendrites
d. Area of the neuron containing the nucleus
e. Composed of axon endings
f. Slender extensions from the cell body that are typical for input
g. Pathway of impulse transmission between the input and output zones
h. A patch of plasma membrane with a great number of gated sodium ion channels

*Matching*

Match each term with its description.

10. _____ neurons [p.423]
11. _____ sensory neurons [p.423]
12. _____ interneurons [p.423]
13. _____ motor neurons [p.423]
14. _____ neuroglial cells [p.423]
15. _____ excitable cells [p.424]
16. _____ action potential [p.424]

a. Neurons and other cells whose plasma membranes show a charge difference that reverses abruptly and briefly in response to outside stimulation
b. Relay signals from interneurons to the body's effectors—muscles and glands—that carry out the specified responses
c. The ever increasing, inward flow of sodium that intensifies as an outcome of its own occurrence; more $Na^+$ flows into the neuron → neuron becomes more positive inside → more gated channels for $Na^+$ open → more $Na^+$ flows into the neuron, and so on
d. The communication units of nervous systems

17. \_\_\_\_ positive feedback [p.424]
18. \_\_\_\_ all-or-nothing event [p.425]
19. \_\_\_\_ stimulus [p.423]

e. A brief charge reversal across a plasma membrane caused by an ever-accelerating flow of ions through channels; sets in motion a series of reversals like itself
f. Located in the spinal cord and brain; receive and process sensory input, and send signals to other neurons
g. Forms of energy, such as light or sound, detected by specific receptors
h. Detect and relay information about stimuli to the spinal cord and brain
i. Diverse cells that metabolically assist, protect, and structurally support neurons
j. The voltage peaks once threshold has been reached; once the positive-feedback cycle starts, nothing stops this full spiking

*Fill in the Blanks*

A (20) _____ [p.426] is a tiny cleft between a neuron's output zone and the input zone of another cell. Some clefts are between two (21) _____ [p.426], and others between a (22) _____ [p.426] and a muscle cell or a gland cell. The (23) _____ [p.426] neuron stores neurotransmitter in vesicles. Arrival of an action potential induces the vesicles to fuse with the plasma membrane and dump molecules of (24) _____ [p.426] into the synaptic cleft. (24) molecules diffuse across the cleft and bind with membrane receptors on the (25) _____ [p.426] cell. Binding induces the formation of a/an (26) _____ [p.426] channel in the (25) cell membrane. Ion flow through the channel is a signal that may (27) _____ [p.426] the (25) cell.

(28) _____ [p.427] signals help drive the membrane toward the threshold.
(29) _____ [p.427] signals have the opposite effect. One of the neurotransmitters, (30) _____ (ACh) [p.426] has excitatory or inhibitory effects on different target cells in the brain, spinal cord, glands, and muscles. For example, motor neurons release ACh at junctions with muscle fiber cells. ACh molecules diffuse across (31) _____ [p.426] and bind with membrane receptors for it on muscle cells. The (32) _____ [p.426] effects can initiate muscle contraction. (33) _____ [p.427] targets the brain and controls mood and memory. Low levels of (33) can lead to (34) _____ [p.427]. (35) _____ [p.427] arouses the body to respond to stress. (36) _____ [p.427] influences learning, fine motor control, and feelings of pleasure. (37) _____ [p.427] is the most common inhibitory signal in the brain. Other modifying substances like (38) _____ [p.427] suppress pain and improve mood.

The process of (39) _____ [p.427] sums all of the signals from different communication lines that reach a neuron's input zone at the same time. Broadly speaking, signals flow from receptor endings of (40) _____ [p.428] neurons to interneurons, then to (41) _____ neurons [p.428]. Billions of (42) _____ [p.428] in your brain engage in something like block parties. In (43) _____ [p.428] circuits, the dendrites and axons of neurons fanning out from one block communicate with other blocks. In (44) _____ [p.428] circuits, signals from many are relayed to just a few. In (45) _____ [p.428] circuits, neurons synapse on themselves, repeating signals like gossip that just won't go away. Such circuits make (46) _____ [p.428] muscles twitch rhythmically while you sleep.

(47) _____ [p.428] are bundles of the long axons of many sensory neurons, motor neurons, or both. Some neuroglial cells wrap like jelly rolls around axons in nerves. They form a (48) _____ [p.428] sheath, an electrical insulator that enhances the propagation of action potentials. In (49) _____ _____ [p.428], an autoimmune disorder, (50) _____ [p.428] blood cells act as if certain proteins on (48) sheaths in the spinal cord are nonself.

(51) _____ [p.429] are automatic motions in response to stimuli. (52) _____ [p.429] neurons synapse directly on motor neurons in the simplest reflex arcs. In the (53) _____ [p.429] reflex, a muscle contracts after gravity or some other load stretches it.

## 25.4. TYPES OF NERVOUS SYSTEMS [pp.430–431]

*Selected Words: radial* symmetry [p.430], bilateral nervous systems [p.430], *central* nervous system [p.431], *peripheral* nervous system [p.431], *afferent* [p.431], *efferent* [p.431]

### Boldfaced Terms

[p.430] nerve net _____

_____

[p.430] ganglion (plural, ganglia) _____

_____

[p.430] cephalization _____

_____

### Choice

Choose from the following:

      a. nerve net    b. bilateral nervous systems    c. evolution of brain and spinal cord

1. _____ Flatworms are the simplest animals having this nervous system. [p.430]
2. _____ Evolved in the earliest fishlike vertebrates from a hollow, tubular nerve cord [p.431]
3. _____ Possessed by radial animals [p.430]
4. _____ Ganglia coordinate signals from paired sensory organs and afford some control over the nerves. [p.430]
5. _____ All interneurons are inside the central nervous system. [p.431]
6. _____ A loose mesh of nerve cells intimately associated with epithelial tissue [p.430]
7. _____ Extends through the body but does not focus information [p.430]
8. _____ A concentration of sensory cells in the body's leading end—not trailing end—was favored. [p.430]
9. _____ The oldest regions still deal with reflex coordination of respiration and other vital functions. [p.431]

# 25.5. THE PERIPHERAL NERVOUS SYSTEM [pp.432–433]
# 25.6. THE CENTRAL NERVOUS SYSTEM [pp.433–436]

***Selected Words:*** *spinal* nerves [p.432], *cranial* nerves [p.432], *fight–flight response* [p.433], *white* matter [p.433], *gray* matter [p.433], *meningitis* [p.433], *paraplegia* [p.433], *tetraplegia* [p.433], forebrain [p.434], midbrain [p.434], hindbrain [p.434], neural tube [p.435], cerebrospinal fluid [p.435], left and right hemispheres [p.435], corpus callosum [p.435], *motor* areas [p.435], primary motor cortex [p.435], premotor cortex [p.435], Broca's area [p.435], *sensory* areas [p.435], *association* areas [p.435]

## *Boldfaced Terms*

[p.432] somatic nerves _____

[p.432] autonomic nerves _____

[p.432] sympathetic neurons _____

[p.433] parasympathetic neurons _____

[p.433] spinal cord _____

[p.434] brain stem _____

[p.434] medulla oblongata _____

[p.434] cerebellum _____

[p.434] pons _____

[p.434] cerebrum _____

[p.434] thalamus _____

[p.434] hypothalamus _____

[p.435] blood–brain barrier _____

[p.435] limbic system _____

*Matching*

Match each term with its description.

1. _____ peripheral nervous system [p.432]
2. _____ somatic nerves [p.432]
3. _____ autonomic nerves [p.432]
4. _____ sympathetic nerves [pp.432–433]
5. _____ parasympathetic nerves [p.433]
6. _____ meningitis [p.433]
7. _____ fight–flight response [p.433]
8. _____ spinal cord [p.433]

a. Spinal and cranial nerves dealing with smooth muscle, cardiac muscle, and glands; they carry signals to and from visceral organs and structures
b. Dominate when the body is not receiving much outside stimulation; these nerves tend to slow down overall activity and divert energy to housekeeping
c. A vital expressway for signals between the peripheral nervous system and the brain; direct reflex connections occur here; signals travel up and down inside white matter; helps control reflexes for limb movements
d. Includes thirty-one pairs of spinal nerves that connect with the spinal cord and twelve pairs of cranial nerves that connect directly with the brain
e. Parasympathetic input slows down, giving way to sympathetic signals that raise your heart rate and blood pressure; you sweat more and breathe faster; adrenal glands secrete epinephrine to put you in a state of arousal
f. Carry signals about moving the head, trunk, and limbs; sensory axons of these nerves deliver messages from receptors in skin, skeletal muscles, and tendons to the central nervous system; their motor axons deliver commands from the brain and spinal cord to skeletal muscles
g. A deadly inflammation of the three coverings around the brain and spinal cord; caused by viral or bacterial infection
h. Dominate in times of sharpened awareness, stress, excitement, or danger; digestion and other basic housekeeping tasks tend to be shelved

*Identification*

Identify the numbered parts of this illustration of the brain; then note, in the parentheses, the function of each part. [p.434]

9. _____ _____ ( );
10. _____ ( );
11. _____ ( );
12. _____ ( );
13. _____ ( ).

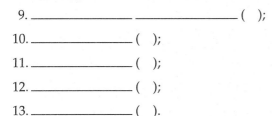

a. Helps control motor skills and posture
b. A coordinating center for sensory input and a relay station for input to the cerebrum
c. Holds reflex centers for respiration, blood circulation, and other vital tasks; can integrate motor responses with many complex reflexes, such as coughing, and it influences sleep
d. The premier center for homeostatic control over the internal environment; central to the behaviors related to internal organ activities, such as thirst, hunger, and sex, and to emotional expression, such as sweating with fear
e. A traffic center for information passing between the cerebellum and higher integrating centers of the forebrain

*Matching*

14. _____ neural tube [p.435]
15. _____ brain stem [p.434]
16. _____ hindbrain [p.434]
17. _____ midbrain [p.434]
18. _____ forebrain [p.434]
19. _____ blood–brain barrier [p.435]
20. _____ motor areas [p.435]
21. _____ sensory areas [p.435]
22. _____ association areas [p.435]
23. _____ limbic system [p.435]

a. Both hemispheres of the cerebrum, the thalamus, and the hypothalamus
b. Protects the brain and spinal cord from harmful substances and exerts some control over which solutes enter cerebrospinal fluid
c. Help us perceive the meaning of specific sensations
d. Medulla oblongata, cerebellum, and pons
e. Persists in vertebrates from the nerve cord of chordate embryos; expands into a brain and spinal cord as the embryo develops
f. Integrate information that brings about our conscious actions
g. Integrates visual input and the motor response in fish and amphibians
h. In the gray layer of the cerebral cortex; affects voluntary activity by controlling skeletal muscle movement
i. Most ancient brain tissue; found in all three brain areas
j. Encircles the upper brain stem; controls emotions and has memory roles; includes the hypothalamus, parts of the thalamus, the amygdala, cingulate gyrus, and hippocampus

## 25.7. DRUGGING THE BRAIN [pp.436–437]

**Selected Words:** psychoactive drugs [p.436], stimulants [p.436], *caffeine* [p.436], *nicotine* [p.436], *cocaine* [p.436], *amphetamines* [p.437], *Ecstasy* [p.437], *crystal meth* [p.437], depressants and hypnotics [p.437], *alcohol* [p.437], *analgesics* [p.437], *endorphins* and *enkephalins* [p.437], *morphine, codeine,* and *heroin* [p.437], *fentanyl* and *oxycodone* [p.437], hallucinogens [p.437], *ketamine* and *PCP* [p.437], marijuana [p.437], *Cannabis* [p.437]

*Boldfaced Term*

[p.436] drug addiction _____

_____

*Choice*

Choose from the following:

      a. stimulants    b. depressants, hypnotics    c. analgesics    d. hallucinogens

1. _____ Depending on the dose and physiological and emotional states, they invite calmness, drowsiness, sleep, coma, or death [p.437]
2. _____ Caffeine, nicotine, and cocaine [p.436]
3. _____ Endorphins and enkephalins are natural types [p.437]
4. _____ Skew sensory perception by interfering with acetylcholine, norepinephrine, or serotonin [p.437]
5. _____ Amphetamines induce massive release of dopamine and norepinephrine; addicts often smoke, snort, inject, or gulp various forms [p.437]
6. _____ Alcohol, or ethyl alcohol, alters cell functions; it is lipid soluble and crosses the blood–brain barrier [p.437]
7. _____ Narcotic forms include morphine, codeine, and heroin [p.437]
8. _____ Synthetic amphetamine or MDMA; users say it enhances sex, trust, and calmness [p.437]
9. _____ Marijuana; like a depressant in low doses; slows down but does not impair motor activity; relaxes the body and elicits a mild euphoria; it also causes disorientation, anxiety bordering on panic, and hallucinations [p.437]
10. _____ Abusers inhale a granular form of cocaine or burn crack cocaine and inhale the smoke; this extremely addictive drug has huge social and economic costs [pp.436–437]

## 25.8. OVERVIEW OF SENSORY SYSTEMS [p.438]
## 25.9. SOMATIC SENSATIONS [p.439]

**Selected Words:** *sensory adaptation* [p.438], *perception* [p.438], *sensation* [p.438], free nerve endings [p.439], Meissner's corpuscle [p.439], bulb of Krause [p.439], Ruffini endings [p.439], Pacinian corpuscles [p.439], *somatic* pain [p.439], *visceral* pain [p.439]

*Boldfaced Terms*

[p.438] stimulus _____

_____

[p.438] thermoreceptors _____

_____

[p.438] mechanoreceptors _____

_____

[p.438] pain receptors _____

[p.438] chemoreceptors _____

[p.438] osmoreceptors _____

[p.438] photoreceptors _____

[p.438] somatic sensations _____

[p.438] special senses _____

[p.439] pain _____

## Fill in the Blanks

A (1) _____ [p.438] is a specific form of energy detected by a sensory receptor.
(2) _____ [p.438] detect forms of mechanical energy. (3) _____ [p.438] detect heat energy. (4) _____ [p.438] receptors detect tissue damage. (5) _____ [p.438] detect chemical energy of specific substances dissolved in the fluid surrounding them. (6) _____ [p.438] detect changes in solute concentration in the surrounding fluid. (7) _____ [p.438] detect visible and ultraviolet light. The brain assesses a stimulus by determining (8) _____ [p.438] nerve pathways are carrying action potentials, the (9) _____ [p.438] of action potentials on each axon in the pathway, and the (10) _____ [p.438] of axons that the stimulus recruited. An animal's brain is prewired, or genetically (11) _____ [p.438], to interpret action potentials in certain ways. Receptors fire action potentials more often when the stimulus is (12) _____ [p.438]. A stronger (13) _____ [p.438] can recruit more sensory receptors than a weaker stimulus is able to do.

In some cases the frequency of action potentials falls or stops even when a stimulus is being maintained at (14) _____ [p.438] strength. A reduced response to a stimulus is called a (15) _____ [p.438] adaptation.

(16) _____ [p.438] is actual understanding of what a stimulus means. A
(17) _____ [p.438] is simply conscious awareness of an external or internal stimulus.

(18) _____ [p.439] sensations arise when signals reach sensory areas of the cerebral cortex, the cerebrum's outermost area of gray matter. You and other mammals discern touch, pressure, cold, warmth, and pain near the body (19) _____ [p.439]. Some encapsulated receptors are common near the body's surface. (20) _____ [p.439] corpuscle adapts very slowly to vibrations of low

frequency. The bulb of (21) _____ [p.439] detects cold. (22) _____ [p.439] endings respond to steady touching and pressure, and high temperatures. (23) _____ [p.439] corpuscles are widely distributed in skin and help you sense fine textures.

(24) _____ [p.439] is a perception of injury to some body region. Sensations of (25) _____ [p.439] pain start with signals from pain receptors in skin, skeletal muscles, tendons, and joints. Sensations of (26) _____ pain [p.439] are associated with internal organs.

### 25.10. THE SPECIAL SENSES [pp.440–445]

**Selected Words:** *vomeronasal organ* [p.440], taste buds [p.440], *umami* [p.440], sclera and cornea [p.441], choroid, ciliary body, iris, and pupil [p.441], aqueous humor [p.441], ciliary muscle [p.442], *nearsightedness* [p.442], *farsightedness* [p.442], *color-blind* [p.443], *red–green color blindness* [p.443], *macular degeneration* [p.443], *cataract* [p.443], *glaucoma* [p.443], *amplitude* [p.444], *frequency* [p.444], *outer* ear [p.444], *middle* ear [p.444], *inner* ear [p.444], *basilar* membrane [p.444], organ of Corti [p.444], *tectorial* membrane [p.444], vestibular apparatus [p.445]

### Boldfaced Terms

[p.440] olfactory receptors _____

_____

[p.440] pheromones _____

_____

[p.440] taste receptors _____

_____

[p.441] vision _____

_____

[p.441] visual field _____

_____

[p.441] lens _____

_____

[p.441] retina _____

_____

[p.442] visual accommodation _____

_____

[p.442] rod cells _____

_____

[p.442] cone cells _____

_____

[p.444] hearing _____

_____

## Choice

Choose from the following:

    a. senses of smell and taste    b. the sense of balance
    c. the sense of hearing    d. the sense of vision

1. _____ The outer is adapted for gathering sounds from the air; the middle amplifies and transmits air waves to the inner. [p.444]
2. _____ The sense begins as olfactory receptors detect water-soluble or volatile (easily vaporized) substances. [p.440]
3. _____ The sense starts at the retina, a tissue of densely packed photoreceptors. [p.441]
4. _____ The pea-sized cochlea, along with the vestibular apparatus, are key components of the inner part. [p.444]
5. _____ The vestibular apparatus and three semicircular canals are involved [p.445]
6. _____ Pheromones, signaling molecules secreted from the exocrine glands of one individual, influence the social behavior of other individuals of its species. [p.440]
7. _____ Requires eyes and image perception in brain centers that receive and interpret patterns of visual stimulation [p.441]
8. _____ Rod cells detect very dim light and cone cells detect bright light. [p.442]
9. _____ Inside your mouth, throat, and especially the upper surface of your tongue, receptors are located in about 10,000 sensory organs. [p.440]
10. _____ The outermost layer consists of the sclera and cornea; the middle layer has a choroid, ciliary body, iris, and pupil; the inner layer includes the retina; the interior has a lens. [p.441]
11. _____ Amplitude and frequency of waves of compressed air [p.444]
12. _____ Axons of these cells relay signals along an olfactory tract for processing in the cerebrum. [p.440]
13. _____ The sense of color begins with red, green, and blue cone cells. [p.442]
14. _____ A thin membrane rapidly vibrates in response to air waves; from this membrane, bones of the middle ear pick up the vibrations and interact to amplify a stimulus by transmitting the force to the oval window. [p.444]

## Self-Quiz

_____ 1. The axon functions as a(n) _____ zone. [p.423]
    a. input
    b. trigger
    c. output
    d. conducting
    e. excitable

_____ 2. The brief charge reversal across a plasma membrane that is caused by an ever accelerating flow of ions through channels is known as a(n) _____ . [p.424]
    a. chemical synapse
    b. action potential
    c. reflex
    d. response
    e. synaptic integration

_____ 3. A _____ is a tiny cleft between a neuron's output zone and the input zone of another cell. [p.426]
   a. chemical synapse
   b. action potential
   c. reflex
   d. response
   e. synaptic integration

_____ 4. In which disease condition do white blood cells act as if certain proteins on myelin sheaths in the spinal cord are nonself? [p.428]
   a. stroke
   b. vertigo
   c. multiple sclerosis
   d. meningitis
   e. color blindness

_____ 5. _____ are the simplest animals having a bilateral nervous system. [p.430]
   a. Sea anemones
   b. Jellyfish
   c. Earthworms
   d. Fish
   e. Flatworms

_____ 6. Parasympathetic nerves _____ . [p.432]
   a. represent a vital expressway for signals between the peripheral nervous system and the brain
   b. tend to slow down overall activity and divert energy to housekeeping
   c. cause the fight–flight response
   d. carry signals about moving the head, trunk, and limbs
   e. dominate in times of sharpened awareness, stress, excitement, or danger

_____ 7. The _____ is a traffic center for information passing between the cerebellum and higher integrating centers of the forebrain. [p.434]
   a. cerebrum
   b. medulla oblongata
   c. brain stem
   d. pons
   e. hypothalamus

_____ 8. The _____ is the main center of homeostatic control of the internal environment. [p.434]
   a. hypothalamus
   b. pons
   c. cerebellum
   d. cerebrum
   e. thalamus

_____ 9. The stimulant _____ fans pleasure by blocking the reabsorption of norepinephrine, dopamine, and other neurotransmitters, so postsynaptic cells aren't released from stimulation. [p.436]
   a. nicotine
   b. amphetamine
   c. cocaine
   d. caffeine
   e. MDMA

_____ 10. The senses of smell and taste depend on _____ . [p.440]
   a. thermoreceptors
   b. mechanoreceptors
   c. osmoreceptors
   d. photoreceptors
   e. chemoreceptors

_____ 11. Rods and cones are located in the _____ . [p.442]
   a. choroids
   b. iris
   c. lens
   d. retina
   e. sclera

_____ 12. The vestibular apparatus functions in the sense of _____ . [p.445]
   a. balance
   b. smell
   c. taste
   d. vision
   e. hearing

# Chapter Objectives/Review Questions

1. Discuss the biochemical and physical dangers of an Ecstasy overdose. [p.422]
2. Identify the structural features and functional zones of a motor neuron; know the functions of each. [p.423]
3. A brief charge reversal caused by an ever accelerating flow of ions through channels across a plasma membrane is a/an _____ _____. [p.424]
4. Explain why, once threshold has been reached, an action potential is an "all or nothing" event. [p.425]
5. A _____ _____ is a tiny cleft between a neuron's output zone and the input zone of another cell. [p.426]
6. Be able to define: *presynaptic cell, post synaptic cell, excitatory signals, inhibitory signals, neurotransmitter,* and *synaptic integration.* [pp.426–427]
7. Describe the path of information flow through the nervous system. [p.428]
8. Define *reflex* and cite an example. [p.429]
9. Radial animals have a _____ _____, a loose mesh of nerve cells intimately associated with epithelial tissue. [p.430]
10. Describe a bilateral nervous system and trace the evolution of the spinal cord and brain. [pp.430–431]
11. Distinguish between somatic and autonomic nerves as well as sympathetic and parasympathetic nerves. [pp.432–433]
12. The _____ _____ is a vital expressway for signals between the peripheral nervous system and the brain. [p.433]
13. Name and describe the components and functions of the major brain subdivisions, hindbrain, midbrain, and forebrain. [p.434]
14. A _____-_____ barrier protects the brain and spinal cord from harmful substances. [p.435]
15. Describe the human cerebrum and its motor, sensory, and association areas. [p.435]
16. The _____ system encircles the upper brain stem; it controls emotions and has roles in memory. [p.435]
17. Define *drug addiction* and be able to list and discuss examples of stimulants, depressants, hypnotics, analgesics, psychedelics, and hallucinogens. [pp.436–437]
18. Be able to list and define the main categories of sensory receptors. [p.438]
19. Describe how sensory receptors convert stimulus energy into action potentials. [p.438]
20. Know the receptors in human skin, their appearance, and the specific function of each; define *somatic* and *visceral* pain. [p.439]
21. Be able to describe the anatomy and functions of the structures concerned with each of the special senses: smell and taste, balance, hearing, and vision. [pp.440–445]

# Integrating and Applying Key Concepts

Monoamine oxidase B (MAO B) is a critical enzyme found in humans. It breaks down the chemicals that allow nerve cells to communicate and regulate blood pressure. New research has shown that cigarette smoke decreases levels of MAO B in kidneys, heart, lungs, and spleen. The damaging effects of cigarette smoke on the lungs is well-established. Speculate on the effects of decreased levels of MAO B in the previously mentioned more peripheral organs.

# 26

# ENDOCRINE CONTROLS

## INTRODUCTION

Chapter 26 describes the human endocrine system including the glands that make it up and the hormones they secrete.

## FOCAL POINTS

- Figure 26.1 [p.449] describes the endocrine glands and their hormones.
- Figure 26.2 [p.451] contrasts the mechanisms of steroid and peptide/protein hormone action.
- Figures 26.3 and 26.4 [p.452, 453] show the hormones secreted by the pituitary.
- Figures 26.6 and 26.8 [p.454, 455] show negative feedback loops.
- Figure 26.9 [p.456] illustrates the pancreatic hormones insulin and glucagon and their antagonistic roles.
- Table 26.3 [p.459] reviews the endocrine glands and their functions.

---

## Interactive Exercises

*Hormones in the Balance* [p.448]

### 26.1. HORMONES AND OTHER SIGNALING MOLECULES [pp.449–451]

**Selected Words:** *endocrine disruptors* [p.448], "target cells" [p.449], *prostaglandins* [p.449], *nitric oxide* [p.449], *signal reception* [p.450], *signal transduction* [p.450], *cellular response* [p.450], *steroid hormones* [p.450], *amine hormones* [p.450], *peptide hormones* [p.450], *protein hormones* [p.450], *androgen insensitivity syndrome* [p.450], *cyclic adenosine monophosphate (cAMP)* [p.450], *kinases* [p.450]

### *Boldfaced Terms*

[p.449] neurotransmitters _____

_____

[p.449] local signaling molecules _____

_____

[p.449] hormones _____

_____

[p.449] endocrine system _____

[p.450] second messenger _____

## Short Answer

1. What are the suspected endocrine disruptor effects of atrazine? [p.448]

## Matching

Choose the most appropriate answer for each term. [p.449]

2. _____ animal hormones
3. _____ neurotransmitters
4. _____ prostaglandins
5. _____ target cell
6. _____ nitric oxide
7. _____ local signaling molecules

a. Released from axon endings of neurons, act swiftly on target cells by diffusing across the tiny synaptic cleft between them
b. Local signaling molecules that affect smooth muscles and inflammation
c. Released by many types of cells; alters chemical conditions in local tissues
d. Local signaling molecule affecting blood pressure and blood distribution
e. Secretory products of endocrine glands, endocrine cells, and certain neurons; they typically reach nonadjacent targets by way of the bloodstream
f. Any cell that has receptors for a signaling molecule

## Choice

For questions 8–17, choose from the following:

a. steroid hormones    b. protein/peptide hormones
c. both steroid and protein/peptide hormones

8. _____ Lipid-soluble molecules derived from cholesterol; testosterone is an example [p.450]
9. _____ Bind to membrane receptors [pp.450–451]
10. _____ One example involves testosterone, abnormal receptors, and a condition called androgen insensitivity syndrome. [p.450]
11. _____ Hormones that often require assistance from second messengers [p.450]
12. _____ Lipid soluble hormones that can diffuse directly across the lipid bilayer of a cell's plasma membrane, then bind to a receptor in the cytoplasm or the nucleus [p.450]
13. _____ A hormone–receptor complex interacts with DNA in a specific gene region; it prevents or stimulates transcription of genes into mRNA; these transcripts are translated into proteins that function as enzymes and other molecules that can carry out a response to the hormonal signal [p.450]

a. steroid hormones   b. protein/peptide hormones
c. both steroid and protein/peptide hormones

14. \_\_\_\_ One type, glucagon, is released when the glucose level in blood declines; it stimulates body cells to convert their glycogen stores to glucose. [p.450]

15. \_\_\_\_ Adenylate cyclase initiates a cascade of reactions by converting ATM to cAMP; cAMP is a signal to start the conversion of many molecules of kinase to its active form; kinases activate other enzymes, and so on until an end reaction converts the glycogen stored in the cell to glucose. [pp.450–451]

16. \_\_\_\_ Others can initiate responses in the target cell by altering the properties of the membrane itself. [p.450]

17. \_\_\_\_ A muscle cell has receptors that bind insulin; this complex induces glucose transporters to move through the cytoplasm and insert themselves into the plasma membrane; such transporters help cells take up glucose faster. [p.450]

*Complete the Table*

18. Complete the table on the next page by matching each gland with its number on this illustration of the endocrine system. Then list the hormones produced by each gland and write a brief statement of hormone action. [p.449]

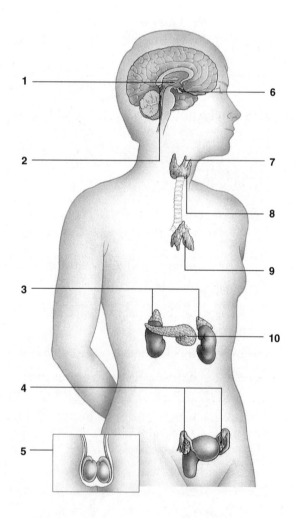

264   Chapter Twenty-Six

| Gland Name | Number on Illustration | Hormones Produced | Hormone Action |
|---|---|---|---|
| a. Hypothalamus | | | |
| b. Pituitary, anterior lobe | | | |
| c. Pituitary, posterior lobe | | | |
| d. Adrenal glands (cortex) | | | |
| e. Adrenal (medulla) | | | |
| f. Ovaries (two) | | | |
| g. Testes (two) | | | |
| h. Thyroid | | | |
| i. Parathyroids (four) | | | |
| j. Pineal | | | |
| k. Thymus | | | |
| l. Pancreatic islets | | | |

## 26.2. THE HYPOTHALAMUS AND PITUITARY GLAND [pp.452–453]

**Selected Words:** *posterior* lobe [p.452], *anterior* lobe [p.452], *pituitary gigantism* [p.453], *pituitary dwarfism* [p.453], *acromegaly* [p.453]

### Boldfaced Terms

[p.452] hypothalamus _____

_____

[p.452] pituitary gland _____

_____

[p.452] releasers _____

_____

[p.452] inhibitors _____

_____

*Identification*

Identify the targets of the posterior and anterior lobes of the pituitary gland shown in these drawings by filling in the blanks; then in the parentheses enter the name(s) (or initials) of the hormone(s) affecting that target.

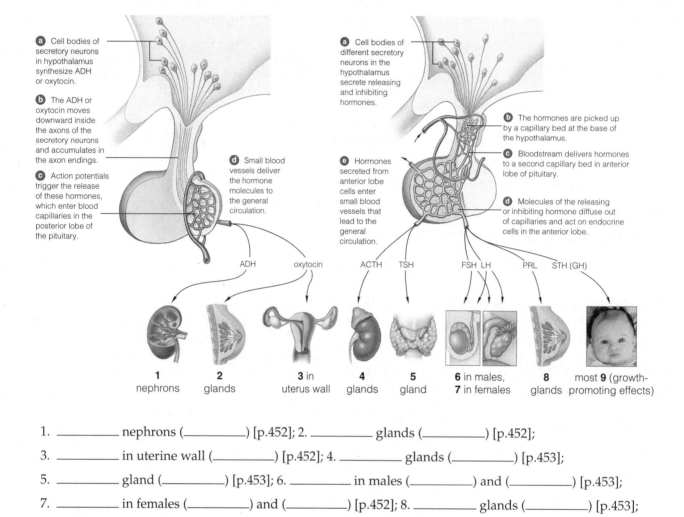

1. _____ nephrons (_____) [p.452]; 2. _____ glands (_____) [p.452];
3. _____ in uterine wall (_____) [p.452]; 4. _____ glands (_____) [p.453];
5. _____ gland (_____) [p.453]; 6. _____ in males (_____) and (_____) [p.453];
7. _____ in females (_____) and (_____) [p.452]; 8. _____ glands (_____) [p.453];
9. most _____ (growth-promoting effects) (_____) [p.453].

*Dichotomous Choice*

Circle the word(s) in parentheses that make each statement correct. [p.452]

10. The (hypothalamus/pituitary gland) is a forebrain center for homeostatic control of the internal environment, the viscera, and emotional states.
11. The (posterior/anterior) lobe of the pituitary stores and secretes two hormones, ADH and oxytocin, that are made by the hypothalamus.
12. The (posterior/anterior) lobe of the pituitary produces and secretes its own hormones that control the release of hormones from other glands.
13. Most hypothalamic hormones with anterior lobe targets are (releasers/inhibitors) which means they stimulate secretion of pituitary hormones.

266  Chapter Twenty-Six

*Matching*

Match each condition to a cause and to its characteristics. [p.453]

14. \_\_\_\_ \_\_\_\_ acromegaly
15. \_\_\_\_ \_\_\_\_ pituitary dwarfism
16. \_\_\_\_ \_\_\_\_ pituitary gigantism

Causes
a. underproduction of somatotropin in childhood
b. overproduction of somatotropin in childhood
c. overproduction of somatotropin in adulthood

Characteristics
d. Affected adults are proportioned like normal adults but larger
e. Connective tissues in the jaws, feet, and hands thicken abnormally; so do some epithelial tissues of the face and tongue
f. Affected adults are proportioned like normal adults but smaller

## 26.3. THYMUS, THYROID, AND PARATHYROID GLANDS [pp.453–454]
## 26.4. ADRENAL GLANDS AND STRESS RESPONSES [p.455]
## 26.5. THE PANCREAS AND GLUCOSE HOMEOSTASIS [pp. 456–457]
## 26.6. HORMONES AND REPRODUCTIVE BEHAVIOR [pp.457–458]

*Selected Words:* "thyroid hormone" [p.453], *goiter* [p.454], *hypothyroidism* [p.454], parathyroid hormone (PTH) [p.454], *rickets* [p.454], cortisol [p.455], aldosterone [p.455], adrenal medulla [p.455], norepinephrine [p.455], epinephrine [p.455], *long-term* stress [p.455], *Cushing syndrome* [p.455], *alpha* cells [p.456], glucagon [p.456], *beta* cells [p.456], insulin [p.456], *delta* cells [p.456], somatostatin [p.456], *diabetes mellitus* [p.457], type 1 diabetes [p.457], type 2 diabetes [p.457], estrogen [p.457], progesterone [p.457], testosterone [p.457], *libido* [p.457], melatonin [p.458], *seasonal affective disorder* [p.458]

*Boldfaced Terms*

[p.453] thymus gland _____

[p.453] thyroid gland _____

[p.454] negative feedback _____

[p.454] parathyroid glands _____

[p.455] adrenal cortex _____

[p.457] puberty _____

[p.458] pineal gland _____

[p.458] biological clock _____

## Fill in the Blanks

Secretion of many hormones, like thyroid hormone, is controlled by a/an (1) _____ _____ _____ [p.453]. In the (1) for thyroid hormone, when blood levels of the hormone rise to a certain point, it inhibits the (2) _____ [p.454] from secreting the thyroid (3) _____ _____ [p.454]. Without (3) the (4) _____ _____ [p.454] cannot secrete the tropic hormone (5) _____ _____ _____ [p.454]. Without (5) the thyroid cannot secrete (6) _____ _____ [p.454]. When levels fall below a certain (7) _____ _____ [p.454] the secretions start again.

## Matching

Match each hormone with its function.

8. _____ aldosterone [p.455]
9. _____ cortisol [p.455]
10. _____ epinephrine [p.455]
11. _____ glucagon [p.456]
12. _____ insulin [p.456]
13. _____ melatonin [p.458]
14. _____ parathyroid hormone [p.454]
15. _____ somatostatin [p.456]
16. _____ testosterone [p.457]
17. _____ thyroid hormone [p.453]

a. regulates metabolic rates
b. increases calcium levels in blood
c. acts on kidneys to regulate sodium and water reabsorption
d. helps maintain glucose levels in the brain by causing breakdown of glycogen, fatty acids, and proteins
e. increases sympathetic stimulation in fight–flight response
f. targets liver to release glucose
g. targets liver and muscle to take up glucose
h. helps control digestion and levels of insulin and glucagon
i. increases libido
j. affects the biological clock

## Short Answer

What gland and what hormone are involved with each of the following problems?

18. goiter [p.454] _____

19. rickets [p.454] _____

20. Cushing syndrome [p.455] _____

21. diabetes mellitus [p.456] _____

22. seasonal affective disorder [p.458] _____

# Self-Quiz

____ 1. Which of the following is *not* associated with peptide/protein hormones? [pp.450–451]
   a. second messengers
   b. androgen insensitivity
   c. cAMP
   d. insulin
   e. glucagon

____ 2. The _____ and the _____ interact as a major neural-endocrine control center that controls activities of other organs, many of which also have endocrine functions. [p.452]
   a. pituitary, hypothalamus
   b. pancreas, hypothalamus
   c. thyroid, parathyroid glands
   d. hypothalamus, posterior pituitary
   e. pituitary, thalamus

____ 3. Neurons of the _____ produce ADH. [p.452]
   a. anterior pituitary lobe
   b. adrenal cortex
   c. posterior pituitary lobe
   d. thyroid
   e. hypothalamus

____ 4. If you were lost in the desert and had no fresh water to drink, the level of _____ in your blood would increase as a means to conserve water. [p.452]
   a. insulin
   b. corticotropin
   c. oxytocin
   d. antidiuretic hormone
   e. salt

For questions 5–8, choose from the following answers:
   a. Estrogen
   b. Parathyroid hormone (PTH)
   c. FSH
   d. Somatotropin
   e. Prolactin

____ 5. _____ stimulates bone cells to release calcium and phosphate and the kidneys to conserve it. [p.454]

____ 6. _____ in females, maintains primary sex organs, influences secondary sexual traits. [p.457]

____ 7. _____ is the hormone associated with pituitary dwarfism, gigantism, and acromegaly. [p.453]

____ 8. _____ is secreted by the pituitary and affects both testes and ovaries.

For questions 9–12, choose from the following answers:
   a. adrenal medulla
   b. adrenal cortex
   c. thyroid
   d. anterior pituitary
   e. posterior pituitary

____ 9. The _____ produces glucocorticoids that help increase the level of glucose in blood. [p.455]

____ 10. The gland that is most closely associated with "fight–flight responses" is the _____ . [p.455]

____ 11. Metabolic rates in warm-blooded animals depend on hormones secreted by the _____ gland. [p.453]

____ 12. Oxytocin, which stimulates contraction of the uterus, is secreted by the _____ gland. [p.452]

____ 13. _____ is the hormone associated with biological clocks. [p.458]
   a. Aldosterone
   b. Oxytocin
   c. Thyroxine
   d. Melatonin
   e. Somatotropin

Endocrine Controls 269

## Chapter Objectives/Review Questions

1. Be able to describe the effects of endocrine disrupters and give an example. [p.448]
2. Give one example that illustrates the effects of *local signaling molecules*. [p.449]
3. Be able to locate and name the major components of the human endocrine system, and state the functions of each component. [Figure 26.1, p.449]
4. Contrast the proposed mechanisms of hormonal action on target cell activities by (a) steroid hormones and (b) peptide hormones that are proteins or are derived from proteins. [pp.450–451]
5. The _____ and the pituitary gland interact closely as a major brain center that controls activities of other organs, many of which have endocrine functions. [p.452]
6. Identify the hormones released from the posterior and anterior lobes of the pituitary, and state their target tissues. [pp.453–454]
7. Most of the hypothalamic hormones with anterior lobe targets are _____; they stimulate secretion of pituitary hormones; but others are _____ that slow down secretion from their targets. [p.452]
8. Pituitary dwarfism, gigantism, and acromegaly are all associated with abnormal secretion of _____ by the pituitary gland. [p.453]
9. In _____ feedback, the rise in the hormone's concentration inhibits its further secretion; with _____ feedback, the rise stimulates it. [pp.453–454]
10. Be able to describe the negative feedback associated with the adrenal cortex and cortisol. [p.455]
11. Describe the role of cortisol in chronic stress, injury, or illness. [p.455]
12. The adrenal _____ contains neurons that secrete epinephrine and norepinephrine. [p.455]
13. Describe the cause of each of the following: simple goiter, hypothyroidism, toxic goiters, and hyperthyroidism. [p.454]
14. Name the glands that secrete PTH, and state the function of this hormone; describe an ailment called rickets, and state its cause. [p.454]
15. Be able to name the hormones secreted by alpha, beta, and delta pancreatic cells; list the effect of each. [p.456]
16. Describe the symptoms of diabetes mellitus, and distinguish between type 1 and type 2 diabetes. [p.457]
17. Be able to relate some of the homeostatic controls over glucose metabolism. [pp.456–457]
18. The pineal gland secretes the hormone _____; relate its action to biological clocks, puberty, jet lag, SAD, and the winter blues. [p.458]

## Integrating and Applying Key Concepts

Current evidence appears to be quite conclusive that humans have contaminated our environment with manmade chemicals. Further, there is growing evidence that these manmade chemicals are disrupting development in frogs, alligators, and other animals. What plan would you propose to provide proof for the endocrine disruption hypothesis? What methods would you suggest for cleansing our environment of manmade chemicals?

# 27

# REPRODUCTION AND DEVELOPMENT

## INTRODUCTION

This chapter looks at the anatomy and function of the human reproductive organs. It also looks at human development from gamete formation and fertilization through birth, aging, and death.

## FOCAL POINTS

- Figure 27.4 [p.464] gives an overview of animal reproduction and development.
- Figure 27.7 [p.467] is a diagram of the reproductive system of the human male.
- Figure 27.9 [p.468] shows sperm formation in humans.
- Figure 27.10 [p.469] diagrams the reproductive system of the human female.
- Figure 27.12 [p.470] illustrates egg maturation and ovulation.
- Figure 27.13 [p.471] outlines menstruation.
- Figure 27.14 [p.472] shows fertilization in humans.
- Figure 27.15 [p.473] compares the effectiveness of various forms of birth control.
- Figure 27.18 [pp.476–477] follows human development from day 1 through day 14.
- Figure 27.19 [p.478] diagrams human placentation.
- Figure 27.20 [p.479] follows human development from day 15 through day 25.
- Figure 27.21 [p.480] shows human development at weeks 4, 5–6, 8, and 16.
- Figure 27.23 [p.483] illustrates labor and delivery.

---

## Interactive Exercises

*Mind-Boggling Births* [p.461]

### 27.1. METHODS OF REPRODUCTION [pp.462–463]
### 27.2. PROCESSES OF ANIMAL DEVELOPMENT [pp.464–466]

**Selected Words:** *parthenogenesis* [p.462], *gamete formation* [p.464], *fertilization* [p.464], *cleavage* [p.464], blastomeres [p.464], blastocoel [p.464], "maternal messages" [p.464], *gastrulation* [p.464], *organ formation* [p.464], *growth and tissue specialization* [p.464]

## Boldfaced Terms

[p.462] sexual reproduction _____

[p.462] asexual reproduction _____

[p.463] yolk _____

[p.464] oocyte _____

[p.464] blastula _____

[p.464] gastrula _____

[p.464] ectoderm _____

[p.464] endoderm _____

[p.464] mesoderm _____

[p.464] selective gene expression _____

[p.464] cell differentiation _____

[p.464] morphogenesis _____

[p.466] embryonic induction _____

[p.466] pattern formation _____

[p.466] morphogens _____

[p.466] master genes _____

[p.466] homeotic genes _____

272   Chapter Twenty-Seven

*Fill in the Blanks*

At the cellular level, (1) _____ [p.462] reproduction utilizes meiosis and gamete formation in two prospective parents. At (2) _____ [p.462], a gamete from one parent fuses with a gamete from the other to form the (3) _____ [p.462] or the first cell of the new individual. During (4) _____ [p.462] reproduction, a single parent produces offspring by one of a variety of mechanisms. New sponges (5) _____ [p.462] from parent sponges and a flatworm dividing into two flatworms represent examples of (4) reproduction. In these cases, all offspring are (6) _____ [p.462] the same as their individual parent, or nearly so. This type of reproduction is useful when gene-encoded traits are highly adapted to a limited and more or less consistent set of (7) _____ [p.462] conditions. Aphids reproduce asexually through (8) _____ [p.462], the development of offspring from (9) _____ _____ [p.462].

Female and male parents bestow different mixes of (10) _____ [p.462] on offspring. The resulting (11) _____ [p.462] in traits improves the odds that some offspring, at least, should survive and reproduce even if conditions change in the environment.

Assuring the survival of at least some (12) _____ [p.463] is also energetically costly. Many species of invertebrates, bony fishes, and frogs release motile sperm, nonmotile eggs, or both into the (13) _____ [p.463]. These animals invest (14) _____ [p.463] in making many gametes, often thousands of them. Nearly all animals on land use (15) _____ [p.463] fertilization, the union of sperm and egg within the female body. They invest metabolic energy to construct elaborate (16) _____ [p.463] organs, such as a penis and a uterus.

Finally, animals set aside energy in forms that can (17) _____ [p.463] the developing individual until it has developed enough to feed itself. Nearly all animal eggs contain (18) _____ [p.463], which is rich in proteins and lipids that nourish embryonic stages. The eggs of some species have much more yolk than others. Sea urchins make tiny (19) _____ [p.463] with little yolk, but produce enormous numbers of them. Mother (20) _____ [p.463] lay notably yolky eggs. The yolk nourishes the (20) (21) _____ [p.463] through an extended period inside an eggshell that forms after fertilization. Your mother put huge demands on her body to protect and nourish you through nine months of (22) _____ [p.463] inside her, from the time you were a nearly (23) _____ [p.463] egg.

## Sequence

Arrange the following events in correct chronological sequence. Write the letter of the first event next to 24, the letter of the second event next to 25, and so on. [p.464]

24. \_\_\_\_
25. \_\_\_\_
26. \_\_\_\_
27. \_\_\_\_
28. \_\_\_\_
29. \_\_\_\_

a. Gastrulation
b. Growth, tissue specialization
c. Fertilization
d. Organ formation
e. Gamete formation
f. Cleavage

## Complete the Table

30. Complete the table below by entering the name of the germ layer (*ectoderm*, *mesoderm*, or *endoderm*) that forms tissues and organs or the systems listed. [p.464]

| Tissues/Organs | Germ Layer |
|---|---|
| Reproductive system | a. |
| Nervous system | b. |
| Gut lining and organs derived from the gut | c. |
| Circulatory system | d. |
| Most of the skin | e. |
| Muscles | f. |
| Excretory system | g. |
| Most of the skeleton | h. |
| Connective tissues of the gut and skin | i. |

## Matching

Choose the most appropriate answer for each term.

31. \_\_\_\_ embryonic induction [p.466]
32. \_\_\_\_ cell differentiation [p.464]
33. \_\_\_\_ information in the egg [p.464]
34. \_\_\_\_ pattern formation [p.466]
35. \_\_\_\_ morphogenesis [p.464]

a. The basis is the selective activation of genes; distinctive cell structures, products, and functions are the outcome
b. A program of orderly changes in the shape, size, and proportions of an embryo results in specialized tissues and early organs; cells divide, grow, migrate, and change; tissues expand and fold and tissues die in controlled ways
c. The developmental fate of one group of embryonic cells is sealed after it is exposed to gene products of other cells in an adjoining tissue
d. Maternal messages in the form of enzymes, mRNA transcripts, and other factors stockpiled in different locations in the cytoplasm
e. Following induction, cells start to form specialized tissues and organs in ordered, spatial patterns; morphogens, master genes, and homeotic genes are involved

## 27.3. REPRODUCTIVE SYSTEM OF HUMAN MALES [pp.467–468]

*Selected Words:* secondary sexual traits [p.467], epididymis [p.467], vasa deferentia (singular, vas deferens) [p.467], seminal vesicles [p.468], prostate gland [p.468], *prostate cancer* [p.468], *testicular cancer* [p.468], prostate-specific antigen (PSA) [p.468], seminiferous tubules [p.468], spermatogonia [p.468], *primary* and *secondary* spermatocytes [p.468], spermatids [p.468], Sertoli cells [p.468], testosterone [p.468], Leydig cells [p.468], luteinizing hormone (LH) [p.468], follicle stimulating hormone (FSH) [p.468]

### Boldfaced Term

[p.467] testes (singular, testis) _____

_____

### Identification

Identify each numbered part of this illustration in the blanks provided. [p. 467]

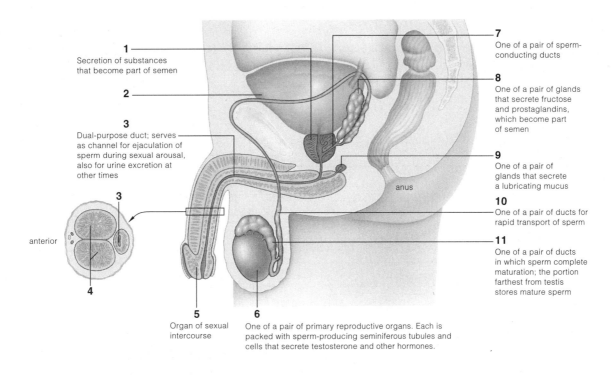

1. _____ _____; 2. _____ _____;
3. _____; 4. _____; 5. _____;
6. _____; 7. _____ _____;
8. _____ _____; 9. _____ _____;
10. _____ _____; 11. _____.

Reproduction and Development  275

## Matching

Match each item with its description. [p.467]

12. \_\_\_\_ bulbourethral gland
13. \_\_\_\_ epididymis
14. \_\_\_\_ prostate
15. \_\_\_\_ scrotum
16. \_\_\_\_ seminal vesicle
17. \_\_\_\_ seminiferous tubule
18. \_\_\_\_ spermatid
19. \_\_\_\_ spermatocyte
20. \_\_\_\_ spermatogonium

a. site of sperm development in the testes
b. immature sperm are stored here and mature
c. an immature pregametic cell that can undergo meiosis to produce more cells
d. a cell that has completed meiosis but has not yet matured into a sperm
e. a sack-like structure that keeps the testes at the right temperature for sperm development
f. a sperm-forming cell during meiosis
g. a gland that secretes signaling molecules and buffers into semen
h. a gland that secretes fructose and prostaglandin into semen
i. a gland that secretes lubricating mucus into semen

## 27.4. REPRODUCTIVE SYSTEM OF HUMAN FEMALES [pp.469–471]

**Selected Words:** *primary* reproductive organs [p.469], Fallopian tubes [p.469], myometrium [p.469], cervix [p.469], *menopause* [p.469], *follicular* phase [p.470], *luteal* phase [p.470], primary oocyte [p.470], zona pellucida [p.471], midcycle LH surge [p.471]

## Boldfaced Terms

[p.469] uterus _____

[p.469] endometrium _____

[p.469] menstrual cycle _____

[p.469] progesterone _____

[p.469] estrogen _____

[p.470] ovulation _____

[p.471] secondary oocyte _____

[p.471] polar bodies _____

[p.471] corpus luteum _____

[p.471] blastocyst _____

## Identification

Identify each numbered part of this illustration in the blanks provided. [p. 469]

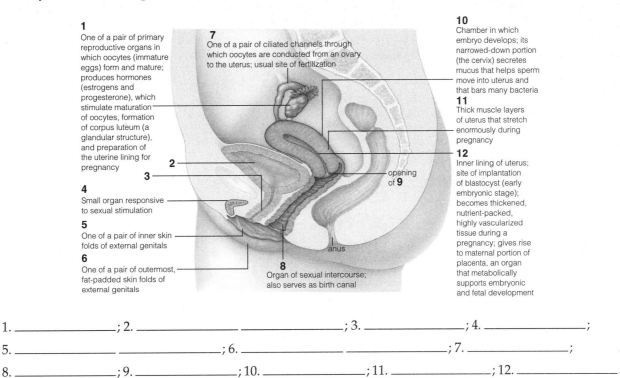

1. _____; 2. _____ _____; 3. _____; 4. _____;
5. _____ _____; 6. _____ _____; 7. _____;
8. _____; 9. _____; 10. _____; 11. _____; 12. _____.

## Matching

Match each term with its description.

13. _____ primary female reproductive organs [p.469]
14. _____ oviducts and uterus [p.469]
15. _____ endometrium [p.469]
16. _____ menstrual cycle [p.469]
17. _____ midcycle surge of LH [p.471]
18. _____ cyclic changes in the ovary [p.471]

a. Triggers ovulation or the release of a secondary oocyte from the ovary
b. A pair of ovaries that produce oocytes and secrete sex hormones
c. Uterine lining consisting of connective tissues, glands, and blood vessels
d. Follicle development with zona pellucida around the primary oocyte; antrum develops in the follicle's cell layer; mature follicle with the secondary oocyte and first polar body, ovulation, and formation of the corpus luteum
e. Pathway for the oocyte to reach the uterus and where fertilization occurs; a hollow, pear-shaped organ where embryos develop
f. Fertile periods are not synchronized with sexual receptiveness; has a follicular phase with endometrium breakdown and regeneration with oocyte maturation, an ovulation phase, and a luteal phase with formation of a corpus luteum

## 27.5. HOW PREGNANCY HAPPENS [pp.472–474]
## 27.6. SEXUALLY TRANSMITTED DISEASES [pp.474–476]

*Selected Words:* coitus [p.472], *erection* [p.472], *ejaculation* [p.472], *orgasm* [p.472], *abstinence* [p.473], *rhythm method* [p.473], *withdrawal* [p.473], *douching* [p.473], *vasectomy* [p.473], *tubal ligation* [p.473], *spermicidal foam* [p.474], *spermicidal jelly* [p.474], *diaphragm* [p.474], *condoms* [p.474], *intrauterine device* (IUD) [p.474], *birth control pills* [p.474], *birth control patch* [p.474], *Depo-Provera* injection [p.474], *Norplant* [p.474], *morning-after pills* [p.474], pelvic inflammatory disease (PID) [p.475], human papillomaviruses (HPV) [p.475], *genital warts* [p.475], *trichomoniasis* [p.475], *chlamydial infection* [p.475], *genital herpes* [p.475], type II *Herpes simplex* virus [p.475], *gonorrhea* [p.475], *syphilis* [p.475], HIV (human immunodeficiency virus) [p.475], AIDS (Acquired Immune Deficiency Syndrome) [p.475]

### Boldfaced Terms

[p.473] ovum (plural ova) _____

_____

[p.474] sexually transmitted diseases, or STDs _____

_____

### Fill in the Blanks

During sexual stimulation in males, a/an (1) _____ [p.472] occurs when more blood enters the penis than exits. A similar enlargement occurs in the female's (2) _____ [p.472]. Continued mechanical stimulation during (3) _____ [p.472], sexual intercourse, often leads to a/an (4) _____ [p.472] which, in males, includes (5) _____ [p.472] which deposits semen in the female's (6) _____ [p.472]. (7) _____ [p.472] in the semen make their way to the female's (8) _____ [p.473] where fertilization usually takes place. Receptors in the head of the (7) bind to the (9) _____ _____ [p.473], a jelly-like layer that surrounds the egg. The sperm release digestive enzymes that allow a single sperm to enter the (10) _____ [p.473]. The (10) is then triggered to complete (11) _____ [p.473] and the two haploid nuclei fuse to form the (12) _____ [p.473].

## Matching

Match each item with its description.

13. _____ abstinence [p.472]
14. _____ rhythm method [p.472]
15. _____ withdrawal [p.472]
16. _____ douching [p.473]
17. _____ vasectomy [p.473]
18. _____ tubal ligation [p.473]
19. _____ spermicidal foam and jelly [p.474]
20. _____ diaphragm [p.474]
21. _____ condoms [p.474]
22. _____ IUD [p.474]
23. _____ birth control pill [p.474]
24. _____ birth control patch [p.474]
25. _____ Depo-Provera and Norplant [p.474]
26. _____ morning-after pills [p.474]

a. A flexible, dome-shaped device inserted into the vagina and positioned over the cervix before intercourse; with spermicide, 84 percent effective
b. Progestin injections or implants that block ovulation; 96 percent effective
c. Cauterizing or cutting and tying off a woman's oviducts; the most effective
d. Chemically rinsing the vagina right after intercourse; unreliable
e. Thin, tight-fitting sheaths worn over the penis during intercourse; with spermicide, up to 95 percent effective
f. Avoiding sex during a woman's fertile period; 74 percent effective
g. Reduces pregnancy odds; Previn suppresses ovulation and blocks corpus luteum secretions; most effective when taken early but some are effective up to five days after intercourse
h. Cutting and tying off each vas deferens of a man; the most effective
i. Orally delivers synthetic estrogens and progesterone-like hormones that block maturation of oocytes and ovulation; 94 percent effective
j. Not engaging in sex at all; the most reliable way to avoid conception
k. Removing the penis from the vagina before ejaculation; a moderately effective method
l. Poisons sperm; chemicals are transferred from an applicator into the vagina before sex; 75 percent effective
m. Applied to skin; delivers the same synthetic hormones as oral contraceptives and blocks ovulation the same way
n. Is placed in the uterus by a physician; may contain progesterone or copper

## Choice

Choose from the following sexually transmitted diseases: [p.475]

a. pelvic inflammatory disease (PID)   b. human papilloma viruses (HPV)   c. trichomoniasis
d. chlamydial infection   e. genital herpes   f. gonorrhea   g. syphilis   h. AIDS

27. _____ Caused by the spirochete, *Treponema pallidum*; eight weeks after infection, a flattened, painless chancre forms; treponemes multiply in the spinal cord, brain, eyes, bones, joints, and mucous membranes; chronic immune reactions may severely damage the brain and spinal cord and lead to general paralysis

28. _____ Primarily a young person's disease; most infected women have no symptoms; untreated men risk inflammation of the epididymes and infertility

29. _____ A secondary outcome of some bacterial STDs; scars the reproductive tract and typically causes chronic pain, infertility, and tubal pregnancies

a. pelvic inflammatory disease (PID)   b. human papilloma viruses (HPV)   c. trichomoniasis
d. chlamydial infection   e. genital herpes   f. gonorrhea   g. syphilis   h. AIDS

30. _____ Infection by HIV leads to this incurable STD; the immune system weakens over time, inviting opportunistic infections; spreads by anal, vaginal, and oral intercourse and intravenous drug use

31. _____ This infection is the most widespread and fastest growing STD in the United States; a few cause genital warts in women on the vagina, cervix, external genitals, and around the anus; in men the growths form on the penis and scrotum

32. _____ Infection is by a flagellated protozoan that causes a yellowish discharge, soreness, and itching of the vagina; infected men are usually symptom free

33. _____ Transmission to new hosts requires direct transmission with the active virus; mucous membranes of the mouth and genitals are vulnerable; painful, small blisters may form on the vulva, cervix, urethra, or anal tissues of infected women; blisters form on the penis and anal tissues of infected men; the virus is reactivated sporadically

34. _____ The bacterium typically enters the body at the mucous membranes of the urethra, cervix, and anal canal during intercourse; an infected female may notice slight vaginal discharge or burning while urinating; oviduct entry causes severe cramps, fever, vomiting, and sterility; in a male, yellow pus oozes from the penis and urination becomes painful and more frequent

## 27.7. HUMAN PRENATAL DEVELOPMENT [pp.476–483]

**Selected Words:** *embryonic* period [p.476], trophoblast [p.477], inner cell mass [p.477], amniotic cavity [p.477], at-home pregnancy tests [p.478], chorionic villi [p.478], "primitive streak" [p.479], *spina bifida* [p.479], *rubella* [p.482], *thalidomide* [p.482], *anti-acne drugs* [p.482], *fetal alcohol syndrome* (FAS) [p.482], secondhand smoke [p.483]

### Boldfaced Terms

[p.476] fetus _____

[p.477] implantation _____

[p.477] amnion _____

[p.477] yolk sac _____

[p.477] allantois _____

[p.477] chorion _____

[p.478] placenta _____

[p.479] somites _____

*Sequence*

Arrange the following human developmental events up to the fetal stage in the proper chronological sequence. Write the letter of the first stage next to 1. The letter of the final process is written next to 12.

1. __e__ [p.477]
2. __j__ [p.477]
3. __l__ [p.477]
4. __h__ [p.477]
5. __k__ [p.477]
6. __a__ [pp.477–478]
7. __d__ [p.478]
8. __i__ [p.479]
9. __f__ [p.479]
10. __g__ [p.479]
11. __b__ [p.479]
12. __c__ [p.480]

a. After endoderm and mesoderm form; pattern formation begins; inductions and interactions among classes of master genes map out basic body plan
b. The blastocyst stops menstruation and avoids sloughing off by secreting HCG (basis of at-home pregnancy tests)
c. By the eighth week, the features define a human fetus; optimal birthing time is 38 weeks after the estimated time of fertilization
d. Formation of extraembryonic membranes, the amnion, yolk sac, chorion, and allantois
e. Zygote
f. Somites and pharyngeal arches form
g. Placenta formation makes possible the exchange of gases and nutrients between the mother and the new individual
h. Implantation; the blastocyst adheres to the uterine lining and invades the mother's tissues; the inner cell mass consists of two flattened cell layers in the shape of a disk
i. Neural tube formation from two folds that will merge
j. Cleavage; genes are now expressed
k. Gastrulation gives rise to primary tissue layers; development of the embryo proper begins
l. A blastocyst with an inner cell mass, the embryonic disk, forms by the fifth day

*Choice*

Choose from the following threats to human development:

a. nutrition   b. risk of infections   c. prescription drugs
d. alcohol   e. cocaine   f. tobacco smoking

13. _____ FAS; symptoms include reduced brain size, a small head size, mental impairment, facial deformities, slow growth, poor coordination, and often heart problems; any amount may harm the fetus. [p.482]

14. _____ Mother must have a well-balanced diet; most fetal organs are highly vulnerable to these deficiencies. [p.482]

15. _____ If used by a pregnant woman, it disrupts the nervous system of her future child. [p.482]

16. _____ None should be used by pregnant women except under close medical supervision; thalidomide and anti-acne drugs are examples of damage potential. [p.482]

17. _____ In cases of exposure to the second-hand type, children were smaller, had twice as many heart abnormalities, and died of more postdelivery complications. [p.483]

18. _____ IgG antibodies in the pregnant woman's blood protect her developing child from all but the most serious bacterial infections; those that are virus-induced diseases can be dangerous in the first six weeks after fertilization. [p.482]

## 27.8. FROM BIRTH ONWARD [pp.483–485]

**Selected Words:** parturition [p.483], *postpartum depression* [p.483], *aging* [p.484], *cancer* cells [p.484], *normal* cells [p.484], *Werner's syndrome* [p.485]

## Boldfaced Terms

[p.483] labor _____

_____

[p.483] oxytocin _____

_____

[p.484] lactation _____

_____

[p.484] prolactin _____

_____

[p.484] telomeres _____

_____

## Matching

Match each term with its description.

1. _____ labor [p.483]
2. _____ oxytocin [p.483]
3. _____ lactation [p.484]
4. _____ prolactin [p.484]
5. _____ CRH [p.483]

a. A hormone that calls for synthesis of enzymes used in milk production; secreted by the anterior lobe of the mother's pituitary gland
b. Secreted by the posterior pituitary; increases uterine contractions during labor
c. The birth process; the amnion typically breaks just prior to birth
d. Contributes to the timing of labor; may also contribute to postpartum depression
e. Milk production

## Fill in the Blanks

As for many species, humans change in size and proportion until reaching (6) _____ [p.484] maturity. The (7) _____ [p.484] period extends from the zygote to the fetus. The (8) _____ [p.484] period extends from the newborn until old age. Postnatal growth is most rapid between the years (9) _____ [p.484] and (10) _____ [p.484]. Not until (11) _____ [p.484] are the bones fully mature. Body tissues gradually deteriorate through (12) _____ [p.484] processes. No known human has lived beyond (13) _____ [p.484] years.

Unlike normal cells, (14) _____ [p.484] cells can continue to divide beyond a cell's normal life span. This is thought to occur because the enzyme (15) _____ [p.485] replaces the (16) _____ [p.485] at the end of the chromosomes that are usually lost during cell divisions. One theory of aging says the cloned sheep, (17) _____ [p.485], suffered age-related problems and premature death because her chromosomes came from a mature donor cell that had already lost (16) and thus could not divide as many times as a normal egg cell. Another theory of aging states that damage (18) _____ [p.485] over time and eventually causes age-related problems.

282 Chapter Twenty-Seven

# Self-Quiz

1. The process of cleavage most commonly produces a(n) _____. [p.464]
   a. zygote
   b. blastula
   c. gastrula
   d. embryo
   e. organ

2. The formation of three primary tissues or germ layers occurs during _____. [p.464]
   a. gastrulation
   b. cleavage
   c. pattern formation
   d. morphogenesis
   e. neural plate formation

3. Muscles differentiate from _____ tissue. [p.464]
   a. ectoderm
   b. mesoderm
   c. endoderm
   d. parthenogenetic
   e. yolky

4. Following embryonic induction, _____ occurs to begin the formation of specialized tissues and organs in ordered, spatial patterns. [p.466]
   a. contact inhibition
   b. ooplasmic localization
   c. cell differentiation
   d. pattern formation
   e. none of the above

5. Master genes are activated by _____. [p.466]
   a. morphogens
   b. agents of cell death
   c. agents of cell differentiation
   d. concentrations of gap gene products
   e. degradable molecules that diffuse as long-range signals

For questions 6 and 7, choose from the following:
   a. Sertoli cells
   b. seminiferous tubules
   c. vas deferens
   d. seminal vesicle
   e. prostate

6. Sperm form in the _____. [p.468]

7. The _____ connects the epididymis on the surface of the testis with the ejaculatory duct. [p.468]

For questions 8–10, choose from the following:
   a. blastocyst
   b. amnion
   c. yolk sac
   d. oviduct
   e. cervix

8. The _____ lies between the uterus and the vagina. [p.469]

9. The _____ is a pathway from the ovary to the uterus. [p.469]

10. The _____ results from the process known as cleavage. [p.464]

For questions 11 and 12, choose from the following answers:
   a. AIDS
   b. Chlamydial infection
   c. Genital herpes
   d. Gonorrhea
   e. Syphilis

11. _____ is a disease caused by *Treponema* that produces a localized ulcer (a chancre). [p.475]

12. _____ is an incurable disease caused by the human immunodeficiency virus. [p.475]

13. The extraembryonic membrane that will enclose the embryo is the _____. [p.477]
    a. chorion
    b. yolk sac
    c. allantois
    d. amnion
    e. implantation membrane

14. Which of the following is *not* associated with the postnatal period? [p.484]
    a. adolescent
    b. fetus
    c. infant
    d. child
    e. pubescent

Reproduction and Development

## Chapter Objectives/Review Questions

1. Be able to state the effects of fertility drugs. [p.462]
2. Understand how asexual reproduction differs from sexual reproduction. Know the costs and benefits associated with having separate sexes. [p.464].
3. The successive cuts of cleavage produce the developmental stage known as a(n) _____. [p.464]
4. Describe early embryonic development and distinguish among the following: gamete formation, fertilization, cleavage, gastrulation, organ formation, and growth and tissue specialization. [p.464]
5. Name each of the three embryonic tissue layers and the organs and systems formed from each. [p.464]
6. Be able to discuss the significance of maternal messages in the oocyte. [p.464]
7. Distinguish between cell differentiation and morphogenesis. [pp.464–466]
8. The emergence of specialized tissues and organs from groups of embryonic cells in predictably ordered, spatial patterns is called _____ _____. [p.466]
9. Name the primary male reproductive organs; follow the path of a mature sperm from the seminiferous tubules to the urethral exit. List every structure encountered along the path, and state the contribution to the nurturing of the sperm. [pp.467–468]
10. Name the four hormones that directly or indirectly control male reproductive function. [p.468]
11. Know all the major structures of the female reproductive tract and state the principal function of each. [pp.469–470]
12. Distinguish the follicular phase of the menstrual cycle from the luteal phase, and explain how the two cycles are synchronized by hormones from the anterior pituitary, hypothalamus, and ovaries. [pp.470–471]
13. State which hormonal event brings about ovulation and which other hormonal events bring about the onset and finish of menstruation. [pp.470–471]
14. Describe the events occurring in sexual intercourse and fertilization. [p.470–473]
15. Identify the three most effective birth control methods used in the United States and the four least effective birth control methods. [pp.473–474]
16. For each STD described, know the causative organism and the symptoms of the disease. [p.475]
17. Describe the events that occur during the first month of human development; include the developed structures of the blastocyst as seen at about 14 days. [pp.476–479]
18. List and describe the four extraembryonic membranes and their functions. [pp.476–477]
19. Characterize the human developmental events occurring from day 15 to day 25. [pp.479–480]
20. Describe the characteristics of the developing human fetus during the second trimester. [pp.480–481]
21. Explain why the mother must be particularly careful of her diet, health habits, and lifestyle during the first trimester after fertilization (especially during the first six weeks). [pp.482–483]
22. Generally, describe the process of labor or delivery. [p.483]
23. List the factors that bring about aging of an animal; compare the programmed life span hypothesis with the cumulative assaults hypothesis. [pp.484–485]

## Integrating and Applying Key Concepts

What rewards do you think a society should give a woman who has at most two children during her lifetime? In the absence of rewards or punishments, how can a society encourage women not to have abortions and yet ensure that the human birth rate does not continue to increase?

# 28

# POPULATION ECOLOGY

## INTRODUCTION

Chapter 28 looks at the characteristics of populations including distribution, density, and life-history patterns. It also applies this to human populations.

## FOCAL POINTS

The introductory section looks at the processes that can occur when a population grows beyond the size that the resources can support.

- Figure 28.6 [p.493] is an idealized S-shaped curve of logistic population growth.
- Table 28.1 [p.495] illustrates a life table for humans in the United States in 2001.
- Figure 28.9 [p.496] contrasts various survivorship curves.
- Figure 28.12 [p.498] is the growth curve for the human population.
- Figure 28.14 [p.500] compares age structure diagrams of several countries with idealized diagrams.
- Figure 28.15 [p.501] illustrates the demographic transition model.

## Interactive Exercises

*The Human Touch* [p.488]

### 28.1. CHARACTERISTICS OF POPULATIONS [pp.489–490]

***Selected Words:*** *pre-reproductive, reproductive,* and *post-reproductive* ages [p.489], *habitat* [p.489], *crude* density [p.489], clumped, nearly uniform, or random distribution patterns [p.489]

### *Boldfaced Terms*

[p.489] ecology _____

[p.489] demographics _____

[p.489] population size _____

[p.489] age structure _____

[p.489] reproductive base _____

[p.489] population density _____

[p.489] population distribution _____

[p.490] quadrats _____

[p.490] capture–mark–recapture methods _____

*Short Answer*

1. Briefly, state the lesson to be learned from the habitation of Easter Island. [p.488]

_____

*Matching*

Match each term with its description.

2. _____ ecology [p.489]
3. _____ demographics [p.489]
4. _____ population size [p.489]
5. _____ age structure [p.489]
6. _____ pre-reproductive, reproductive, and post-reproductive ages [p.489]
7. _____ reproductive base [p.489]
8. _____ population density [p.489]
9. _____ population distribution [p.489]
10. _____ crude density [p.489]
11. _____ clumped, nearly uniform, or random [p.489]
12. _____ quadrats [p.490]
13. _____ capture–mark–recapture method [p.490]

a. Includes pre-reproductive and reproductive age categories
b. A measured number of individuals in some specified area
c. Sampling areas of the same size and shape, such as rectangles, squares, and hexagons
d. The number of individuals in some specified area or volume of a habitat
e. The number of individuals in each of several to many age categories
f. The vital statistics of the population
g. An example of age categories in a population
h. Capturing mobile individuals and marking them in some fashion and recapturing them after some period of time
i. The systematic study of how organisms interact with one another and their physical and chemical environment
j. Possible distribution patterns shown by a population
k. The general pattern in which individuals are dispersed in a specified area
l. The number of individuals that represent the population's gene pool

286   Chapter Twenty-Eight

*Population Density Problem*

14. In a capture–mark–recapture experiment, 20 squirrels were captured and marked. A week later, 25 squirrels were captured and it was noted that 5 of those had been previously marked. What is the estimated population size? [p.490]

## 28.2. POPULATION SIZE AND EXPONENTIAL GROWTH [pp.491–492]
## 28.3. LIMITS ON THE GROWTH OF POPULATIONS [pp.492–494]

**Selected Words:** *capita* [p.491], J-shaped curve [p.492], *actual* rate of increase [p.492], *sustainable* supply of resources [p.493], S-shaped curve [p.494], *bubonic plague* [p.494], *pneumonic plague* [p.494]

### Boldfaced Terms

[p.491] immigration _____

[p.491] emigration _____

[p.491] migration _____

[p.491] zero population growth _____

[p.491] per capita _____

[p.491] $r$ _____

[p.492] exponential growth _____

[p.492] doubling time _____

[p.492] biotic potential _____

[p.493] limiting factor _____

[p.493] carrying capacity _____

[p.493] logistic growth _____

[p.494] density-dependent controls _____

_____

[p.494] density-independent factors _____

_____

## Fill in the Blanks

Population size increases as a result of births and (1) _____ [p.491]—the arrival of new residents from other populations of the species. Population sizes decrease as a result of deaths and (2) _____ [p.491], whereby individuals permanently move out of the population. Population size also changes on a predictable basis as a result of daily or seasonal events called (3) _____ [p.491]. However, this is a recurring round trip between two regions, so these transient effects need not be initially considered in a study of population size. If it is assumed that immigration is balancing emigration over time, the effects of both on population size may be ignored. This allows the definition of (4) _____ _____ [p.491] growth as an interval during which the number of births is balanced out by the number of deaths. The population size is stabilized and has no overall increase or decrease.

Births, deaths, and other variables that might affect population size can be measured as (5) _____ _____ [p.491] rates, or rates per individual. *Capita* means (6) _____ [p.491], as in (6) counts. Visualize 2,000 mice living in a cornfield. If the female mice collectively give birth to 1,000 mice per month, the birth rate would be (7) _____ [p.491] per mouse, per month. If 200 of the 2,000 mice die in that interval, the death rate would be (8) _____ [p.491] per mouse per month.

If it is assumed that the birth rate and death rate remain constant, both can be combined into a single variable—the net (9) _____ [p.491] per individual per unit (10) _____ [p.491], or $r$. For the mouse population in the example above, the value of $r$ is (11) _____ [p.491] per mouse per month. This example provides a method of representing population growth per unit of time by the equation (12) _____ [p.491].

If the monthly increases in the individuals in a population are plotted against time, one finds a graph line in the shape of a (13) _____ [p.492]. Then one knows that the (14) _____ [p.492] growth is being tracked. The length of time it takes for a population to double in size is its (15) _____ [p.492] time. When a population displays its (16) _____ _____ [p.492] in terms of growth, it is the maximum rate of increase per individual under ideal conditions.

## Problems

For questions 17–18, consider the equation $G = rN$, where $G$ = the population growth rate per unit time, $r$ = the net population growth rate per individual per unit time, and $N$ = the number of individuals in the population. [p.491]

17. Five thousand rats live in the streets of a city. The rats give birth to 2,000 rats every month. Every month 490 of the 5,000 rats die. Calculate the following:

    a. birth rate = _____

    b. death rate = _____

    c. the value of $r$ = _____

18. If a society decides it is necessary to lower its value of $N$ through reproductive means because supportive resources are dwindling, it must lower either its net reproduction per individual per unit or its _____ _____ _____.

## Fill in the Blanks

Most of the time, (19) _____ [p.492] circumstances keep any population from fulfilling its biotic potential. Place a lone bacterial cell in a culture flask containing culture medium with glucose and other nutrients. At first, the growth pattern seems to be (20) _____ [p.493]. Then growth (21) _____ [p.493], and the population size is rather stable. Following the stable period, the population size plummets until all the bacterial cells are (22) _____ [p.493]. What happened? As the population grew ever larger, it used up more and more (23) _____ [p.493]. Nutrient scarcity became an (24) _____ [p.493] signal for cells to stop dividing; the cells starved to death.

Any essential resource that is in short supply is a (25) _____ _____ [p.493] on population growth. Examples are food, mineral ions, refuge from predators, living space, and even a pollution-free habitat. One factor alone often puts the brakes on (26) _____ [p.493] growth. Adding a new supply of nutrients to the exponentially growing (27) _____ [p.493] population would not keep it from crashing. Like all other organisms, bacteria produce (28) _____ [p.493] wastes that pollute their environment and stop their further (29) _____ [p.493] growth.

A (30) _____ [p.493] supply of resources will determine the population's size. (31) _____ _____ [p.493] is the maximum number of individuals of a population or species that a given environment can sustain indefinitely. By the (32) _____ _____ [p.493] pattern, a small population starts growing slowly in size, then it grows rapidly, and finally its size levels off once the carrying capacity is reached. The pattern plots out as a(n) (33) _____ [p.493]-shaped curve. This curve is an approximation of what actually goes on in nature. If population growth overshoots carrying capacity, the (34) _____ [p.494] rate skyrockets, and the (35) _____ [p.494] rate plummets. These two responses drive the number of individuals down to the carrying capacity or lower.

## Choice

Choose from the following [p.494]:

       a. density-dependent controls    b. density-independent factors

36. _____ A sudden freeze

37. _____ Overcrowding

38. _____ Bubonic plague, pneumonic plague

39. _____ Application of pesticides in your backyard

40. _____ Predators, parasites, and pathogens

41. _____ Droughts, floods, and earthquakes

42. _____ Deforestation

## 28.4. LIFE HISTORY PATTERNS [pp.495–498]

**Selected Words:** life tables [p.495], "survivorship" schedule [p.495], *Type I* curves [p.496], *Type II* curves [p.496], *Type III* curves [p.496]

### Boldfaced Terms

[p.495] life history pattern _____

_____

[p.495] cohort _____

_____

[p.496] survivorship curve _____

_____

### Matching

Match each term with its description.

1. _____ life history pattern [p.495]
2. _____ cohort [p.495]
3. _____ survivorship schedule [p.495]
4. _____ survivorship curve [p.496]
5. _____ type I curves [p.496]
6. _____ type II curves [p.496]
7. _____ type III curves [p.496]

a. A graph line that emerges when ecologists plot a cohort's age-specific survival in a habitat
b. Reflect high survivorship until fairly late in life, then a large increase in deaths
c. A group of individuals, tracked from the time of birth until the last one dies; data is gathered such as the number of offspring born to individuals in each age interval
d. Signify a death rate that is highest early on; typical of species that produce many small offspring and do little, if any, parenting
e. For each species, a set of adaptations that influence survival, fertility, and age at first reproduction
f. Reflect a fairly constant death rate at all ages; typical of organisms just as likely to die of disease at any age, such as lizards, small mammals, and large birds
g. May show, for example, the number of individuals reaching some specified age ($x$)

*True–False*

If the statement is true, write a T in the blank. If the statement is false, correct it by changing the underlined word(s) and writing the correct word(s) in the answer blank. [p.497]

_____ 8. Many diverse experiments support the hypothesis that natural selection <u>has not</u> been a factor in vertebrate coevolution.

_____ 9. Guppy populations deal with predators, especially pike-cichlids and <u>killifish</u>.

_____ 10. Smaller killifish prey on smaller, immature guppies but not on the larger adults; <u>larger</u> pike-cichlids live in other streams and prey on larger and sexually mature guppies.

_____ 11. Reznick and Endler hypothesized that predation is a selective agent acting on <u>killifish</u> life history patterns.

_____ 12. In pike-cichlid streams, guppies grow <u>slower</u> and their body size is smaller at maturity, compared to guppies in killifish-dominated streams.

_____ 13. Guppies hunted by pike-cichlids reproduce earlier in life, produce far <u>more</u> offspring, and they do so earlier in life.

_____ 14. Following laboratory experiments, the researchers concluded the differences between guppies preyed upon by different predators have <u>an environmental</u> foundation.

_____ 15. After study, and as predicted, body sizes of the guppy lineage subjected to killifish predation over time were <u>smaller</u> at maturity; guppies raised with pike-cichlids tended to mature earlier.

_____ 16. The control study that was initiated when Reznick and Endler first visited Trinidad showed that the <u>pike-cichlid</u> predator had influenced the body size, frequency of reproduction, and other aspects of guppy life history patterns.

_____ 17. Later laboratory experiments involving two generations of guppies confirmed that the differences studied in guppies have a genetic basis, one of the requirements for the occurrence of <u>genetic drift</u>.

## 28.5. HUMAN POPULATION GROWTH [pp.498–501]

**Selected Words:** *preindustrial* stage [p.500], *transitional* stage [p.500], *industrial* stage [p.501], *postindustrial* stage [p.501]

**Boldfaced Terms**

[p.500] total fertility rate (TFR) _____

_____

[p.500] demographic transition model _____

_____

## Short Answer

1. List the three possible reasons humans began sidestepping natural population controls. [pp.498–499]

2. Explain why continued human population growth cannot continue indefinitely. [p.499]

3. What is meant by TFR? [p.500]

4. Compare the TFRs on a worldwide basis for 1950 and 2003. What is the current replacement-level fertility for developed countries and for developing countries? [p.500]

## Matching

Match each age structure diagram with its description. [p.500]

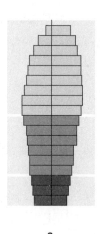
a

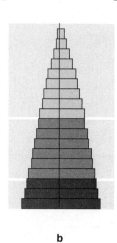
b

c

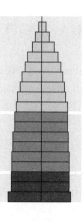
d

5. _____ zero growth
6. _____ rapid growth
7. _____ negative growth
8. _____ slow growth

*Choice*

Using the age structure diagrams on the previous page, choose the diagram that best fits the countries given below. [p.500]

9. \_\_\_\_ United States and Canada

10. \_\_\_\_ India and Mexico

*Sequence*

Arrange the following stages of the demographic transition model in correct chronological sequence.

11. \_\_\_\_
12. \_\_\_\_
13. \_\_\_\_
14. \_\_\_\_

a. Industrial stage: population growth slows dramatically and industrialization is in full swing; a slowdown starts mainly because people want to move from the country to cities, and urban people want smaller families [p.501]
b. Preindustrial stage: living conditions are the harshest before technological and medical advances become widespread; birth and death rates are high, so the growth rate is low [p.500]
c. Postindustrial stage: population growth rates become negative; the birth rate falls below the death rate, and population size slowly decreases [p.501]
d. Transitional stage: industrialization begins, food and health care improve; death rates drop, birth rates stay high in agricultural societies [pp.500–501]

*Fill in the Blanks*

The United States represents (15) _____ [p.501] percent of the world population. The United States produces (16) _____ [p.501] percent of the world's goods and services. It uses (17) _____ [p.501] percent of the world's processed minerals and a big portion of nonrenewable energy sources. People of the United States generate at least (18) _____ [p.501] percent of all pollution and trash.

By contrast, India produces only about (19) _____ [p.501] percent of all goods and services and uses (20) _____ [p.501] percent of available minerals and nonrenewable energy resources. It generates only about (21) _____ [p.501] percent of the total pollution and waste products.

*Environmental Impact Problem*

22. The average person in the United States produces how many times more pollution and waste products than the average person in India? _____ [p.501]

## Self-Quiz

\_\_\_\_ 1. The number of individuals that contribute to a population's gene pool is _____ . [p.489]
a. the population density
b. the population growth
c. the population birth rate
d. the population size
e. the population distribution

\_\_\_\_ 2. How are the individuals in a population *most often* dispersed? [p.489]
a. clumped
b. very uniform
c. nearly uniform
d. random
e. very random

_____ 3. Assuming immigration is balancing emigration over time, _____ may be defined as an interval in which the number of births is balanced by the number of deaths. [p.491]
   a. the lack of a limiting factor
   b. exponential growth
   c. saturation
   d. zero population growth
   e. logistic growth

_____ 4. The population growth rate (G) is equal to the _____ net population growth rate per individual (r) and number of individuals (N). [p.491]
   a. sum of
   b. product of
   c. doubling of
   d. difference between
   e. tripling of

_____ 5. Assuming the birth and death rate remain constant, both can be combined into a single variable, r, or _____ . [p.491]
   a. the per capita rate
   b. the minus migration factor
   c. the number of individuals
   d. the net reproduction per individual per unit

_____ 6. A population that is growing exponentially in the absence of limiting factors can be illustrated accurately by a(n) _____ . [p.492]
   a. S-shaped curve
   b. J-shaped curve
   c. curve that terminates in a plateau phase
   d. tolerance curve
   e. resource curve

_____ 7. Which of the following is *not* characteristic of logistic growth? [p.493]
   a. S-shaped curve
   b. leveling off growth as carrying capacity is reached
   c. unrestricted growth
   d. slow growth of a low-density population followed by rapid growth
   e. limiting factors

_____ 8. The maximum number of individuals of a population (or species) that a given environment can sustain indefinitely defines _____ . [p.493]
   a. the carrying capacity of the environment
   b. exponential growth
   c. the doubling time of a population
   d. density-independent factors
   e. logistic growth

_____ 9. The survivorship curve typical of industrialized human populations is type _____ . [p.496]
   a. I
   b. II
   c. III
   d. both I and II
   e. both I and III

_____ 10. The beginning of industrialization, a rise in food production, improvement in health care, rising birth rates, and declining death rates describe the _____ stage of the demographic transition model. [pp.500–501]
   a. preindustrial
   b. transitional
   c. industrial
   d. postindustrial
   e. includes both preindustrial and transitional

# Chapter Objectives/Review Questions

1. Be able to define: *ecology, demographics, population size, age structure, reproductive base, population density, population distribution, crude density, quadrats,* and *capture–mark–recapture method.* [p.490]
2. List and describe the three patterns of dispersion illustrated by populations in a habitat. [p.489]
3. Distinguish immigration from emigration; define *migration* and *zero population growth.* [p.491]
4. In the equation $G = rN$, as long as $r$ holds constant, any population will show _____ growth; be able to calculate a population growth rate. [p.490]

5. _____ _____ is the maximum rate of increase per individual under ideal conditions. [p.492]
6. List several examples of limiting factors and explain how they influence population curves. [p.493]
7. _____ _____ is the maximum number of individuals of a population or species that a given environment will sustain indefinitely. [p.493]
8. Tell how the pattern of logistic growth shows how carrying capacity can affect population size. [pp.493–494]
9. Define *density-dependent control* and *density-independent* factors of populations; cite two examples of each. [p.494]
10. Each species has a _____ _____ pattern, or a set of adaptations that influence survival, fertility, and age at first reproduction; each pattern reflects the individual's schedule of reproduction. [p.495]
11. Life insurance companies and ecologists track a _____, a group of individuals from the time of birth until the last one dies. [p.495]
12. Explain how the construction of life tables and survivorship curves can be useful to humans in managing the distribution of scarce resources. [pp.495–496]
13. After consideration of the research results obtained on guppies and their predators by Reznick and Endler, provide an explanation for these differences. [p.497]
14. Be able to list three possible reasons why growth of the human population is out of control. [pp.498–499]
15. Understand the age structure diagrams for populations growing at different rates. [p.500]
16. List and describe the four stages of the demographic transition model. [pp.500–501]
17. Be able to generally compare the implications of the resource consumption of India and the United States. [p.501]

## Integrating and Applying Key Concepts

Following your study of this chapter and with consideration of what you have learned about the problems facing planet Earth, prepare a list of positive changes you could make in your own living habits. These changes would help solve the serious problems we face as a human population. Share your knowledge and your list with family and friends and request that they each prepare a similar list. Above all, encourage everyone to *act* on the personal changes that are proposed.

# 29

# COMMUNITY STRUCTURE AND BIODIVERSITY

## INTRODUCTION

The chapter begins with a discussion of communities and the interactions of populations within them. Chapter 29 also covers succession, exotic species, patterns of diversity, and conservation.

## FOCAL POINTS

- Figures 29.4 and 29.5 [pp.506, 507] describe interspecific competition.
- Figure 29.6 [p.507] looks at resource partitioning.
- Figures 29.10, 29.11, 29.12, and 29.30 [pp.510, 511, 525] illustrate camouflage and mimicry.
- Figure 29.13, 29.14, and 29.15 [p.512] show parasitism.
- Figures 29.16 and 29.17 [pp.513, 514] illustrate ecological succession.
- Figures 29.22 and 29.23 [pp.518, 519] look at patterns of species diversity.
- Figure 29.27 [p.522] maps out areas of extreme vulnerability on the globe.
- Figure 29.28 [p.523] illustrates strip logging.

## Interactive Exercises

*Fire Ants in the Pants* [p.503]

### 29.1. WHICH FACTORS SHAPE COMMUNITY STRUCTURE? [pp.504–505]

*Selected Words:* fire ants [p.503], *fundamental* niche [p.504], *realized* niche [p.504]

*Boldfaced Terms*

[p.504] habitat _____

_____

[p.504] community _____

_____

[p.504] niche _____

[p.504] commensalism _____

[p.504] mutualism _____

[p.504] interspecific competition _____

[p.504] predation _____

[p.504] parasitism _____

[p.504] symbiosis _____

[p.505] coevolution _____

## Short Answer

1. Describe the biological controls being enlisted by ecologists in an attempt to control *Solenopsis*, the genus name of the dreadful fire ant. [p.503]

## Fill in the Blanks

Every community has a characteristic structure, which arises as an outcome of five factors. First, (2) _____ [p.504] and topography interact to dictate temperatures, rainfall, type of soil, and other habitat conditions. Second, the kinds and quantities of (3) _____ [p.504] and other resources that become available through the year affect which species live there. Third, (4) _____ [p.504] traits of each species help it (5) _____ [p.504] and exploit specific resources on the habitat. Fourth, species (6) _____ [p.504] by competition, predation, and mutually beneficial actions. Fifth, the overall pattern of population sizes affects (7) _____ [p.504] structure.

Community Structure and Biodiversity  297

*Matching*

Match each item with its description. [p.504]

8. _____ habitat
9. _____ niche
10. _____ fundamental niche
11. _____ realized niche
12. _____ indirect interactions
13. _____ commensalism
14. _____ mutualism
15. _____ intraspecific competition
16. _____ predation
17. _____ parasitism
18. _____ symbiosis

a. An interaction that directly helps one species but does not affect the other much, if at all
b. Disadvantages flow both ways between species
c. This one has some constraining factors and does shift in large and small ways over time, as individuals respond to a mosaic of changes
d. Generally means "living together"; commensalism, mutualism, and parasitism are all cases
e. Possess physical and chemical features, such as temperature, and an array of species; an organism's "address"
f. An interaction that directly benefits one species, the predator
g. The one that might prevail in the absence of competition and other factors that can constrain how individuals get and use resources
h. The distinct sum of an organism's activities and relationships as it goes about getting and using the resources required for survival and reproduction; the "profession" an organism has
i. An interaction that directly benefits one species, the parasite
j. Five categories having different effects on population growth
k. Where benefits flow both ways between the interacting species

## 29.2. MUTUALLY BENEFICIAL INTERACTIONS [p.505]
## 29.3. COMPETITIVE INTERACTIONS [pp.506–507]

**Selected Words:** *obligatory* mutualism [p.505], *intraspecific* competition [p.506], *interspecific* competition [p.506], *exploitative* competition [p.506], *scramble* competition [p.506], *Paramecium* [p.506]

*Boldfaced Terms*

[p.506] competitive exclusion _____

_____

[p.507] resource partitioning _____

_____

*Complete the Table*

1. Complete the table by describing how each of the organisms listed is intimately dependent on the other for survival and reproduction in an obligatory mutualistic symbiotic interaction. [p.505]

| Organism | Dependency |
|---|---|
| a. Yucca moth | |
| b. Yucca plant | |

*Choice*

Choose from the following:

      a. intraspecific competition    b. interspecific competition
      c. competitive exclusion    d. resource partitioning

2. \_\_\_\_ Two species of *Paramecium* are grown in the same culture but cannot coexist indefinitely. [p.506]
3. \_\_\_\_ Two species of *Paramecium* are grown in the same culture and can coexist indefinitely. [p.506]
4. \_\_\_\_ Bristly foxtail grasses, Indian mallow plants, and smartweed plants coexist in the same habitat. [p.507]
5. \_\_\_\_ Nine species of chipmunks live in different habitats on the slopes of the Sierra Nevada. [p.506]
6. \_\_\_\_ A male mockingbird chases away all other male mockingbirds from its territory. [p.506]

### 29.4. PREDATOR–PREY INTERACTIONS [pp.508–511]

*Selected Words:* type I, II, and III responses (models) [p.508], *three-level* interaction [p.509], prey defenses [p.509]

*Boldfaced Terms*

[p.508] predators _____

[p.508] prey _____

[p.509] warning coloration _____

[p.510] mimicry _____

[p.510] camouflage _____

*Matching*

Match each of the following with its term. (*Note*: The same letter may be used more than once, but use only one letter per blank.)

      a. predator–prey interactions    b. camouflage    c. warning coloration
      d. mimicry    e. last-chance trick    f. predator responses to prey

1. \_\_\_\_ Cornered earwigs, skunks, and stink beetles producing awful odors [p.510]
2. \_\_\_\_ Canadian lynx and snowshoe hare [pp.508–509]
3. \_\_\_\_ Chameleons who can blend with their background and hold themselves motionless for a long time [p.511]

Community Structure and Biodiversity

a. predator–prey interactions    b. camouflage    c. warning coloration
d. mimicry    e. last-chance trick    f. predator responses to prey

4. \_\_\_\_ Grasshopper mice plunging the noxious chemical-spraying tail end of their beetle prey into the ground to feast on the head end [p.510]
5. \_\_\_\_ Aggressively stinging yellow jackets are the likely model for nonstinging edible wasps. [p.510]
6. \_\_\_\_ Polar bears on snow, striped tigers crouched in tall-stalked grasses, and scorpionfish well hidden on the seafloor, and the preying mantis [p.511]
7. \_\_\_\_ Dangerous or repugnant species such as brightly colored poisonous frogs of the genus *Dendrobates* [p.509]
8. \_\_\_\_ Resemblance of *Lithops*, a desert plant, to a small rock [p.510]
9. \_\_\_\_ Yellow-banded wasps or an orange-patterned monarch butterfly [p.509]
10. \_\_\_\_ Baboon on the run turns to give canine tooth display to a pursuing leopard. [p.510]
11. \_\_\_\_ Least bittern with coloration similar to that of surrounding withered reeds [p.510]

## 29.5. PARASITES AND PARASITOIDS [pp.511–512]

*Selected Words:* social parasites [p.512], biological controls [p.512]

### Boldfaced Terms

[p.511] parasites _____

[p.511] vectors _____

[p.512] parasitoids _____

### Fill in the Blanks

(1) _____ [p.511] have pervasive influences on populations. They drain (2) _____ [p.511] from hosts, render hosts more vulnerable to (3) _____ [p.511], and are less attractive to potential mates. Some parasitic infections cause (4) _____ [p.511]. Some can shift the ratio of (5) _____ [p.511] males to females. In such ways, parasitic infections lower (6) _____ [p.511] rates, raise (7) _____ rates [p.511], and intervene in intraspecific and interspecific competition.

Sometimes the gradual drain of nutrients during a parasitic infection indirectly leads to (8) _____ [p.511]. The host can become so weakened that it can't fight off (9) _____ [p.511] infections. In evolutionary terms, killing a host is not good for a parasite's (10) _____ [p.511] success. Usually, only two types of interactions kill the host. First, the parasite may attack a (11) _____ [p.511] host, which has no coevolved defenses against it. Second, the host may be supporting too many active, individual (12) _____ [p.511] at the same time.

*True–False*

If the statement is true, write a T in the blank. If the statement is false, correct it by changing the underlined word(s) and writing the correct word(s) in the answer blank. [p.512]

_____ 13. Many <u>vertebrates</u> such as viruses, bacteria, protozoans, sporozoans, flatworms, roundworms, fleas, ticks, mites, and lice are parasites.

_____ 14. Dodder and mistletoe are types of <u>predatory</u> plants.

_____ 15. Cuckoos and North American cowbirds alter the <u>digestive</u> behavior of another species to complete their life cycle.

_____ 16. Because parasites and parasitoids can help control the population growth of other species, many are raised commercially and then selectively released as <u>biological controls</u>.

## 29.6. CHANGES IN COMMUNITY STRUCTURE OVER TIME [pp.513–514]

*Selected Terms:* pioneer species [p.513], *climax* community [p.513], *facilitation* [p.514], *inhibition* [p.514], *tolerance* [p.514]

*Boldfaced Terms*

[p.513] ecological succession _____

_____

[p.513] primary succession _____

_____

[p.513] secondary succession _____

_____

[p.514] intermediate disturbance hypothesis _____

_____

*Choice*

Choose from the following:

    a. primary succession    b. secondary succession    c. intermediate disturbance hypothesis

1. ____ A process that begins when pioneer species colonize a barren habitat, such as a new volcanic island [p.513]

2. ____ When trees fall in tropical forests, small gaps open in the dense tree canopy and more light reaches patches on the forest floor; then conditions favor growth of previously suppressed small trees and germination of pioneers or shade-intolerant species. [p.514]

3. ____ Included are lichens and mosses [p.513]

4. ____ A disturbed area within a community recovers. [pp.513–514]

5. ____ This pattern is common in abandoned fields, burned forests, and volcanic disturbances. [p.513]

6. ____ Many types are mutualists with nitrogen-fixing bacteria and outcompete earlier plants in nitrogen-poor habitats. [p.513]

a. primary succession    b. secondary succession    c. intermediate disturbance hypothesis

7. \_\_\_\_ The type that would occur when land is exposed by the retreat of a glacier [p.513]
8. \_\_\_\_ A freeze kills some plant species in an area and allows others to grow. [p.514]

## 29.7. FORCES CONTRIBUTING TO COMMUNITY INSTABILITY [pp.515–517]

*Selected Words:* **jump** dispersal [p.516], kudzu [p.516], *myxomatosis* [p.517]

### Boldfaced Terms

[p.515] keystone species _____

[p.516] geographic dispersal _____

[p.516] exotic species _____

### Matching

Match each item with its description.

1. \_\_\_\_ community stability [p.515]
2. \_\_\_\_ keystone species [p.515]
3. \_\_\_\_ geographic dispersal [p.516]
4. \_\_\_\_ jump dispersal [p.516]
5. \_\_\_\_ exotic species [p.516]

a. A rapid geographic dispersal mechanism, as when an insect might travel in a ship's cargo hold from an island to the mainland
b. An outcome of forces that have come into an uneasy balance
c. Resident of an established community that has moved from its home range and successfully taken up residence elsewhere; *Caulerpa*, European rabbits, and kudzu are examples
d. A dominant species that can shape community structure; the periwinkle is an example
e. Residents of established communities move out from their home range and successfully take up residence elsewhere; permanently insinuate themselves into a new community

## 29.8. PATTERNS OF SPECIES DIVERSITY [pp.518–519]

*Selected Words:* island patterns [p.518], biodiversity [p.519]

### Boldfaced Terms

[p.518] distance effect _____

[p.519] area effect _____

## Fill in the Blanks

The most striking pattern of biodiversity corresponds to distance from the (1) _____ [p.518]. For most groups of plants and animals, on land and in the seas, the number of coexisting species is greatest in the (2) _____ [p.518], and it systematically declines toward the poles. Tropical latitudes intercept more incoming (3) _____ [p.518] of higher intensity. As one outcome, (4) _____ [p.518] availability tends to be greater and more reliable in the tropics than elsewhere. All year long, different (5) _____ [p.518] species in humid tropical forests construct new leaves, flowers, and fruit. Year in and year out, they support many (6) _____ [p.518] of diverse herbivores, nectar foragers, and fruit eaters. Species diversity might be self-(7) _____ [p.518]. The diversity of (8) _____ [p.518] species in tropical forests is much greater than in comparable forests at higher latitudes. When more plant species compete and coexist, more species of (9) _____ [p.518] similarly compete and coexist, partly because no single (9) can overcome all of the chemical defenses of all the different plants. Thus, more (10) _____ [p.518] and (11) _____ [p.518] species tend to evolve in response to a diversity of prey and hosts. The same applies to diversity on tropical (12) _____ [p.518].

## Problems

After studying the distance effect, the area effect, and species diversity patterns as related to the equator, answer the following questions.

13. There are two islands (b and c) of the same size and topography that are equidistant from the African coast (a), as shown in the illustration below. Which will have the higher species diversity values? _____ [pp.518–519]

14. The two islands d and c are the same size and topography. Which will have the higher species diversity? _____ [pp.518–519]

15. The two islands c and e are the same distance from the coast and the equator but e is 10 times larger than c. Which will have the higher species diversity values? _____ [pp.518–519]

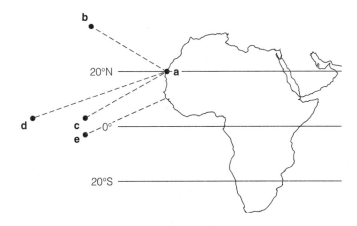

## 29.9. CONSERVATION BIOLOGY [pp.519–523]

**Selected Words:** K–T asteroid impact [p.519], *endemic* species [p.520], species introductions [p.520], overharvesting [p.520]

### Boldfaced Terms

[p.520] biodiversity _____

_____

[p.520] endangered species _____

_____

[p.520] habitat loss _____

_____

[p.520] indicator species _____

_____

[p.521] conservation biology _____

_____

[p.521] hot spots _____

_____

[p.521] ecoregion _____

_____

[p.523] strip logging _____

_____

[p.523] riparian zone _____

_____

### Matching

Match each term with its description.

1. _____ K–T asteroid impact [p.519]
2. _____ sixth major extinction event [p.520]
3. _____ endangered species [p.520]
4. _____ endemic [p.520]
5. _____ habitat loss [p.520]
6. _____ species introductions [p.520]
7. _____ conservation biology [p.521]
8. _____ hot spots [p.521]

a. Includes systematic surveys of all biological diversity, an attempt to decipher the evolutionary and ecological origins of diversity, and an attempt to identify methods to maintain and use biodiversity for the good of human populations
b. The physical reduction in suitable places to live, as well as the loss of suitable habitat through chemical pollution
c. A broad land or ocean region defined by climate, geography, and producer species
d. Any endemic species that is highly vulnerable to extinction

9. _____ ecoregion [p.521]
10. _____ strip logging [pp.522–523]
11. _____ riparian zone [p.523]

e. A relatively narrow corridor of vegetation along a stream or a river; a defense against flood damage; shade cast by the canopy of taller shrubs and trees helps conserve water; offers wildlife food, shelter, and shade
f. Delivered the *coup de grace* to some lineages although some dinosaur lineages became extinct during the 10 million years that preceded it
g. A newer method of sustainable development; a profitable cycle of logging
h. A species that originates in one region and is found nowhere else
i. Human activities
j. Another major threat to endemic species; a key factor in almost 70 percent of the cases where endemic species are being driven to extinction
k. Habitats of a large number of species that are found nowhere else and are in greatest danger of extinction

## Self-Quiz

_____ 1. _____ have physical and chemical features, such as temperatures, and an array of species. [p.504]
   a. Habitats
   b. Niches
   c. Interactions
   d. Ecosystems
   e. Climates

_____ 2. A lopsided interaction that directly benefits one species but does not affect the other much, if at all, is _____. [p.504]
   a. commensalism
   b. competitive exclusion
   c. predation
   d. mutualism
   e. parasitism

_____ 3. The relationship between the yucca plant and the yucca moth that pollinates it is best described as _____. [p.505]
   a. camouflage
   b. commensalism
   c. competitive exclusion
   d. obligatory mutualism
   e. mimicry

_____ 4. _____ is represented by foxtail grass, mallow plants, and smartweed because their root systems subdivide different areas of the soil in a field. [p.507]
   a. Succession
   b. Resource partitioning
   c. A climax community
   d. A disturbance
   e. Competitive exclusion

_____ 5. A baboon turns abruptly and displays its long, sharp canines to a pursuing leopard. This is an example of _____. [p.510]
   a. warning coloration
   b. mimicry
   c. camouflage
   d. last-chance trick
   e. obligatory mutualism

_____ 6. When an inexperienced predator attacks a yellow-banded wasp, the predator receives the pain of a stinger and will not attack again. This is an example of _____. [p.510]
   a. mimicry
   b. camouflage
   c. a prey defense
   d. warning coloration
   e. both c. and d.

Community Structure and Biodiversity 305

_____ 7. _____ is among the pervasive influences that parasites have on populations. [p.511]
   a. Draining host nutrients
   b. Predator vulnerability
   c. Sterility
   d. Death
   e. all of the above

_____ 8. During the process of community succession, _____. [p.513]
   a. pioneer populations colonize a barren habitat
   b. pioneers include lichens and mosses
   c. pioneers improve conditions for other species
   d. pioneers set the stage for their own replacement
   e. all of the above

_____ 9. The _____ is an example of a keystone species. [p.515]
   a. kudzu vine
   b. sea star
   c. mussel
   d. European rabbit
   e. barnacle

_____ 10. The most striking patterns of species diversity on land and in the seas relate to the _____. [p.518]
   a. distance effect
   b. area effect
   c. immigration rate for new species
   d. distance from the equator
   e. resource availability

## Chapter Objectives/Review Questions

1. Discuss possible controls for the two species of invading Argentine fire ants. [p.503]
2. List the five factors that shape the structure of a biological community. [p.504]
3. A _____ niche is the one that might prevail in the absence of competition and other factors that can constrain how individuals get and use resources. [p.504]
4. Be able to define the following terms and cite examples: *habitat, niche, realized niche, indirect interactions, commensalism, mutualism, interspecific competition, predation, parasitism,* and *symbiosis*. [p.504]
5. Describe and cite examples of intraspecific competition, interspecific competition, competitive exclusion, and resource partitioning. [pp.506–507]
6. Describe the interactions of predator–prey, warning coloration, mimicry, camouflage, moment-of-truth defenses, and predator responses to prey; cite examples. [pp.508–511]
7. List pervasive influences that parasites have on populations, the kinds of parasites, and tell how parasites may be potential biological control agents. [pp.511–512]
8. Be able to describe the patterns of ecological succession; include the following terms: pioneer species, climax community, and primary and secondary succession. [pp.513–514]
9. List and explain examples of forces leading to community instability. [pp.515–517]
10. Describe mainland, marine, and island patterns of diversity. [pp.518–519]
11. Be able to define the following terms and cite examples: *K–T impact, endangered species, habitat loss, species introductions, conservation biology, hot spots, ecoregion, bioeconomic analysis, sustainable development, strip logging,* and *riparian zone*. [pp.519–523]

## Integrating and Applying Key Concepts

Is there a *fundamental niche* that is occupied by humans? If you think so, describe the minimal abiotic and biotic conditions required by populations of humans in order to live and reproduce. If you do not think so, state why. (Note that *"thrive"* and *"be happy"* are not criteria.)

# 30

# ECOSYSTEMS

## INTRODUCTION

Chapter 30 looks at ecosystems and the energy and nutrient flow through them.

## FOCAL POINTS

- Figure 30.3 [p.528] illustrates a simple food chain.
- Figure 30.4 [p.529] shows how simple food chains fit together in complex food webs.
- Figure 30.7 [p.531] diagrams biomass and energy pyramids for Silver Springs, FL.
- Figure 30.9 [p.532] looks at the hydrologic cycle.
- Figure 30.12 [pp.534–535] illustrates the carbon cycle.
- Figure 30.13 [p.536] describes the greenhouse effect.
- Figures 30.14 and 30.15 [p.537] graph changes in greenhouse gases and temperature over time.
- Figure 30.16 [p.538] diagrams the nitrogen cycle.
- Figure 30.18 [p.540] shows the phosphorus cycle.

## Interactive Exercises

*Bye-Bye Bayou* [p.526]

### 30.1. THE NATURE OF ECOSYSTEMS [pp.527–529]

*Selected Words:* global interactions [p.526], *herbivores* [p.527], *carnivores* [p.527], *parasites* [p.527], *omnivores* [p.527], *scavenger* [p.527]

*Boldfaced Terms*

[p.527] ecosystem _____

[p.527] primary producers _____

[p.527] consumers _____

Ecosystems  307

[p.527] detritivores _____

_____

[p.527] decomposers _____

_____

[p.528] trophic levels _____

_____

[p.528] food chain _____

_____

[p.528] food webs _____

_____

[p.529] grazing food web _____

_____

[p.529] detrital food web _____

_____

## Short Answer

1. What effects will global warming have on coastal states? [p.526]

_____

_____

_____

_____

## Matching

Match each term with its description. [p.527]

2. _____ consumers
3. _____ herbivores
4. _____ carnivores
5. _____ parasites
6. _____ decomposers
7. _____ detritivores
8. _____ omnivores
9. _____ scavenger
10. _____ ecosystem

a. A type of consumer that dines on animals, plants, fungi, protistans, and even bacteria
b. A type of consumer that eats a living host's tissues but usually does not kill it
c. An array of organisms and their physical environment, all interacting through a one-way flow of energy and a cycling of raw materials
d. A type of consumer that eats flesh
e. Feed on tissues of other organisms; contains several sub-categories
f. Heterotrophs that ingest decomposing organic matter; examples are crabs and earthworms
g. A type of consumer that eats plants
h. An animal consumer that ingests dead plants, animals, or both all of the time or some of the time; vultures, termites, and many beetles are examples
i. Feed on the organic products and the remains of all organisms

*Fill in the Blanks*

All the organisms of an (11) _____ [p.527] can be classified by their functional roles in a hierarchy of feeding relationships called (12) _____ [p.528] levels. A (13) _____ [p.528] is a straight-line sequence of steps by which energy stored in autotroph tissues enters higher trophic levels. In reality, it is more accurate to think of food chains cross-connecting with one another, as a (14) _____ _____ [p.528]. In most cases, (15) _____ [p.528] that producers initially captured passes through no more than (16) _____ or _____ [p.528] trophic levels. All of the heat losses from an ecosystem represent a one-way flow of (17) _____ [p.528] out of the ecosystem. There are two categories of food webs. In a (18) _____ [p.529] food web, energy flows from photoautotrophs to herbivores, and then through (19) _____ [p.529]. By contrast, in a (20) _____ [p.529] food web, energy flows from photoautotrophs through detritivores and (21) _____ [p.529].

## 30.2. BIOLOGICAL MAGNIFICATION IN FOOD WEBS [p.530]

*Selected Words:* DDT [p.530], Rachel Carson [p.530], *Silent Spring* [p.530]

*Boldfaced Term*

[p.530] biological magnification _____
_____

*Short Answer*

1. Define *biological magnification,* cite an example of it, and tell what person was important in identifying this concept in nature. [p.530]
_____
_____
_____
_____
_____
_____

## 30.3. STUDYING ENERGY FLOW THROUGH ECOSYSTEMS [p.531]

*Selected Words:* ecological pyramid [p.531], *gross* primary production [p.531], *net* primary production [p.531]

*Boldfaced Terms*

[p.531] biomass pyramid _____
_____

[p.531] energy pyramid _____

[p.531] primary productivity _____

[p.531] net ecosystem production _____

## Matching

Choose the most appropriate answer for each term. [p.531]

1. _____ primary productivity
2. _____ gross primary production
3. _____ net amount
4. _____ net ecosystem production
5. _____ ecological pyramid
6. _____ energy pyramids

a. All trapped energy
b. Show how usable energy diminishes as it flows through ecosystems over time
c. All energy stored in growing plants
d. A figure arrived at by subtracting the energy used by plants and soil organisms from the gross primary production
e. Rate at which producers get and store an amount of energy in their tissues during a specified interval
f. A method ecologists use to represent trophic structure; primary producers are the base

## 30.4. GLOBAL CYCLING OF WATER AND NUTRIENTS [pp.532–533]

*Selected Words:* atmospheric cycles [p.532], sedimentary cycles [p.532]

### Boldfaced Terms

[p.532] biogeochemical cycle _____

[p.532] hydrologic cycle _____

[p.532] watershed _____

[p.533] salinization _____

[p.533] groundwater _____

[p.533] desalinization _____

## Complete the Table

1. Complete the following table by summarizing the functions of the three types of biogeochemical cycles. [p.532]

| Biogeochemical Cycle | General Function(s) |
|---|---|
| a. Hydrologic cycle | |
| b. Atmospheric cycles | |
| c. Sedimentary cycles | |

## Matching

Choose the most appropriate answer for each term.

2. _____ hydrologic cycle [p.532]
3. _____ watershed [p.532]
4. _____ effects of deforestation [p.533]
5. _____ desalinization [p.533]
6. _____ heavy irrigation [p.533]
7. _____ salinization [p.533]
8. _____ groundwater contamination [p.533]
9. _____ water pollution [p.533]
10. _____ water wars [p.533]

a. Caused by evaporation in regions where soil drains poorly; can stunt growth of crop plants, eventually kill them, and decrease yields
b. Accounts for two-thirds of the fresh water used by human populations; can itself change the land's suitability for agriculture
c. Conflicts arising from reductions in water flow, pollution, and silt buildup in aquifers, rivers, and lakes; regional, national, and global planning for the future is overdue
d. Water evaporates into the lower atmosphere, stays aloft as vapor, clouds, and ice crystals, then falls mainly as rain and snow; influenced by ocean circulation and wind patterns; moves nutrients into and out of ecosystems
e. Groundwater receiving toxic chemicals leached from some landfills, hazardous waste dumps, wells that store hazardous chemicals, and underground tanks used to store gasoline, oil, and solvents; groundwater becomes unreclaimable
f. Any region where precipitation becomes funneled into a single stream or river
g. Inputs of sewage, animal wastes, and toxic chemicals from power-generating plants and factories make water unfit to drink; runoff from fields puts sediments and pesticides into water along with other nutrients such as phosphate
h. May disrupt nutrient availability for an entire ecosystem; causes a shift in nutrient outputs
i. Removal of salt from seawater; costly fuel energy drives the processes

## 30.5. CARBON CYCLE [pp.534–535]
## 30.6. GREENHOUSE GASES, GLOBAL WARMING [pp.536–537]

*Selected Words:* carbon–oxygen cycle [p.534], peat [p.535], ancient aquatic ecosystems [p.535], Carboniferous swamp [p.535], greenhouse gases [p.536], pollutants [p.536]

## Boldfaced Terms

[p.534] carbon cycle _____

_____

[p.536] greenhouse effect _____

_____

[p.536] global warming _____

_____

### Choice

For questions 1–13, choose from the following:

        a. carbon cycle    b. global warming

1. _____ Consider air pollutants such as carbon dioxide, nitrogen oxides, chlorofluorocarbons, ozone, and other photochemical oxidants that form through chemical interactions involving the sun's rays [p.537]
2. _____ Photosynthesizers incorporate carbon atoms into organic compounds [p.534]
3. _____ Carbon–oxygen cycle, an alternate name [p.534]
4. _____ Gases impede the escape of longer, infrared wavelengths from the Earth into space [p.536]
5. _____ Carbon reservoirs → atmosphere and oceans → through organisms → carbon reservoirs; global movement [pp.534–535]
6. _____ The "greenhouse gases" [p.536]
7. _____ In ancient aquatic ecosystems, carbon incorporation occurred in staggering numbers of shells and other hard body parts of foraminiferans and other species. [p.535]
8. _____ Carbon dioxide is the most abundant form of carbon in the atmosphere; carbon dissolved in ocean water is mainly combined with oxygen as bicarbonate or carbonate. [p.534]
9. _____ Carbon enters the atmosphere by aerobic respiration, fossil fuel burning, and volcanic eruptions. [p.534]
10. _____ Warming of Earth's lower atmosphere due to accumulation of certain gases [p.536]
11. _____ The burning of fossil fuels releases carbon that was incorporated into organic compounds by plants of Carboniferous forests; this may release more carbon than can be cycled naturally to the ocean's storage reservoirs. [p.535]
12. _____ Polar ice is melting and glaciers are retreating; the sea level may have risen as much as twenty centimeters (eight inches) in the past century. [p.536]
13. _____ Atmospheric $CO_2$ is at its highest level since 420,000 years ago; it may well be the highest for the past 20 million years. [p.536]

## 30.7. NITROGEN CYCLE [pp.538–539]

**Selected Words:** nitrogen-fixing bacteria [p.538], nodules [p.538], legumes [p.538], acid rain [p.539]

### Boldfaced Terms

[p.538] nitrogen cycle _____

_____

[p.538] nitrogen fixation _____

[p.539] ammonification _____

[p.539] nitrification _____

[p.539] denitrification _____

[p.539] ion exchange _____

## Matching

Match each term with its description.

1. _____ atmospheric nitrogen [p.538]
2. _____ nitrogen-fixing bacteria [p.538]
3. _____ nitrogen fixation [p.538]
4. _____ nitrogenous waste and remains of organisms [p.539]
5. _____ clover, beans, peas, and other legumes [p.539]
6. _____ denitrification [p.539]
7. _____ nitrogen cycle description [pp.538–539]
8. _____ nitrogen scarcity [p.539]
9. _____ human impact on the nitrogen cycle [p.539]
10. _____ ion exchange [p.539]

a. Symbionts with nitrogen-fixing bacteria that form root nodules
b. Ammonium, nitrite, and nitrate that form during the nitrogen cycle are very vulnerable to leaching; nitrogen fixation comes at metabolic cost to plants
c. Degraded by bacteria and fungi; they use part of the released proteins and amino acids; plants and nitrifying bacteria use some of the leftover ammonia; bacteria strip electrons, leaving nitrite that other bacteria use to form nitrate that is taken up by plants
d. Occurs in the atmosphere (largest reservoir); only certain bacteria, volcanic action, and lightning can convert $N_2$ into forms that can enter food webs
e. A few species of soil bacteria that convert $N_2$ to ammonia ($NH_3$), which becomes ionized (as ammonium $NH_4^+$) in their cytoplasm; cyanobacteria are typically capable of this metabolism
f. Ions dissociate from soil particles, and other dissolved ions replace them; change in the composition of soil water disrupts the number and kinds of ions in exchangeable form, the types that plants take up; this can harm plants after heavy fertilizer applications
g. Triple covalent bonds join its two atoms ($N_2$), and very few organisms have the metabolic equipment to break them
h. A wasteful process by which certain bacteria convert nitrate or nitrite to $N_2$ and $N_2O$ (nitrous oxide), much of which escapes
i. A huge amount of nitrogen is lost through deforestation and grassland conversion for agriculture; this is countered by crop rotation and heavy fertilizer applications
j. Introduction of excess nitrogen into ecosystems through sewage and burning fossil fuels in power plants and vehicles

### 30.8. PHOSPHORUS CYCLE [pp.540–541]

*Boldfaced Terms*

[p.540] phosphorus cycle _____
_____

[p.541] eutrophication _____
_____

*Fill in the Blanks*

In the (1) _____ [p.540] cycle, this element passes quickly through food webs as it moves from land to ocean sediments, then slowly back to land. The Earth's crust is the largest (2) _____ [p.540], as it is for other minerals. In rock formations, phosphorus occurs mainly as phosphate (3) _____ [p.540]. Weathering and soil erosion puts phosphates into streams and rivers, then (4) _____ [p.540] sediments. The ions slowly accumulate with other minerals, mainly on the submerged shelves of continents, and form (5) _____ [p.540] deposits. Millions of years go by. Where movements of crustal plates uplift part of the (6) _____ [p.540], phosphates become exposed on drained land surfaces. In time, weathering and erosion release phosphates from exposed rocks once again.

All organisms use the phosphate (7) _____ [p.540] to make phospholipids, NADPH, ATP, nucleic acids, and other compounds. Plants take up dissolved phosphates from soil (8) _____ [p.540]. Phosphates then pass up the food chain. All animals excrete phosphates as a (9) _____ [p.540] in urine and feces. Bacterial and fungal (10) _____ [p.540] in soil release phosphates, then plants take them up again in this rapid cycle.

The (11) _____ [p.540] cycle helps move most minerals through ecosystems. Of all minerals, phosphorus is the most prevalent (12) _____ [p.540] factor in all natural ecosystems. Some (13) _____ [p.540] activities are lowering the levels of phosphorus in natural ecosystems. This is especially so for tropical and subtropical regions of (14) _____ [p.540] countries. In developed countries, soils are now overloaded with phosphorus after years of heavy applications of (15) _____ [p.540]. Dissolved phosphorus that enters streams, rivers, lakes, and estuaries can promote dense (16) _____ [p.541] blooms. In many freshwater (17) _____ [p.541], nitrogen isn't a limiting factor on algal growth because of the abundance of nitrogen-fixing bacteria. (18) _____ [p.541] is. When decomposers act on the remains of algal blooms, they deplete water of (19) _____ [p.541], which kills fishes and other organisms.

(20) _____ [p.541] refers to the nutrient enrichment of any ecosystem that is naturally low in (21) _____ [p.541]. This is a natural process, but phosphorus inputs accelerate it.

# Self-Quiz

____ 1. An array of organisms and their physical environment, interacting through a one-way flow of energy and a cycling of materials, is a(n) _____. [p.527]
   a. population
   b. community
   c. ecosystem
   d. biosphere
   e. food web

____ 2. _____ consume only dead or decomposing particles of organic matter. [p.527]
   a. Herbivores
   b. Parasites
   c. Detritivores
   d. Carnivores
   e. Omnivores

____ 3. The members of feeding relationships are structured in a hierarchy, the steps of which are called _____. [p.527]
   a. organism levels
   b. energy source levels
   c. eating levels
   d. trophic levels
   e. energy transfers

____ 4. A straight-line sequence of who eats whom in an ecosystem is sometimes called a(n) _____. [p.528]
   a. trophic level
   b. food chain
   c. ecological pyramid
   d. food web

____ 5. In a grazing food web, the primary consumers are _____. [p.529]
   a. herbivores
   b. carnivores
   c. detritivores
   d. decomposers
   e. both a. and c.

____ 6. _____ refers to an increase in concentration of a nondegradable (or slowly degradable) substance in organisms as it is passed along food chains. [p.530]
   a. Ecosystem modeling
   b. Nutrient input
   c. Biogeochemical cycle
   d. Biological magnification
   e. Gross primary production

____ 7. In the carbon cycle, carbon enters the atmosphere through _____. [p.534]
   a. carbon dioxide fixation
   b. aerobic respiration, burning, and volcanic eruptions
   c. oceans and accumulation of plant biomass
   d. release of greenhouse gases
   e. both b. and d. are correct

____ 8. Which of the following is not considered a greenhouse gas? [p.536]
   a. carbon dioxide
   b. chlorofluorocarbons
   c. methane
   d. nitrogen
   e. water vapor

____ 9. Ecosystems lose nitrogen to the air naturally by the activities of _____. [p.539]
   a. ammonifying bacteria
   b. denitrifying bacteria
   c. nitrogen-fixing bacteria
   d. ion exchange bacteria
   e. nitrifying bacteria

____ 10. _____ refers to the nutrient enrichment of any ecosystem that is naturally low in nutrients, a natural process. [p.541]
   a. Nitrification
   b. Ion exchange
   c. Ammonification
   d. Eutrophication
   e. Carbon fixation

## Chapter Objectives/Review Questions

1. Name and define the principal participants of an ecosystem. [p.527]
2. A(n) _____ is an array of organisms and their physical environment, all interacting through a flow of energy and a cycling of materials. [p.527]
3. Explain why nutrients can be completely recycled but energy cannot. [p.527]
4. Members of an ecosystem fit somewhere in a hierarchy of energy transfers (feeding relationships) called _____ levels. [p.528]
5. Distinguish between food chains and food webs. [p.528]
6. In most cases, _____ that producers initially captured passes through no more than four or five trophic levels. [p.528]
7. Compare grazing food webs with detrital food webs. Present an example of each. [p.529]
8. Describe how DDT damages ecosystems; discuss biological magnification. [p.530]
9. Distinguish between primary productivity, gross primary productivity, and the net amount. [p.531]
10. Ecologists often represent the trophic structure as an ecological _____. [p.531]
11. Briefly, summarize the hydrologic cycle, atmospheric cycles, and sedimentary cycles. [p.532]
12. Define *desalinization, salinization,* and *groundwater.* [p.533]
13. The carbon cycle traces carbon movement from reservoirs in the _____ and oceans, through organisms, then back to reservoirs. [pp.534–535]
14. Certain gases cause heat to build up in the lower atmosphere, a warming action known as the _____ effect. [p.536]
15. Describe the chemical events that occur during nitrogen fixation, degrading of nitrogenous wastes, and ammonia production. [pp.538–539]
16. Be able to discuss various aspects of the human impact on the nitrogen cycle. [p.539]
17. Be able to trace the major steps in the phosphorus cycle. [pp.540–541]
18. _____ is the name for nutrient enrichment of an ecosystem that is naturally low in nutrients. [p.541]

## Integrating and Applying Key Concepts

In 1971, *Diet for a Small Planet* was published. Frances Moore Lappé, the author, felt that people in the United States of America wasted protein and ate too much meat. She said, "We have created a national consumption pattern in which the majority, who can pay, overconsume the most inefficient livestock products [cattle] well beyond their biological needs (even to the point of jeopardizing their health), while the minority, who cannot pay, are inadequately fed, even to the point of malnutrition." Cases of marasmus (a nutritional disease caused by prolonged lack of food calories) and kwashiorkor (caused by severe, long-term protein deficiency) have been found in Nashville, Tennessee, and on an Indian reservation in Arizona, respectively. Lappé's partial solution to the problem was to encourage people to get as much of their protein as possible directly from plants and to supplement that with less meat from the more efficient converters of grain to protein (chickens, turkeys, and hogs) and with seafood and dairy products.

Most of us realize that feeding the hungry people of the world is not just a matter of distributing the abundance that exists—it is also a matter of political, economic, and cultural factors. Yet, it is still valuable to consider applying Lappé's idea to our everyday living. Devise two full days of breakfasts, lunches, and dinners that would enable you to exploit the lowest acceptable trophic levels to sustain yourself healthfully.

# 31

# THE BIOSPHERE

## INTRODUCTION

This chapter looks at several problems affecting the biosphere. It also looks at climate and the land biomes and aquatic provinces.

## FOCAL POINTS

- Figure 31.1 [p.544] shows the global air circulation pattern.
- Figure 31.2 [p.545] diagrams world temperature zones and how they are affected by the Earth's orbit.
- Figure 31.3 [p.545] looks at the ozone layer.
- Figure 31.5 [p.546] maps out acidic precipitation in the United States.
- Figure 31.7 [p.548] correlates ocean currents to climate zones.
- Figure 31.8 [p.549] shows the effects of topography on air moisture.
- Figure 31.9 [pp.550–551] is a world map of biogeographic realms and biomes.
- Figures 31.11, 31.13, 31.14, 31.15, 31.16, 31.17, and 31.18 [pp.552–556] illustrate various biomes.
- Figure 31.20 [p.557] shows lake zonation.
- Figures 31.21, 31.22, 31.23, 31.24, and 31.25 [pp.558–561] describe marine zones and provinces.
- Figures 31.26 and 31.27 [pp.561–562] relate upwellings and downwellings to El Niño.

## Interactive Exercises

*Surfers, Seals, and the Sea* [p.543]

### 31.1. AIR CIRCULATION AND CLIMATES [pp.544–547]

**Selected Words:** El Niño [p.543], weather [p.544], topography [p.544], "ozone layer" [p.545], "ozone hole" [p.545], chlorofluorocarbons (CFCs) [p.545], *industrial* smog [p.546], *photochemical* smog [p.546], nitric oxide [p.546], peroxyacyl nitrates (PANs) [p.546], sulfur dioxides [p.546], "photovoltaic cells" [p.547], *solar–hydrogen energy* [p.547], *wind farms* [p.547], "visual pollution" [p.547]

**Boldfaced Terms**

[p.543] biosphere _____

[p.544] climate _____

[p.544] temperature zones _____

[p.546] thermal inversion _____

[p.546] acid rain _____

*Short Answer*

1. Briefly, describe the apparent cause for the terrible storms and destruction of life and property that occurred in the winter of 1997–1998. [p.543]

2. What are the factors that affect a region's climate? [p.544]

*Matching*

Match each term with its description.

3. _____ ozone layer [p.544]
4. _____ temperature zones [pp.544–545]
5. _____ ozone thinning [p.545]
6. _____ CFCs [p.545]
7. _____ thermal inversion [p.546]
8. _____ industrial smog [p.546]
9. _____ photochemical smog [p.546]
10. _____ acid rain [p.546]
11. _____ solar–hydrogen energy and wind farms [p.547]

a. When weather conditions trap a layer of cool, dense air under a warm air layer
b. Major ozone eaters, once used as coolants in refrigerators and air conditioners
c. Defined by latitudinal variations in solar heating, combined with modified patterns of air circulation
d. Unlike fossil fuels, they are unlimited energy resources
e. 0 to 100 times more acidic as rainwater; causes great damage to many materials, disrupts organism physiology, the chemistry of ecosystems, and biodiversity
f. Located at between 17 and 27 kilometers above sea level, where these molecules are most concentrated
g. Occurs in cold, wet winters as a gray haze over industrialized cities that burn coal and other fossil fuels
h. A seasonal event, vast, and once called an "ozone hole"
i. Occurs in large cities in warm climates, forms a brown haze especially in natural basins; nitric oxide is the main pollutant from vehicle exhaust

# 31.2. THE OCEAN, LANDFORMS, AND CLIMATES [pp.548–549]

*Selected Words:* currents [p.548], *topography* [p.549], *monsoons* [p.549], mini-monsoons [p.549]

## Boldfaced Term
[p.549] rain shadow _____

_____

## Fill in the Blanks

When (1) _____ [p.548], the vast volume of ocean water expands, where (2) _____ [p.548], it shrinks. Sea level is (3) _____ [p.548] at the equator than it is at either pole. The volume of water in that "slope" is enough to get surface waters moving in response to (4) _____ [p.548], most often towards the poles. Along the way, water (5) _____ [p.548] air parcels above it.

Rapid mass flows of water, or (6) _____ [p.548], arise from the tug of trade winds and westerlies. Their direction and properties are influenced by the Earth's (7) _____ [p.548], the distribution of land masses, and shapes of oceanic basins. The water circulates (8) _____ [p.548] in the Northern Hemisphere, and it circulates (9) _____ [p.548] in the Southern Hemisphere.

Swift, deep, and narrow currents of nutrient-poor waters move parallel with the (10) _____ [p.548] coast of continents.

Pacific Northwest coasts are cool, foggy, and mild in the (11) _____ [p.548] because winds nearing the coast give up heat to cold coastal water, which the California Current is moving down toward the equator. Baltimore and Boston are muggy in the summer because air above the Gulf Stream gains (12) _____ [p.548] and moisture, which southerly and easterly winds deliver to those cities. Winters are milder in London and Edinburgh than they are in Ontario because, at the same latitude, the North Atlantic Current picks up (13) _____ [p.548] water from the Gulf Stream. As it flows past northwestern Europe, it gives up (14) _____ [p.548] energy to the prevailing winds.

(15) _____ [p.549] refers to a region's physical features, such as elevation. A (16) _____ _____ [p.549] is a semiarid or arid region of sparse rainfall on the side of high mountains not facing the (17) _____ [p.549]. (18) _____ [p.549] are air circulation patterns with effects on continents north or south of warm oceans. Converging trade winds and the equatorial sun cause intense heating and (19) _____ [p.549]. Recurring coastal breezes are like mini- (20) _____ [p.549].

The Biosphere 319

## 31.3. REALMS OF BIODIVERSITY [pp.550–556]

***Selected Words:*** chaparral [p.550], dry shrublands [p.552], dry woodlands [p.552], grasslands [p.552], *shortgrass* and *tallgrass* prairie [p.553], Dust Bowl [p.553], savannas [p.553], *monsoon* grasslands [p.553], *evergreen broadleafs* [p.554], *deciduous broadleafs* [p.554], tropical rain forest [p.554], monsoon forests [p.554], temperate deciduous forests [p.554], coniferous forest [p.554], boreal forests [p.554], montane coniferous forests [p.554], temperate rain forests [p.554], pine barrens [p.554], *taigas* [p.554], tundra [p.556], *arctic* tundra [p.556], *alpine* tundra [p.556]

### Boldfaced Terms

[p.550] biogeographic realms _____

_____

[p.550] biomes _____

_____

[p.551] hot spots _____

_____

[p.552] desertification _____

_____

[p.556] permafrost _____

_____

### Choice

For questions 1–10, choose from the following:

      a. dry shrublands    b. dry woodlands    c. grasslands
          d. savannas    e. monsoon grasslands

1. _____ North America's main types are prairies. [p.552]
2. _____ Broad belts of grasslands with a smattering of shrubs and trees; extended seasonal droughts [p.553]
3. _____ Steinbeck's *Grapes of Wrath* and James Michener's *Centennial* speak eloquently of the Dust Bowl's impact on people. [p.553]
4. _____ Form in southern Asia where heavy rains alternate with a dry season; dense stands of tall, coarse grasses form, then die back and often burn during the dry season [p.553]
5. _____ Local names for this biome include fynbos and chaparral; shrubs have highly flammable leaves and quickly burn to the ground [p.552]
6. _____ Dominant trees can be tall but do not form a dense canopy; includes eucalyptus woodlands of southwestern Australia and oak woodlands of California and Oregon [p.552]
7. _____ Grazing and burrowing species are the dominant animals; their activities combined with periodic fires help keep shrublands and forests from encroaching on the fringes. [p.552]

8. _____ Rainfall is sparse and fast-growing grasses dominate; where rainfall is higher, these areas grade into tropical woodlands with low trees and shrubs and tall, coarse grasses [p.553]

9. _____ Annual rainfall of 25–100 centimeters prevents deserts from forming. [p.552]

10. _____ Fast-growing grasses dominate where rainfall is sparse, but acacia and other shrubs grow where there is slightly more moisture. [p.553]

For questions 11–20, choose from the following:

    a. tropical rain forests    b. other broadleaf forests    c. coniferous forests
    d. arctic tundra    e. alpine tundra

11. _____ In the Northern Hemisphere, montane coniferous forests extend southward through the great mountain ranges; spruces and firs dominate in the north and at higher elevations; they give way to firs and pines in the south and at lower elevations [p.555]

12. _____ Highly productive forest; decomposition and mineral cycling are notably rapid given the hot, humid climate [p.554]

13. _____ Deciduous broadleaf forests, monsoon forests, temperate deciduous forests [p.554]

14. _____ Boreal forests or taiga [p.554]

15. _____ Extremely cold, plants grow rapidly in the nearly continuous sunlight of a brief growing season; lichens and compact, hardy, shallow-rooted plants are a base for food webs [p.556]

16. _____ Southern pine forests dominate the coastal plains of the southern Atlantic and Gulf states. [p.555]

17. _____ Even in summer, shaded patches of snow persist; no permafrost [p.556]

18. _____ Most have thickly cuticled, needle-shaped leaves and recessed stomata, adaptations that help conserve water [p.554]

19. _____ Permafrost and peat bogs [p.556]

20. _____ Palmettos grow below loblolly and other pines in the Deep South. [p.555]

## 31.4. THE WATER PROVINCES [pp.557–561]

**Selected Words:** littoral [p.557], limnetic [p.557], profundal [p.557], *phytoplankton* [p.557], *zooplankton* [p.557], *thermocline* [p.557], *oligotrophic* lakes [p.558], *eutrophic* lakes [p.558], estuaries [p.558], mangrove wetlands [p.559], rocky shores [p.559], *upper* littoral [p.559], *midlittoral* [p.559], *lower* littoral [p.559], *sandy* shores [p.559], coral reefs [p.559], *coral bleaching* [p.560], *pelagic* province [p.560], neritic zone [p.560], oceanic zone [p.560], *benthic* province [p.560], phytoplankton "pastures" [p.560], hydrothermal vents [p.560]

### Boldfaced Terms

[p.557] spring overturn _____

_____

[p.557] fall overturn _____

_____

[p.558] eutrophication _____

[p.560] marine snow _____

[p.561] upwelling _____

## Identification

Identify each numbered part of this diagram. [p.557]

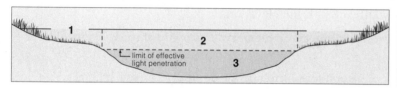 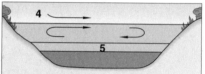

1. _____; 2. _____; 3. _____;
4. _____; 5. _____.

## Choice

Choose from the following:

    a. lake ecosystems     b. wetlands and the intertidal zone     c. rocky and sandy coastlines
         d. coral reefs     e. open ocean     f. hydrothermal vents

6. _____ Wave-resistant formations that consist of accumulated remains of countless marine organisms [pp.559–560]

7. _____ Estuaries depend on inflows of unpolluted fresh water; they are declining because of fresh water diversion upstream and polluted water flooding in. [p.558]

8. _____ Spring overturn and fall overturn [p.557]

9. _____ Three zones are present: upper littoral, midlittoral, and lower littoral. [p.559]

10. _____ Dinoflagellates live as symbionts in tissues of most of these organisms. [p.560]

11. _____ Food webs begin with marine snow. [p.560]

12. _____ Near-freezing water seeps into these fissures in the seafloor and becomes heated to extremely high temperatures. [p.560]

13. _____ A thermocline forms by midsummer; an abrupt drop in temperature at this midlayer blocks vertical mixing. [p.557]

14. _____ In tidal flats at tropical latitudes, nutrient-rich mangrove wetlands are found. [p.558]

15. _____ Cyanobacteria, green algae, and diatoms dominate its phytoplankton; rotifers and copepods are common in zooplankton. [p.557]

16. _____ Waves and currents continually rearrange stretches of loose sediments, the sandy and muddy shores. [p.559]

17. _____ The sides are honeycombed, with cell-sized chambers; the protection provided by the chambers may have promoted molecular self-assembly into the first living cells. [pp.560–561]

18. _____ The benthic province and the pelagic province with its neritic and oceanic zones [p.560]

19. _____ Oligotrophic and eutrophic types; eutrophication [p.558]

## Identification

Identify each numbered item in this illustration. Then, in the parentheses, match each term with its description. [p.561]

20. _____ province ( )

21. _____ province ( )

22. _____ zone ( )

23. _____ zone ( )

a. The full volume of ocean water
b. All water above the continental shelves
c. Includes sediments and rocks of the ocean bottom; begins with continental shelves and extends to deep-sea trenches
d. Water of the ocean basins

## 31.5. APPLYING KNOWLEDGE OF THE BIOSPHERE [pp.562–563]

*Selected Words:* La Niña [p.563], cholera [p.563]

### Boldfaced Terms

[p.562] downwelling _____

[p.562] El Niño _____

[p.562] ENSO _____

The Biosphere 323

*Matching*

Choose the most appropriate answer for each term.

1. \_\_\_\_\_ upwelling [p.562]
2. \_\_\_\_\_ downwelling [p.562]
3. \_\_\_\_\_ El Niño [p.562]
4. \_\_\_\_\_ ENSO [p.562]
5. \_\_\_\_\_ between ENSOs [p.562]
6. \_\_\_\_\_ during an ENSO [p.562]
7. \_\_\_\_\_ La Niña [p.563]
8. \_\_\_\_\_ Rita Colwell [p.563]

a. El Niño Southern Oscillation; massive dislocated rainfall patterns; seesawing of atmospheric pressure in the western equatorial Pacific; largest reservoir of warm water and air in the world; source of heavy rainfall that releases enough heat to drive the world's air circulation system
b. Surface winds that prevail in the western equatorial Pacific pick up speed and "drag" the surface waters east
c. Occurs every three to seven years; driven eastward into Peru's coastline, the nutrient-poor water is forced downward to displace cooler waters of the Humboldt Current, which prevents upwelling
d. Seesaw weather oscillation following six to eighteen months of El Niño
e. Warm currents of a downwelling usually arrive at Peru around Christmas; fishermen gave this event a name meaning "the little one," a reference to baby Jesus
f. Along western coasts, cold, deep, often nutrient-rich water moving vertically up in response to friction and Earth's rotational force, prevailing north winds slowly deflect a mass of cold water westward
g. The warm reservoir and heavy rain associated with it moves to the west
h. Correlated cholera outbreaks with El Niño episodes; the dormant stage of *V. cholerae* resides in copepods, previously unknown

# Self-Quiz

\_\_\_\_\_ 1. The distribution of different types of ecosystems is influenced by _____. [pp.544–545]
   a. global air circulation patterns and landforms
   b. modified patterns of air circulation
   c. surface ocean currents
   d. rainfall patterns at different latitudes
   e. all of the above

\_\_\_\_\_ 2. Photochemical smog is characterized by _____. [p.546]
   a. a gray haze
   b. heavy burning of fossil fuels
   c. a brown haze and nitric oxide
   d. formation over natural basins
   e. automobile exhausts

\_\_\_\_\_ 3. _____ are air circulation patterns that influence the continents north or south of warm oceans; land heats intensely, forming low pressure that draws in moisture-laden air from the ocean; air circulating north and south gives rise to alternating wet and dry seasons. [p.549]
   a. Geothermal ecosystems
   b. Upwellings
   c. Taigas
   d. Monsoons
   e. Downwellings

\_\_\_\_\_ 4. A broad belt of grasslands with a smattering of shrubs and trees is known as a _____. [p.553]
   a. warm desert
   b. savanna
   c. tundra
   d. taiga
   e. grassland

_____ 5. _____ form at latitudes of about 30° north and south, have low humidity, high evaporation rates, and the ground cools fast at night. [p.552]
   a. Shrublands
   b. Savannas
   c. Taigas
   d. Deserts
   e. Tropical rain forests

_____ 6. In tropical rain forests, _____ . [p.554]
   a. there is heavy and frequent rainfall
   b. litter does not accumulate
   c. soils are weathered and humus-deficient, and have poor nutrient reservoirs
   d. decomposition and mineral cycling are rapid
   e. all of the above

_____ 7. A lake's _____ zone ranges from open sunlit water to depths where light and photosynthesis are not much to speak of; it has phytoplankton and zooplankton. [p.557]
   a. epileptic
   b. limnetic
   c. littoral
   d. profundal
   e. neritic

_____ 8. The lake's upper layer cools, sinks, and the thermocline vanishes, lake water mixes vertically, and again dissolved oxygen moves down and nutrients move up. This describes the _____ . [p.557]
   a. spring overturn
   b. summer overturn
   c. fall overturn
   d. winter overturn
   e. oligotrophic lake

_____ 9. Chemoautotrophic bacteria are the primary producers for _____ . [p.560]
   a. hydrothermal vent communities
   b. desert communities
   c. lake communities
   d. coniferous forest communities
   e. coral reef communities

_____ 10. Rita Colwell is noted for _____ . [p.563]
   a. explaining the cause of ENSOs
   b. the discovery of upwelling
   c. the discovery of downwelling
   d. correlation of cholera epidemics with El Niño episodes and the discovery of a copepod reservoir
   e. discovering an abundance of copepods in phytoplankton

## Chapter Objectives/Review Questions

1. Briefly, describe the causes and effects of the El Niño event occurring in the winter of 1997–1998. [p.543]
2. _____ means average weather conditions, such as temperature, humidity, wind speed, cloud cover, and rainfall, over time. [p.544]
3. State the reason that most forms of life depend on the ozone layer. [p.545]
4. Be able to describe the causes of global air circulation patterns. [pp.544–545]
5. Describe how the tilt of the Earth's axis affects annual variation in the amount of incoming solar radiation. [p.545]
6. Be able to define *temperature zones, ozone thinning, CFCs, chlorine monoxide, methyl bromide, thermal inversion, industrial smog, photochemical smog, nitric oxide, sulfur dioxides, acid rain, solar–hydrogen energy,* and *wind farms*. [pp.544–547]
7. Which way an ocean current flows depends on prevailing winds, the Earth's _____ , gravity, the shapes of ocean basins, and the presence of land masses. [p.548]
8. _____ refers to a region's physical features, such as elevation. [p.549]
9. Define *rain shadow, monsoons,* and *mini-monsoons*. [p.549]
10. A _____ is characterized by habitat conditions and community structure, such as endemic species. [p.550]
11. Define *desertification*. [p.552]

The Biosphere 325

12. Be able to characterize the location, climate, and the dominant species found in the following: dry shrublands, dry woodlands, grasslands, shortgrass and tallgrass prairies, savannas, monsoon grasslands, evergreen and deciduous broadleaf forests, evergreen conifers, monsoon forests, temperate deciduous forests, boreal forests, montane coniferous forests, temperate rain forests, taigas, arctic tundra, and alpine tundra. [pp.552–556]
13. A _____ is a body of standing fresh water. [p.557]
14. Be able to describe the littoral, limnetic, and profundal zones of a lake. [p.557]
15. Define *phytoplankton, spring overturn, thermocline, fall overturn, oligotrophic lakes, eutrophic lakes,* and *eutrophication.* [pp.557–558]
16. Define *estuary;* characterize the conditions and organisms found there. [p.558]
17. Name and describe the three zones of rocky shores. [p.559]
18. Tell how coral reefs form and describe their value. [pp.559–560]
19. Describe the benthic and pelagic provinces, the neritic and oceanic zones of the ocean; define *marine snow.* [p.560]
20. State the significance of ocean upwelling and downwelling. [pp.561–562]
21. Describe conditions of ENSO occurrence and how this phenomenon interrelates ocean surface temperatures, the atmosphere, and the land. [p.562]
22. Describe the work of scientist Rita Colwell and how she was able to relate El Niño events to the incidence of cholera epidemics. [p.563]

## Integrating and Applying Key Concepts

Can you suggest practical methods of keeping Earth's biomes healthy while providing at least the minimal needs of *all* Earth's residents (not just humans)? If so, outline the requirements of such a system and devise a way in which it could be established.

# 32

# BEHAVIORAL ECOLOGY

## INTRODUCTION
The final chapter looks at animal behavior comparing instinctive with learned behavior in various organisms.

## FOCAL POINTS
- Figure 32.4 [p.569] shows imprinting in geese.
- Figure 32.6 [p.571] describes the adaptive value of nest decorating in starlings.
- Figures 32.7 and 32.8 [pp.572, 573] show communication displays.
- Figure 32.16 [p.578] illustrates specialization in eusocial insects.

---

## Interactive Exercises

*My Pheromones Made Me Do It* [p.566]

### 32.1. SO WHERE DOES BEHAVIOR START? [pp.567–571]

**Selected Words:** *behavioral* responses [p.567], *intermediate* response [p.567], *sign stimuli* [p.568], *genetically based capacity to learn* [p.570], *social experience* [p.570], *reproductive success* [p.570], *adaptive behavior* [p.570], *social behavior* [p.570], *selfish behavior* [p.570], *altruism* [p.570]

### Boldfaced Terms

[p.566] pheromone _____

[p.568] instinctive behavior _____

[p.568] fixed action pattern _____

[p.569] learned behavior _____

[p.569] imprinting _____

[p.570] natural selection _____

## Short Answer

1. Describe the mechanism that explains the behavior of Africanized honeybees. [p.566]

## Matching

Match each term with its description.

2. _____ evidence of the genetic basis of behavior [p.567]
3. _____ intermediate response [p.568]
4. _____ hormonal guidance of behavior [pp.567–568]
5. _____ instinctive behavior [p.567]
6. _____ learned behavior [p.569]
7. _____ imprinting [p.569]
8. _____ adaptive value of behavior [p.570]

a. Time-dependent form of learning; an example is young geese
b. Behavior performed without having first been learned by actual experience in the environment; examples are snake tongue-flicking and strikes of newborn garter snakes, cuckoo birds laying eggs in nests of other species with young cuckoos rolling eggs of natural-born offspring out of the nest, and the instinctive smile of newborn human infants
c. Using natural selection theory to develop and test possible explanations of why a behavior has endured; this helps explain how some actions bestow reproductive benefits that offset the reproductive costs (disadvantages) associated with them; examples are lions hunting food, dispersing lemmings, and starlings including sprigs of greenery in their nests to repel mites
d. Term applied to genetically based behavioral reactions of hybrid offspring; Arnold's hybrid garter snakes displayed an intermediate response to slug chunks and odors
e. Demonstrated by the effects of oxytocin and the matings of prairie voles
f. Reported by Stevan Arnold and his experiments with garter snakes and banana slugs
g. Animals processing information about experiences and using it to change or vary responses to stimuli; an example is bumblebee avoidance by toads once stung on their tongues

## 32.2. COMMUNICATION SIGNALS [pp.572–573]

*Selected Words:* *signaling* pheromones [p.572], *priming* pheromones [p.572], *composite* signal [p.572], ritualized behavior [p.573], *illegitimate receiver* [p.573], *illegitimate signaler* [p.573]

### Boldfaced Term

[p.572] communication signals _____

_____

### Choice

Choose from the following:

      a. communication signals      b. communication displays
      c. illegitimate signaler        d. illegitimate receiver

1. _____ Assassin bugs hook dead and drained termite bodies on their dorsal surfaces and acquire the odor of their victims; this deception allows assassin bugs to hunt termite victims more easily. [p.573]
2. _____ Acoustical signals from male birds, frogs, grasshoppers, whales, and other animals make sounds that attract females; prairie dogs bark, wolves howl, and kangaroo rats drum their feet on the ground when advertising territory possession. [p.572]
3. _____ The play bow of dogs and wolves [p.572]
4. _____ Bombykol molecules (signaling pheromones) released by female silk moths serve as sex attractants. [p.572]
5. _____ Normal movements may be exaggerated or frozen; feathers, manes, claws, and other body parts are notably enlarged, patterned, or colored; ritualization of bird courtship [p.573]
6. _____ Termites act defensively when detecting scents from invading ants whose scent signals are meant to elicit cooperation from other ants. [p.573]
7. _____ A dominant male baboon's "yawn" that exposes large canines [p.572]
8. _____ A volatile odor, a priming pheromone, in the urine of certain male mice triggers and enhances estrus in female mice. [p.572]
9. _____ After finding a source of pollen or nectar, a foraging honeybee returns to its colony (a hive) and performs a complex dance; other bees follow and maintain contact with the dancer, a tactile display. [p.573]
10. _____ Tungara frogs issue two "come hither" calls to females, one simple, the other complex. [p.573]
11. _____ The tungara frog calls also mean "dinner here" to fringed-lipped bats. [p.573]
12. _____ A lion emits a spine-tingling roar to keep in touch with its pride or to threaten rivals. [p.572]
13. _____ Ears laid back against the head of a zebra convey hostility, but ears pointing up convey its absence; a zebra with laid-back ears isn't too riled up when its mouth is open just a bit, but when the mouth is gaping, watch out. This is a composite signal. [p.572]

## 32.3. MATES, OFFSPRING, AND REPRODUCTIVE SUCCESS [pp.574–575]

*Selected Words:* sexual selection [p.574], "nuptial gift" [p.574], elk [p.575], parental care [p.575]

*Complete the Table*

1. Enter the common names of the animals that fit the text examples of mating and parenting behaviors.

| Animal Examples | Descriptions of Sexual Selection |
|---|---|
| [p.574] a. | Females choose males that offer superior food; a male kills a moth or some other insect, then releases a sex pheromone to attract females to his nuptial gift; only after she has been eating for about five minutes or so does she accept sperm from her partner. |
| [p.574] b. | Females shop around for good-looking males. |
| [pp.574–575] c. | Males congregate in a communal display ground called a lek; each male stakes out a few square meters as his territory; females are attracted to the lek to observe male displays and usually select and mate with only one male; many females choose the same stand-out male, so most of the males never mate. |
| [p.575] d. | Females of some species cluster in defendable groups when they are sexually receptive; males compete for access to the clusters; combative males are favored. |
| [p.575] e. | Extended parental care improves the likelihood that the current generation of offspring will survive; this behavior comes at a reproductive cost to the adults. |

## 32.4. COSTS AND BENEFITS OF SOCIAL GROUPS [pp.576–577]

*Selected Words:* cooperative predator avoidance [p.576], cooperative hunting [p.576]

*Boldfaced Term*

[p.576] selfish herd _____
_____

*Matching*

Match each term with its description.

1. _____ cooperative predator avoidance [p.576]
2. _____ the selfish herd [p.576]
3. _____ cooperative hunting [pp.576–577]
4. _____ dominance hierarchies [p.577]
5. _____ cost and benefits of social behavior [p.577]

a. Many predatory mammals, including wolves, lions, and wild dogs, live in social groups and cooperate in this behavior; this group behavior seems to be less successful than solitary ventures and so this group living is explained by factors other than success in this behavior
b. By their position in a group, some individuals form a living shield against predation on others; a simple society formed by reproductive self-interest; male bluegill sunfishes compete for safer nesting sites in the center of the colony
c. In most environments, costs outweigh benefits; cormorants and many other seabirds form enormous colonies but all must compete for a share of the same ecological pie; large groups are attractive to predators and crowding invites parasites and diseases readily transmitted from host to host; another cost is the risk of being killed or exploited by others in the group; breeding pairs of herring gulls cannibalize a neighbor's eggs, and any wandering chicks
d. A cooperative response reduces the risk to all in a group; examples are vervet monkeys, meerkats, prairie dogs, and many other mammals that give alarm calls; Australian sawfly caterpillars live in clumps or branches so all are aware of a disturbance and, as an example, Great Tits prefer to eat one caterpillar at a time rather than attack a clump
e. Wolf packs have one dominant male that breeds with a dominant female while other pack members are nonbreeding relatives that hunt, bring food to the den, and guard the young; baboons live in large groups and high-ranking females displace lower-ranked females from food, water, and grooming opportunities

## 32.5. WHY SACRIFICE YOURSELF? [pp.578–579]

*Selected Words:* eusocial insects [p.578], altruistic [p.579]

*Boldfaced Term*

[p.579] inclusive fitness _____

_____

*Fill in the Blanks*

None of the (1) _____ [p.579] individuals of a honeybee hive, termite colony, or naked mole-rat clan can contribute genes to the next generation. So how are the (2) _____ [p.579] that underlie altruistic behavior perpetuated? By a theory of (3) _____ _____ [p.579] based on William Hamilton's work, genes associated with caring for one's (4) _____ [p.579] might be favored in some situations.

Behavioral Ecology   *331*

A sexually reproducing, diploid parent caring for offspring is not helping exact (5) _____ [p.579] copies of itself. Each of its gametes, and each of its offspring, inherits (6) _____-_____ [p.579] of its genes. Other individuals of the social group with the same ancestors also share genes with their (7) _____ [p.579]. Two siblings (brothers or sisters) are as genetically similar as a parent is to one of its offspring. Nephews and nieces have inherited about (8) _____-_____ [p.579] of their uncle's genes.

Thus, nonbreeding workers in such societies may indirectly promote their (9) "_____-_____" [p.579] genes through exceptional altruistic behavior directed toward their relatives. All individuals in honeybee, termite, and ant colonies are members of an (10) _____ [p.579] family. So are the naked mole-rats. H. Kern Reeve and his colleagues used DNA (11) _____ [p.579] to determine the extent of relatedness among the members of a naked mole-rat clan. He discovered that all the individuals from the same clan are very close (12) _____ [p.579]. He also found out that they are very different (13) _____ [p.579] from members of other clans. Each clan was highly (14) _____ [p.579] through generations of brother–sister, mother–son, and father–daughter matings among breeding individuals that can live for a long time.

In Damaraland mole-rats, with a social organization similar to their naked cousins, the nonbreeders of both sexes cooperatively assist one breeding pair and DNA studies have shown that breeding pairs are usually (15) _____ [p.579]. Other factors such as an arid environment and patchy food supplies may select for mole-rat eusocial (16) _____ [p.579], that is, their cooperative burrow digging and food searches.

## 32.6. A LOOK AT PRIMATE SOCIAL BEHAVIOR [p.580]
## 32.7. HUMAN SOCIAL BEHAVIOR [p.581]

*Selected Words:* "fishing sticks" [p.580], *autism* [p.581]

*Choice*
Choose from the following: [p.580]

        a. chimpanzees    b. bonobos    c. both a. and b.

1. _____ Females are receptive to sex at any time, not just during their fertile cycle.
2. _____ Adult males are related and females are not.
3. _____ Sex is a premier binding force in their social life.
4. _____ Females spend time together and form strong bonds.
5. _____ Males cooperate to hunt monkeys and to attack neighboring groups; they also attack and kill infants.
6. _____ Females are often unrelated and interact little with one another.
7. _____ Females have sexual organs that let them copulate while facing a partner.

8. \_\_\_\_ Infanticidal behavior has never been observed among the males.
9. \_\_\_\_ They often walk on two legs.
10. \_\_\_\_ By sexual attractiveness and direct solicitations, females build strong alliances with many males.

---

## Self-Quiz

\_\_\_\_ 1. A young toad flips its sticky-tipped tongue and captures a bumblebee that stings its tongue; in the future, the toad leaves bumblebees alone. This is _____ . [p.569]
   a. instinctive behavior
   b. a fixed reaction pattern
   c. altruistic
   d. learned behavior
   e. imprinting

\_\_\_\_ 2. Newly hatched goslings follow any large moving objects to which they are exposed shortly after hatching; this is an example of _____ . [p.569]
   a. reproductive success
   b. imprinting
   c. piloting
   d. migration
   e. instinctive behavior

\_\_\_\_ 3. _____ provides an example of an illegitimate signaler. [p.573]
   a. A soldier termite killing an ant on cue
   b. An assassin bug with acquired termite odor
   c. A termite pheromone alarm signal
   d. The "yawn" of a dominant male baboon
   e. A tactile display by bees

\_\_\_\_ 4. The claiming of the more protected central locations of the bluegill colony by the largest, most powerful males suggests _____ . [p.576]
   a. cooperative predator avoidance
   b. the selfish herd
   c. a huge parent cost
   d. self-sacrificing behavior
   e. parental care

\_\_\_\_ 5. A chemical odor in the urine of male mice triggers and enhances estrus in female mice. The source of stimulus for this response is a _____ . [p.572]
   a. generic mouse pheromone
   b. signaling pheromone
   c. priming pheromone
   d. cue from female mice
   e. composite signal

\_\_\_\_ 6. Bombykol released by a female silk moth summons males from kilometers away. This is an example of a(n) _____ signal. [p.572]
   a. priming pheromone
   b. visual
   c. acoustical
   d. tactile
   e. signaling pheromone

\_\_\_\_ 7. Male birds sing to secure territories and attract a mate. This is an example of a(n) _____ signal. [p.572]
   a. chemical
   b. visual
   c. acoustical
   d. tactile
   e. pheromone

\_\_\_\_ 8. Australian sawfly caterpillars live in clumps on branches; when something disturbs their clump, they collectively react. This is an example of _____ . [p.576]
   a. a selfish herd
   b. cooperative predator avoidance
   c. self-sacrificing behavior
   d. a dominance hierarchy
   e. cooperative hunting

_____ 9. Caring for nondescendant relatives favors the genes associated with helpful behavior and is classified as _____ . [p.579]
   a. dominance hierarchy
   b. indirect selection
   c. altruism
   d. predator avoidance
   e. both b. and c.

_____ 10. When H. Kern Reeve studied the genetic relationship in naked mole-rat clans, he discovered _____ . [p.579]
   a. dominance hierarchy
   b. indirect selection
   c. that all those from the clan are very close relatives
   d. altruism
   e. very few individuals in the same clan are related

_____ 11. From the following characteristics, choose the one that is *not* characteristic of bonobos? [p.581]
   a. adult males are related and females are not
   b. females spend time together and form strong bonds
   c. exhibit bipedalism
   d. infanticidal behavior
   e. ability to copulate facing a partner

## Chapter Objectives/Review Questions

1. Describe the chemical signal mechanism utilized by Africanized bees. [p.566]
2. What role does oxytocin have in the sexual behavior of prairie voles? [pp.567–568]
3. Compare learned and instinctive behaviors and give an example of each. [pp.567–570]
4. Be able to define *reproductive success, adaptive behavior, social behavior, selfish behavior,* and *altruism.* [p.570]
5. Be able to explain why starlings festoon their nests with sprigs of greenery. How does this have adaptive value? [pp.570–571]
6. Play bows by dogs and wolves, a baboon "yawn," ritualization of courtship displays, and bee tactile communication are all examples of _____ displays. [pp.572–573]
7. When termites detect ant scents meant for other ants and kill ants, the termites are said to be _____ receivers of a signal meant for individuals of a different species. [p.573]
8. Describe how the two calls of male tungara frogs are interpreted by female frogs and fringe-lipped bats. [p.573]
9. Assassin bugs covered with termite scent are able to use deception to hunt termite victims more easily and as such are acting as _____ signalers. [p.573]
10. Describe the sexual selection displayed by female hangingflies and female fiddler crabs. [p.574]
11. Competition for ready-made _____ favors highly combative male lions, sheep, elk, elephant seals, and bison. [p.575]
12. If females are fighting over males, then males probably provide some service beyond sperm delivery, such as _____ care. [p.575]
13. Vervet monkeys, meerkats, and prairie dogs sound _____ calls. [p.576]
14. Explain how a selfish herd, such as exhibited by bluegills, illustrates reproductive self-interest. [p.576]
15. Describe the dominance hierarchies exhibited by wolf packs and baboons living in large groups. [p.577]
16. Explain how honeybee colonies, termites, and naked mole-rats illustrate altruism. [p.579]
17. Explain the theory of inclusive fitness based on the work of William Hamilton. [p.579]

18. Describe the primate behavior patterns of chimpanzees and bonobos. [p.580]
19. Summarize what is known about human pheromones; explain why infanticide is sometimes natural. [p.581]

## Integrating and Applying Key Concepts

Think about communication signals that humans use, and list them. Do you believe a dominance hierarchy exists in human society? Think of examples.

# ANSWERS

## Chapter 1  Invitation to Biology

*What Am I Doing Here?* [p.1]

**1.1. LIFE'S LEVELS OF ORGANIZATION** [pp.2–3]
1. The study of biology can deepen your perspective on life and sharpen how you interpret the natural world, including human nature. Any of your personal reasons may be acceptable. 2. d; 3. c; 4. f; 5. h; 6. e; 7. j; 8. k; 9. b; 10. i; 11. a; 12. g; 13. biosphere; 14. ecosystem; 15. community; 16. population; 17. multicelled organism; 18. organ system; 19. organ; 20. tissue; 21. cell; 22. molecule; 23. atom.

**1.2. OVERVIEW OF LIFE'S UNITY** [pp.4–5]
1. DNA; 2. proteins; 3. structural; 4. enzymes; 5. molecules; 6. parents; 7. Inheritance; 8. Reproduction; 9. development; 10. metabolism; 11. b; 12. c; 13. a; 14. b; 15. c; 16. producers; 17. environment; 18. biosphere; 19. changes; 20. responses; 21. receptors; 22. stimuli; 23. bloodstream; 24. internal; 25. pancreas; 26. sugar; 27. normal; 28. homeostasis.

**1.3. IF SO MUCH UNITY, WHY SO MANY SPECIES?** [pp.6–7]

**1.4. AN EVOLUTIONARY VIEW OF DIVERSITY** [p.8]
1. f; 2. d; 3. b; 4. c; 5. a; 6. e; 7. a; 8. b; 9. b; 10. b; 11. b; 12. a; 13. b.

**1.5. THE NATURE OF BIOLOGICAL INQUIRY** [p.9]
**1.6. THE POWER OF EXPERIMENTAL TESTS** [pp.10–11]
**1.7. THE SCOPE AND LIMITS OF SCIENCE** [p.12]
1. g; 2. a; 3. d; 4. b; 5. e; 6. c; 7. f; 8. experiment; 9. prediction; 10. theory; 11. prediction; 12. belief; 13. scientific theory; 14. variable; 15. test predictions; 16. cause and effect; 17. sampling error; 18. quantitative; 19. subjective; 20. supernatural; 21. conviction (faith).

*Self-Quiz*
1. a; 2. b; 3. b; 4. c; 5. b; 6. c; 7. d; 8. a; 9. d; 10. b.

## Chapter 2  Molecules of Life

*Science or Supernatural?* [p.15]

**2.1. ATOMS AND THEIR INTERACTIONS** [pp.16–17]
**2.2. BONDS IN BIOLOGICAL MOLECULES** [pp.18–19]

1.

a. hydrogen   b. carbon   c. nitrogen   d. chlorine

2.
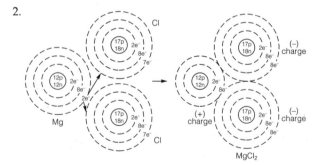

3. a, c; 4. b; 5. c; 6. a, b; 7. Hydrogen bonds typically form between different parts of proteins and other large

molecules that have been folded into three-dimensional shapes. They also form between two molecules, as when they hold the nucleotide strands of DNA together.

**2.3. WATER'S LIFE-GIVING PROPERTIES [pp.20–21]**
**2.4. ACIDS AND BASES [pp.22–23]**
1. hydrophilic; 2. hydrophobic; 3. hydrogen bonds; 4. heat reservoir; 5. evaporation; 6. cools; 7. cohesion; 8. hydroxide; 9. pH; 10. acidic; 11. basic (alkaline); 12. tenfold; 13. buffer systems.

**2.5. MOLECULES OF LIFE—FROM STRUCTURE TO FUNCTION [pp.23–25]**
1. organic; 2. hydrocarbons; 3. functional; 4. carbon; 5. four; 6. sugars; 7. fatty; 8. acids; 9. nucleotides; 10. monomers; 11. enzymes; 12.a. hydrolysis; b. condensation.

**2.6. THE TRULY ABUNDANT CARBOHYDRATES [pp.26–27]**
1. d; 2. h; 3. b; 4. c; 5. g; 6. f; 7. e; 8. i; 9. a.

**2.7. GREASY, FATTY—MUST BE LIPIDS [pp.28–29]**
1.a. unsaturated (circle one double bond); b. saturated; c. unsaturated (circle three double bonds); 2.a. phospholipid; b. sterol; c. fatty acid; d. triglyceride (triacyl glycerol); 3. b; 4. d; 5. e; 6. b; 7. e; 8. c; 9. e; 10. d; 11. e; 12. a; 13. b; 14. b; 15. d; 16. b.

**2.8. PROTEINS—DIVERSITY IN STRUCTURE AND FUNCTION [pp.30–31]**
**2.9. WHY IS PROTEIN STRUCTURE SO IMPORTANT? [pp.32–33]**
1. a. coiled secondary structure; b. tertiary structure; c. quaternary structure; d. tertiary structure; e. linear primary structure; f. sheet-like secondary structure; 2. One mutation in hemoglobin puts the amino acid valine, not glutamate, at the sixth position of the beta chain. Under oxygen tension, the red blood cells carrying the mutated hemoglobin take on a sickle shape and clump together. These cells rupture easily, remnants clog and rupture capillaries and disrupt normal blood circulation, thus starving tissues of oxygen. This has significant negative consequences for the body's tissues and organs; 3. transport, structure/shape, movement, enzymes, receptors, hormones, defense.

**2.10. NUCLEOTIDES AND THE NUCLEIC ACIDS [pp.34–35]**
1. Three complete nucleotides are present; 2. b; 3. a; 4. d; 5. f; 6. c; 7. e.

*Self-Quiz*
1. a; 2. d; 3. c; 4. b; 5. a; 6. d; 7. a; 8. c; 9. c; 10. c.

# Chapter 3   HOW CELLS ARE PUT TOGETHER

*Animalcules and Cells Fill'd With Juices* [p.38]

**3.1. WHAT IS "A CELL"? [p.39]**
**3.2. MOST CELLS ARE *REALLY SMALL* [pp.40–41]**
1. Galileo Galilei; 2. Robert Hooke; 3. Antoni van Leeuwenhoek; 4. Robert Brown; 5. Theodor Schwann; 6. Rudolf Virchow; 7. All organisms consist of one or more cells. The cell is the smallest unit with the properties of life. Only a living cell can give rise to other cells.; 8. It is a comparison of the size of the cell's surface with its volume. For good diffusion of nutrients and wastes, larger cells need to be longer, more flattened or more frilled to increase the surface-to-volume ratio.; 9. a; 10. c; 11. b; 12. d; 13. a; 14. plasma membrane; 15. ribosomes; 16. magnetic; 17. 4 ($2^2$), 8 ($2^3$).

**3.3. THE STRUCTURE OF CELL MEMBRANES [pp.42–43]**
**3.4. A CLOSER LOOK AT PROKARYOTIC CELLS [pp.44–45]**
**3.5. A CLOSER LOOK AT EUKARYOTIC CELLS [pp.45–49]**

**3.6. WHERE DID ORGANELLES COME FROM? [pp.50–51]**
1. Phospholipids; 2. fatty acid tails; 3. head; 4. lipid bilayer; 5. proteins, glycolipids, and steroids; 6. hydrogen bonds; 7. fluid mosaic; 8. water-soluble; 9. proteins; 10. Proteins; 11. a; 12. c; 13. a; 14. b; 15. c; 16. a; 17. b; 18. b; 19. b; 20. c; 21. c; 22. c; 23. c; 24. a; 25. b; 26. c; 27. e; 28. a; 29. f; 30. d; 31. b; 32. d; 33. w; 34. c; 35. a, y; 36. k, t; 37. u; 38. m, x; 39. e, n; 40. f, o; 41. h, q; 42. l, v; 43. i, r; 44. j, s; 45. Bacteria ingested by ancestors of eukaryotic cells were not digested but continued to survive inside the ancestral eukaryote. They developed a mutualistic relationship that eventually became so interdependent that they became incapable of independent life. Both mitochondria and chloroplasts are self replicating and have their own circular DNA and ribosomes.

**3.7. THE DYNAMIC CYTOSKELETON [pp.52–53]**
**3.8. CELL SURFACE SPECIALIZATIONS [pp.54–56]**
1. cytoskeleton; 2. protein; 3. Microtubules; 4. chromosomes; 5. Colchicine; 6. taxol; 7. Microfilaments; 8. plasma membrane; 9. muscle; 10. Intermediate

filaments; 11. cancer; 12. Flagella (Cilia); 13. cilia (flagella); 14. microtubules; 15. motor proteins; 16. pseudopod; 17. outside; 18. T; 19. Pectin; 20. inside; 21. T; 22. b; 23. c; 24. a.

*Self-Quiz*
1. d; 2. c; 3. c; 4. c; 5. b; 6. c; 7. c; 8. a; 9. d; 10. c.

# Chapter 4  How Cells Work

*Beer, Enzymes, and Your Liver* [p.58]

### 4.1. INPUTS AND OUTPUTS OF ENERGY [pp.59–60]
1. liver; 2. acetate; 3. 12; 4. 5; 5. 1½; 6. alcoholic cirrhosis; 7. alcoholic hepatitis; 8. Binge drinking; 9. heart; 10. c; 11. b; 12. a; 13. b; 14. b; 15. b; 16. Continued input of the sun's energy allows living organisms to maintain order; 17. liver damage, accidents, date rape, unprotected sex, death.

### 4.2. INPUTS AND OUTPUTS OF SUBSTANCES [p.61]
### 4.3. HOW ENZYMES MAKE SUBSTANCES REACT [pp.62–64]
1. reactants (substrates); 2. products; 3. Enzymes; 4. cofactors; 5. reversible; 6. chemical equilibrium; 7. metabolic pathways; 8. biosynthetic; 9. degradative; 10. antioxidant; 11. Inhibitors; 12. feedback inhibition; 13. d; 14. c; 15. b; 16. a; 17. e; 18. functional group transfers, electron transfers, condensation, cleavage; 19. helping substrates get together, orienting substrates in favorable positions, shutting out water, inducing changes in enzyme shape.

### 4.4. DIFFUSION AND METABOLISM [pp.65–66]
### 4.5. WORKING WITH AND AGAINST DIFFUSION [pp.66–67]
### 4.6. WHICH WAY WILL WATER MOVE? [pp.68–69]
### 4.7. CELL BURPS AND GULPS [p.70]
1. Can diffuse: carbon dioxide, oxygen, water. Cannot diffuse: calcium ion, glucose, protein; 2. d; 3. c.; 4. Only certain things can pass across the membrane. It maintains the balance between what is in the cell and what is outside.; 5. steepness of gradient, heat, particle size, electric gradient, pressure gradient; 6. facilitated diffusion; 7. Active transport; 8. energy; 9. sodium; 10. potassium; 11. b; 12. c; 13. a; 14. a; 15. b; 16. c; 17. c; 18. c; 19. c; 20. c; 21. hydrostatic; 22. turgor; 23. exocytosis; 24. endocytosis; 25. phagocytosis.

*Self-Quiz*
1. a; 2. a; 3. d; 4. d; 5. a; 6. d; 7. b; 8. c; 9. a; 10. c.

# Chapter 5  Where It Starts—Photosynthesis

*Sunlight and Survival* [p.73]

### 5.1. THE RAINBOW CATCHERS [pp.74–75]
1. autotrophs; 2. heterotrophs; 3. photosynthesis; 4. chlorophyll *a*; 5. green; 6. green; 7. carotenoids; 8. yellow; 9. anthocyanins; 10. Phycobilins; 11. bacteriorhodopsin; 12. antenna; 13. higher; 14. ATP; 15. $12 H_2O + 6 CO_2 \rightarrow 6 O_2 + C_6H_{12}O_6 + 6 H_2O$; 16.a. twelve; b. carbon dioxide; c. oxygen; d. glucose; e. six; 17. two outer; 18. thylakoid; 19. stroma; 20. sunlight; 21. water; 22. carbon dioxide; 23. light-dependent; 24. ATP; 25. NADPH; 26. light-independent; 27. glucose; 28. $O_2$; 29. $H_2O$ (metabolic water); 30. Oxygen, a product of photosynthesis, started accumulating long ago in what had been an oxygen-free atmosphere. This changed the atmosphere and the life forms. The global rise in free oxygen favored the evolution of aerobic respiration, the most efficient way for releasing energy from organic compounds. In the evolutionary view, photosynthesis is the main reason you and every animal are around today, breathing in oxygen that keeps you alive.

### 5.2. LIGHT-DEPENDENT REACTIONS [pp.76–77]
1. light-harvesting complex; 2. photosystem; 3. electron; 4. electron transfer chain; 5. $H^+$ (hydrogen ion); 6. thylakoid; 7. lipid; 8. ATP synthase; 9. ADP; 10. ATP; 11. $NADP^+$; 12. NADPH; 13. water; 14. oxygen; 15. ATP; 16. cyclic.

### 5.3. LIGHT-INDEPENDENT REACTIONS [pp.78–79]
### 5.4. PASTURES OF THE SEAS [p.80]
1. $CO_2$ (f); 2. PGA (h); 3. ATP (a); 4. NADPH (g); 5. PGAL (d); 6. glucose (b); 7. ATP (c); 8. RuBP (e); 9. c; 10. a; 11. b (c); 12. c; 13. b; 14. a; 15. c; 16. c; 17. b (c); 18. a; 19 bacteria (protists); 20. protists (bacteria); 21. 45 billion; 22. $CO_2$; 23. global warming; 24. food webs.

*Self-Quiz*
1. d; 2. c; 3. d; 4. c; 5. d; 6. d; 7. e; 8. e; 9. a; 10. c.

# Chapter 6  How Cells Release Chemical Energy

*When Mitochondria Spin Their Wheels* [p.82]

**6.1. OVERVIEW OF ENERGY-RELEASING PATHWAYS** [pp.83–84]
1. Defective mitochondria; 2. glycolysis; 3. cytoplasm; 4. glucose; 5. pyruvate; 6. two; 7. mitochondrion; 8. thirty-six; 9. Krebs cycle; 10. carbon dioxide; 11. NADH (FADH$_2$); 12. FADH$_2$ (NADH); 13. two; 14. electron transfer phosphorylation; 15. thirty-two; 16. oxygen; 17. electrons; 18. hydrogen; 19. water; 20. glucose + 6 O$_2$ → 6 CO$_2$ + 6 H$_2$O; 21. one molecule of glucose plus six molecules of oxygen (in the presence of appropriate enzymes) yield six molecules of carbon dioxide plus six molecules of metabolic water.

**6.2. GLYCOLYSIS—GLUCOSE BREAKDOWN STARTS** [pp.84–85]
1. g; 2. c; 3. h; 4. a; 5. e; 6. b; 7. d; 8. f.

**6.3. SECOND AND THIRD STAGES OF AEROBIC RESPIRATION** [pp.86–88]
1. inner mitochondrial; 2. outer mitochondrial; 3. inner; 4. outer; 5. pyruvate; 6. carbon dioxide; 7. Coenzyme A; 8. acetyl–CoA; 9. ATP; 10. oxaloacetate; 11. NADH; 12. NADH; 13. FAD; 14. NADH; 15. FADH$_2$; 16. hydrogen; 17. inner; 18. outer; 19. ATP synthases; 20. ATP; 21. ADP; 22. P$_i$; 23. Free oxygen helps keep transport chains free for operation. It withdraws spent electrons at the end of the chains and combines with H$^+$ to form water, a by-product; 24. Thirty-two ATP molecules typically form in the third stage of aerobic respiration. The net harvest, including the four ATP from preceding stages, is thirty-six ATP from one molecule of glucose.

**6.4. ANAEROBIC ENERGY-RELEASING PATHWAYS** [pp.88–89]
1. c; 2. a; 3. b; 4. c; 5. b; 6. c; 7. a; 8. c; 9. a; 10. c; 11. a; 12. b; 13. a; 14. c; 15. b.

**6.5. ALTERNATIVE ENERGY SOURCES IN THE BODY** [pp.89–91]
**6.6. CONNECTIONS WITH PHOTOSYNTHESIS** [pp.91–92]
1. f; 2. c; 3. f; 4. e; 5. g; 6. g; 7. g; 8. c; 9. f; 10. c; 11. a; 12. d; 13. g; 14. b; 15. a; 16. h; 17. Put simply, photosynthesizing organisms produce energy-rich glucose and oxygen for themselves and for heterotrophs that must utilize aerobic respiratory pathways to produce ATP molecules. All organisms that utilize respiratory pathways produce carbon dioxide as a by-product, and this is a raw material for photosynthesizing organisms to produce more glucose and oxygen. Do not make the mistake of thinking that photosynthesizing organisms are only capable of photosynthetic pathways. They too must utilize the respiratory pathways to produce ATP molecules (ATP is often referred to as the "energy currency of life").

*Self-Quiz*
1. d; 2. b; 3. d; 4. a; 5. c; 6. e; 7. e; 8. b; 9. d; 10. e; 11. d.

# Chapter 7  How Cells Reproduce

*Henrietta's Immortal Cells* [p.94]

**7.1. OVERVIEW OF CELL DIVISION MECHANISMS** [p.95]
**7.2. INTRODUCING THE CELL CYCLE** [p.96]
1. HeLa cells are tumor cells taken from and named for a cancer patient, Henrietta Lacks, in 1951. Henrietta Lacks died (at age thirty-one) two months following her diagnosis of cancer. HeLa cells have continued to divide in culture and are used for cancer research in laboratories all over the world. The legacy of Henrietta Lacks continues to benefit humans everywhere. 2. h; 3. j; 4. g; 5. b; 6. k; 7. d; 8. a; 9. i; 10. c; 11. e; 12. f; 13. G1 interval; 14. S interval; 15. G2 interval; 16. prophase; 17. metaphase; 18. anaphase; 19. telophase; 20. cytoplasmic division; 21. interphase; 22. mitosis; 23. daughter cells; 24. 15; 25. 22; 26. 14; 27. 13; 28. 21; 29. 20; 30. 21.

**7.3. MITOSIS MAINTAINS THE CHROMOSOME NUMBER** [pp.97–99]
**7.4. DIVISION OF THE CYTOPLASM** [pp.100–101]
1. interphase—daughter cells (f); 2. anaphase (a); 3. late prophase (g); 4. metaphase (d); 5. cells at interphase (e); 6. early prophase (c); 7. transition to metaphase (b); 8. telophase (h); 9. a; 10. b; 11. a; 12. a; 13. a; 14. b; 15. b; 16. b; 17. a.

**7.5. MEIOSIS AND SEXUAL REPRODUCTION** [pp.101–103]
**7.6. HOW MEIOSIS PUTS VARIATION IN TRAITS** [p.104–105]
1. a; 2. a; 3. b; 4. a; 5. b; 6. a; 7. b; 8. a; 9. b; 10 b; 11. two; 12. homologous; 13. haploid; 14. sister chromatids; 15. two; 16. chromosome; 17. one; 18. alleles; 19. four; 20. haploid; 21. interphase preceding meiosis I;

22. meiosis II; 23. twenty-three; 24. e ($2n = 2$); 25. d ($2n = 2$); 26. b ($2n = 2$); 27. a ($n = 1$); 28. c ($n = 1$), 29. Crossing over in prophase I and the metaphase I alignments.

### 7.7. FROM GAMETES TO OFFSPRING [pp.106–107]
### 7.8. THE CELL CYCLE AND CANCER [pp.108–109]
1. b; 2. c; 3. a; 4. c; 5. b; 6. b; 7. b; 8. b; 9. c; 10. b; 11. 2 (2n); 12. 5 (n); 13. 4 (n); 14. 1 (2n); 15. 3 (n); 16. a; 17. b; 18. e; 19. d; 20. c; 21. Fertilization also adds to variation among offspring. During prophase I, an average of two or three crossovers take place in each human chromosome. Random positioning of pairs of paternal and maternal chromosomes at metaphase I results in one of millions of possible chromosome combinations in each gamete. Of all male and female gametes produced, which two get together is a matter of chance. The sheer numbers of combinations that can exist are staggering. 22. First, cancer cells grow and divide abnormally. Second, cancer cells have profoundly altered cytoplasm and plasma membranes. Third, cancer cells have a weakened capacity for adhesion. Fourth, cancer cells usually have a lethal effect.

*Self-Quiz*
1. a; 2. d; 3. c; 4. e; 5. c; 6. d; 7. a; 8. a; 9. c; 10. b; 11. b.

---

## Chapter 8   Observing Patterns in Inherited Traits

*Menacing Mucus* [p.112]

### 8.1. TRACKING TRAITS WITH HYBRID CROSSES [pp.113–117]
1. genotype: 1/2 $Tt$; 1/2 $tt$, phenotype: 1/2 tall; 1/2 short; 2. genotype: ¼ WW, ½ Ww, ¼ ww; phenotype ¾ purple, ¼ white; 3. Both white parents are heterozygous and had one white (dominant) and one black (recessive) allele. You would expect ¾ white and ¼ black from such a cross; 4.a. 9/16; pigmented eyes, right-handed; b. 3/16 pigmented eyes, left-handed; c. 3/16 blue-eyed, right-handed; d. 1/16 blue-eyed, left-handed.

### 8.2. NOT-SO-STRAIGHTFORWARD PHENOTYPES [pp.117–119]
1. Genotypes: 1/4 $I^A I^A$; 1/4 $I^A I^B$; 1/4 $I^A i$; 1/4 $I^B i$, phenotypes: 1/2 A; 1/4 AB; 1/4 B; 2. Genotypes: 1/4 $I^A I^B$; 1/4 $I^B i$; 1/4 $I^A i$; 1/4 $ii$, phenotypes: 1/4 AB; 1/4 B; 1/4 A; 1/4 O; 3. Genotypes: 1/4 $I^A I^A$; 1/2 $I^A I^B$; 1/4 $I^B I^B$, phenotypes: 1/4 A; 1/2 AB; 1/4 B; 4.a. phenotype: all pink; genotype: all $RR''$; b. phenotypes: 1/4 red; 1/2 pink; 1/4 white; genotypes: 1/4 RR; 1/2 RR'; 1/4 R'R'; 5. 3/8 black; 1/2 yellow; 1/8 brown; 6. The genotype of the male parent is $RrPp$ and the genotype of the female parent is $rrpp$. The offspring are 1/4 walnut comb, $RrPp$; 1/4 rose comb, $Rrpp$; 1/4 pea comb, $rrPp$; 1/4 single comb, $rrpp$.

### 8.3. COMPLEX VARIATIONS IN TRAITS [pp.120–122]
1. b (a); 2. a, b; 3. a; 4. b; 5. a; 6. a.

### 8.4. THE CHROMOSOMAL BASIS OF INHERITANCE [pp.123–124]
### 8.5. IMPACT OF CROSSING OVER ON INHERITANCE [p.125]
1. wild-type; 2. homologous; 3. cell division; 4. sons; 5. mothers; 6. daughters; 7. *SRY* is the master gene for male sex determination and its expression leads to the formation of testes, which are the primary male reproductive organs. 8. d.

### 8.6. HUMAN GENETIC ANALYSIS [pp.126–127]
### 8.7. EXAMPLES OF HUMAN INHERITANCE PATTERNS [pp.127–129]
1. The woman's mother is heterozygous normal, *Gg*; the woman is also heterozygous normal, *Gg*. The galactosemic man, gg, has two heterozygous normal parents, *Gg*. The two normal children are heterozygous normal, *Gg*; the galactosemic child is *gg*; 2. Assuming the father is heterozygous with Huntington's disorder and the mother normal, the chances are .50 that the son will develop the disease; 3. If only male offspring are considered, the probability is .50 that the couple will have a color-blind son; 4. The probability is that 1/2 of the sons will have hemophilia; the probability is 0 that a daughter will express hemophilia; the probability is that 1/2 of the daughters will be carriers; 5. If the woman marries a normal male, the chance that her son would be color blind is .50; if the woman marries a color-blind male, the chance that her son would be color blind is .25.

### 8.8. STRUCTURAL CHANGES IN CHROMOSOMES [p.130]
### 8.9. CHANGE IN THE NUMBER OF CHROMOSOMES [pp.131–133]
1. duplication; 2. inversion; 3. deletion (number 3); 4. translocation; 5. deletion; 6. All gametes will be abnormal; 7. About half of all flowering plant species are polyploids but this condition is lethal for humans; 8. b; 9. c; 10. d; 11. a; 12. e.

### 8.10. SOME PROSPECTS IN HUMAN GENETICS [pp.133–134]
1. e; 2. b; 3. d; 4. c; 5. a.

*Self-Quiz*
1. d; 2. c; 3. b; 4. e; 5. a; 6. d; 7. b; 8. a; 9. c; 10. b.

*Integrating and Applying Key Concepts*
1. Yes, the husband has a case because he could not have supplied either of his daughter's recessive genes. His only X chromosome bears the gene for normal iris, the dominant gene.
2. Couple 1 child Y; Couple 2 child Z; Couple 3 child X.

# Chapter 9  DNA Structure and Function

*Here Kitty, Kitty, Kitty, Kitty, Kitty* [p.137]

**9.1. THE HUNT FOR FAME, FORTUNE, AND DNA** [pp.138–139]
1. Surviving clones typically have health problems. Like Dolly, many become overweight as they age. Others are unusually large from birth or have some enlarged organs. Cloned mice have lung and liver problems, and almost all die prematurely. Cloned pigs have heart problems, they limp, and one never did develop an anus or a tail; 2. d; 3. a; 4. f; 5. e; 6. b; 7. c.

**9.2. DNA STRUCTURE AND FUNCTION** [pp.140–142]
1. A five-carbon sugar called deoxyribose, a phosphate group, and one of the four nitrogen-containing bases (A, T, G, or C); 2. guanine (pu); 3. cytosine (py); 4. adenine (pu); 5. thymine (py); 6. deoxyribose; 7. phosphate; 8. purine; 9. pyrimidine; 10. purine; 11. pyrimidine; 12. nucleotide.

**9.3. DNA REPLICATION AND REPAIR** [pp.142–143]
1. replication; 2. Enzymes; 3. strand; 4. nucleotides; 5. T; 6. C; 7. double helix; 8. saved (conserved); 9. semi-conservative; 10. polymerases; 11. phosphate; 12. energy; 13. ligases; 14. double helix; 15. DNA polymerases; 16. mutation.
17. T- A T -A
    G- C G -C
    A- T A -T
    C- G C -G
    C- G C -G
    C- G C -G

**9.4. USING DNA TO CLONE MAMMALS** [pp.143–144]
1. a; 2. b; 3. b; 4. c; 5. b; 6. b.

*Self-Quiz*
1. d; 2. a; 3. b; 4. a; 5. a; 6. c; 7. e; 8. d; 9. d; 10. c.

# Chapter 10  Gene Expression and Control

*Ricin and Your Ribosomes* [p.146]

**10.1. MAKING AND CONTROLLING THE CELL'S PROTEINS** [p147]
**10.2. HOW IS RNA TRANSCRIBED FROM DNA?** [pp.147–149]
1. plutonium; 2. botulism toxin; 3. castor oil; 4. ribosomes; 5. proteins; 6. RNA; 7. nucleus; 8. mRNA (rRNA, tRNA); 9. rRNA (mRNA, tRNA); 10. tRNA (mRNA, rRNA); 11. RNA polymerase; 12. promoter; 13. mRNA codes for the actual sequence of amino acids in the protein, rRNA is a part of the ribosome, tRNA transports amino acids to the ribosome; 14. DNA has the sugar deoxyribose while RNA contains ribose. In DNA, T is the complement of A while in RNA, U is A's complement. DNA is usually double stranded while RNA is usually single stranded. DNA is for information storage in the nucleus/nucleoid. RNA is for information use in the cytoplasm. 15. It must be "capped," have a poly-A tail added, and have introns removed and the remaining exons spliced together.
16. Exons can be spliced together in different ways. This is called alternative splicing.

**10.3. DECIPHERING mRNA** [pp.149–151]
**10.4. FROM mRNA TO PROTEIN** [pp.152–153]
1. UAGGUCAU; 2. triplet; 3. 64; 4. 61; 5. AUG; 6. UAA; 7. UAG; 8. UGA; 9. 1; 10. UGG; 11. 3; 12. AUU, AUC, AUA; 13. 6; 14. UCU, UCC, UCA, UCG, AGU, AGC; 15. T; 16. mitochondria; 17. first and second; 18. one mRNA connected to several ribosomes; 19. e; 20. c; 21. g; 22. b; 23. h; 24. a; 25. i; 26. d; 27. f.

**10.5. MUTATED GENES AND THEIR PROTEIN PRODUCTS** [pp.154–155]
**10.6. CONTROLS OVER GENE EXPRESSION** [pp.156–159]
1. b; 2. a; 3. c; 4. a; 5. a; 6. When they insert in a gene, they can disrupt the information blocking or altering the expression of the gene. 7. hormones; 8. signaling molecules; 9. negative; 10. promoter; 11. RNA polymerase; 12. Operators; 13. repressors; 14. allolactose; 15. shape; 16. operon; 17. lactase; 18. lactose intolerance; 19. specialized (differentiated); 20. cell differentiation; 21. 5, 10; 22. transcriptional; 23. transcriptional; 24. transcriptional; 25. translational (transcript processing); 26. posttranslational; 27. transcript processing; 28. transcript processing; 29. They are "master genes" that cause body parts to form in a basic body plan. 30. one; 31. Barr bodies; 32. mosaic; 33. anhidrotic ectodermal dysplasia; 34. calico; 35. female.

*Self-Quiz*
1. b; 2. a; 3. d; 4. a; 5. d; 6. c; 7. c; 8. a; 9. c; 10. a.

## Chapter 11  Studying and Manipulating Genomes

*Golden Rice or Frankenfood?* [p.162]

**11.1. A MOLECULAR TOOLKIT** [pp.163–164]
**11.2. HAYSTACKS TO NEEDLES** [pp.165–166]
**11.3. DNA SEQUENCING** [p.167]
**11.4. FIRST JUST FINGERPRINTS, NOW DNA FINGERPRINTS** [p.168]
**11.5. TINKERING WITH THE MOLECULES OF LIFE** [pp.169–170]
**11.6. PRACTICAL GENETICS** [pp.171–173]
**11.7. WEIGHING THE BENEFITS AND RISKS** [pp.173–174]

1. 124 million; 2. blindness; 3. beta carotene; 4. genetic engineering; 5. Frankenfood; 6. b; 7. a; 8. c; 9. b; 10. c; 11. c; 12. b; 13. g; 14. f; 15. b; 16. e; 17. d; 18. a; 19. c; 20. c; 21. a; 22. e; 23. b; 24. d; 25. DNA is heated to form single strands. Primers are mixed in. Cooling the mixture promotes base pairing of primers and DNA fragments. DNA polymerase synthesizes a new strand starting at the primers. The process is repeated. 26. primers; 27. Heat-tolerant; 28. *Thermus aquaticus;* 29. DNA polymerase; 30. free nucleotides; 31. DNA; 32. doubles; 33. Researchers mix DNA with primers, nucleotides, DNA polymerase, and fluorescent nucleotides which terminate synthesis. This produces fragments of different lengths, each tagged at the end. Fragments are separated by size using electrophoresis. As fragments reach the end of the gel, a fluorescent reader determines which fluorescent nucleotide is present and thus which nucleotide is next in the sequence. 34. T; 35. Ninety-nine; 36. tandem repeats; 37. size or length.

1. recombinant DNA; 2. Human Genome Project; 3. 99; 4. structural genomics; 5. comparative genomics; 6. gene therapy; 7. g; 8. a, b; 9. b; 10. c; 11. d; 12. e; 13. f; 14. a; 15. c; 16. c; 17. mice; 18. somatotropin; 19. cystic fibrosis; 20. collagen; 21. immune cells; 22. SCID-X1; 23. nine; 24. leukemia (cancer); 25. Xenotransplantation; 26. immunity-inducing; 27. Making humans "better" by replacing "unfavorable" genes with "desirable" ones; 28. Pig cells were engineered not to have the gene Ggta1, which codes for placement of a cell surface sugar that the human immune system responds strongly to. Organs from pigs lacking two copies of the gene would not stimulate the human immune system as strongly as organs from normal pigs.

*Self-Quiz*
1. c; 2. b; 3. a; 4. c; 5. c; 6. d; 7. d; 8. b; 9. d; 10. c.

## Chapter 12  Processes of Evolution

*Rise of the Super Rats* [p.176]

**12.1. EARLY BELIEFS, CONFOUNDING DISCOVERIES** [pp.177–180]
**12.2. THE NATURE OF ADAPTATION** [pp.181–182]
1. g; 2. k; 3. i; 4. l; 5. h; 6. m; 7. j; 8. f; 9. a; 10. d; 11. b; 12. n; 13. e; 14. c; 15. Tomato species respond differently to salty water. *Solanum lycopersicon,* the grocery store tomato, wilts quickly in a salty environment. The Galápagos tomato (*S. cheesmanii*) survives and reproduces in seawater-washed soils. Its salt tolerance is a heritable adaptation because gene transfers from the wild species to the commercial one yield a small, edible hybrid. The hybrid tolerates irrigation water that is two parts fresh and one part salty. It may grow in salty croplands where fresh water is scarce; llama hemoglobin and dromedary camel hemoglobin are better than ours at latching on to oxygen. They pick up oxygen in the lungs

far more efficiently. The two animals are closely related but live in different environments. Superficially, at least, the oxygen-binding affinity of llama hemoglobin appears to be an adaptation to thin air at high altitudes. Apparently though, it is not. If the trait arose in a shared ancestor to llamas and camels, then in what respect has it been adaptive at *low* elevations when it's adaptive at *high* elevations? Their most recent ancestors lived in very different environments with different oxygen concentrations. The gene may have proven to be adaptive during a long-term shift in climate, such as the alternately warm and cool temperatures of the Eocene. Hemoglobin's oxygen-binding capacity does go down as temperatures go up. Perhaps the effects of the mutant gene were neutral at first?

### 12.3. INDIVIDUALS DON'T EVOLVE, POPULATIONS DO [pp.182–183]
### 12.4. WHEN IS A POPULATION *NOT* EVOLVING? [p.184]

1. M; 2. P; 3. M; 4. B; 5. P; 6. population; 7. qualitative; 8. polymorphism; 9. quantitative; 10. pool; 11. alleles; 12. phenotype; 13. frequencies; 14. genetic equilibrium; 15. evolving; 16. Microevolution; 17. mutation; 18. lethal; 19. neutral; 20. Natural; 21. No mutation, infinitely large population isolated from others of its species, random mating with respect to the alleles being studied, and all individuals survive and reproduce equally;

22 a. homozygous dominant = $p^2 \times 200 = (0.8)^2 \times 200 = 0.64 \times 200 = 128$ individuals; b. $q = (1.00 - p) = 0.20$; homozygous recessive = $q^2 \times 200 = (0.2)^2 \times 200 = (0.04) \times (200) = 8$ individuals; c. heterozygotes = $2pq \times 200 = 2 \times 0.8 \times 0.2 \times 200 = 0.32 \times 200 = 64$ individuals. Check: $128 + 8 + 64 = 200$; 23. If $p = 0.70$, since $p + q = 1$, $0.70 + q = 1$; then $q = 0.30$, or 30 percent; 24. If $p = 0.60$, since $p + q = 1$, $0.60 + q = 1$; then $q = 0.40$; thus, $2pq = 0.48$, or 48 percent.

### 12.5. NATURAL SELECTION REVISITED [pp.185–188]
### 12.6. MAINTAINING VARIATION IN A POPULATION [pp.188–189]

1.a. directional; b. disruptive; c. stabilizing; 2. c; 3. a; 4. a; 5. b; 6. b; 7. c; 8. b; 9. a; 10. a; 11. b; 12. a.

### 12.7. GENETIC DRIFT—THE CHANCE CHANGES [pp.190–191]
### 12.8. GENE FLOW—KEEPING POPULATIONS ALIKE [pp.191–192]

1. Genetic drift; 2. small; 3. Fixation; 4. homozygous; 5. bottleneck; 6. founder; 7. Inbreeding; 8. genetic diversity; 9. Ellis–van Creveld; 10. emigration; 11. immigration; 12. gene flow.

*Self-Quiz*
1. e; 2. c; 3. d; 4. c; 5. c; 6. b; 7. e; 8. e; 9. b; 10. c.

---

## Chapter 13  Evolutionary Patterns, Rates, and Trends

*Measuring Time* [p.194]

### 13.1. FOSSILS—EVIDENCE OF ANCIENT LIFE [pp.195–196]
### 13.2. DATING PIECES OF THE PUZZLE [pp.196–198]

1. g; 2. k; 3. i; 4. h; 5. c; 6. a; 7. l; 8. e; 9. j; 10. b; 11. f; 12. d; 13. b; 14. e; 15. c; 16. a; 17. d; 18. c; 19. b; 20. e; 21. d; 22. a; 23. d; 24. d.

### 13.3. EVIDENCE FROM BIOGEOGRAPHY [pp.199–200]
### 13.4. MORE EVIDENCE FROM COMPARATIVE MORPHOLOGY [pp.201–202]

1. uniformity; 2. time; 3. jigsaw puzzle; 4. Pangea; 5. north-south; 6. crust; 7. plates; 8. plate tectonics; 9. fossils; 10. heavy; 11. small; 12. Gondwana; 13. evolution; 14. convergence; 15. divergence; 16. d; 17. c; 18. a; 19. e; 20. b.

### 13.5. EVIDENCE FROM PATTERNS OF DEVELOPMENT [pp.203–204]
### 13.6. EVIDENCE FROM DNA, RNA, AND PROTEINS [pp.204–205]

1. D; 2. D; 3. B; 4. B; 5. D; 6. D; 7. B.

### 13.7. REPRODUCTIVE ISOLATION, MAYBE NEW SPECIES [pp.205–207]
### 13.8. INTERPRETING THE EVIDENCE: MODELS FOR SPECIATION [pp.208–211]

1. c; 2. a; 3. e; 4. d; 5. b; 6. f; 7. d; 8. e; 9. a; 10. c; 11. b.

### 13.9. PATTERNS OF SPECIATION AND EXTINCTIONS [pp.211–212]
### 13.10. ORGANIZING INFORMATION ABOUT SPECIES [pp.213–215]

1. h; 2. f; 3. g; 4. d; 5. b; 6. i; 7. c; 8. a; 9. j; 10. e; 11. kingdom: Plantae; phylum: Anthophyta; class: Dicotyledonae; order: Asterales; family: Asteraceae; genus: *Archibaccharis*; (specific epithet): *linearilobа*; 12. sharks, crocodiles, birds, and mammals; 13. crocodiles, birds, and mammals; 14. crocodiles and birds; 15. birds; 16. crocodiles.

*Self-Quiz*
1. e; 2. c; 3. a; 4. b; 5. f; 6. d; 7. f; 8. d; 9. b; 10. e; 11. c; 12. a.

# Chapter 14  EARLY LIFE

*Looking for Life in All the Odd Places* [p.218]

**14.1. ORIGIN OF THE FIRST LIVING CELLS**
   [pp.219–223]
1. hot; 2. Yellowstone; 3. 80; 4. 176; 5. Extreme thermophiles; 6. 110; 7. 230; 8. nanobes; 9. 170; 10. 338; 11. electron; 12. four; 13. hydrogen (nitrogen); 14. nitrogen (hydrogen); 15. carbon monoxide (carbon dioxide); 16. carbon dioxide (carbon monoxide); 17. salty waters; 18. 200; 19. a; 20. f; 21. e; 22. c; 23. b; 24. d; 25. Archean; 26. bacteria; 27. archaea; 28. eukaryotes; 29. cyclic; 30. stromatolites; 31. noncyclic; 32. oxygen; 33. aerobic respiration; 34. chemical origin.

**14.2. WHAT ARE EXISTING PROKARYOTES LIKE?**
   [pp.224–228]
1. b; 2. d; 3. a; 4. c; 5. c(b); 6. b(a); 7. f(e); 8. d(c); 9. g(f); 10. e(d); 11. a(g); 12. a. wetland mud, guts of mammals and termites, hydrothermal vents; b. strip electrons from hydrogen or ethanol, produce methane, strict anaerobes; c. extreme halophiles; d. can produce ATP by aerobic metabolism or anaerobically by photosynthesis; e. extreme thermophiles; f. hot springs and hydrothermal vents; g. freshwater ponds; h. photoautotrophs, some fix nitrogen; i. *Rhizobium* (some spirochetes); j. associated with roots of certain plants; k. *Lactobacillus*; l. *Borrelia burgdorferi*; m. inside ticks and several mammals including humans; n. spoiled food and grain; o. causes botulism.

**14.3. THE CURIOUSLY CLASSIFIED PROTISTS**
   [pp.228–234]
1. a, b; 2. b, f, g; 3. b, f; 4. d, f; 5. d, e; 6. c; 7. f, g; 8. a, b, f, h; 9. a, e; 10. c; 11. c; 12. c; 13. b; 14. b; 15. d; 16. b; 17. d; 18. a; 19. c; 20. d; 21. d.

**14.4. THE FABULOUS FUNGI** [pp.234–238]
1. b; 2. a; 3. a; 4. b; 5. c; 6. c; 7. a; 8. a; 9. c; 10. a; 11. b.

**14.5. VIRUSES, VIROIDS, AND PRIONS** [pp.239–241]
**14.6. EVOLUTION AND INFECTIOUS DISEASES**
   [p.241]
1. proteins and nucleic acids; 2. T; 3. only a few types of; 4. lysogenic; 5. herpes simplex virus; 6. proteins; 7. d; 8. a; 9. e; 10. b; 11. c; 12. b; 13. c; 14. a; 15. a,c; 16. a,b,c; 17. a; 18. b; 19. c; 20. c; 21. c; 22. a.

*Self-Quiz*
1. d; 2. d; 3. c; 4. c; 5. c; 6. c; 7. b; 8. c; 9. a; 10. c; 11. c; 12. a; 13. d; 14. b; 15. d; 16. d.

*Chapter Objectives/Review Questions*
13. a. 2; b. 9; c. 6; d. 7; e. 5; f. 10; g. 11; h. 1; i. 4; j. 3; k. 12; l. 8.

# Chapter 15  Plant Evolution

*Beginnings, and Endings* [p.244]

**15.1. PIONEERS IN A NEW WORLD** [pp.245–247]
1. g; 2. i; 3. h; 4. c; 5. a; 6. b; 7. d; 8. e; 9. j; 10. f; 11. nonvascular; 12. Vascular; 13. seedless; 14. gymnosperms; 15. seed-bearing; 16. angiosperms; 17. spores; 18. megaspores; 19. gametophytes; 20. microspores; 21. pollen grains.

**15.2. THE BRYOPHYTES—NO VASCULAR TISSUES**
   [pp.247–248]
1. T; 2. lack; 3. rhizoids; 4. T; 5. sporophyte; 6. acidic; 7. T; 8. diploid, sporophyte; 9. sporophyte, diploid, gametophyte, haploid; 10. Meiosis, haploid spores; 11. gametophyte, haploid; 12. gametophyte; 13. gametophyte; 14. haploid sperms; 15. haploid eggs; 16. 15.

**15.3. SEEDLESS VASCULAR PLANTS** [pp.248–250]
1. b; 2. a; 3. d; 4. c; 5. c; 6. d; 7. b; 8. c; 9. c; 10. b; 11. b; 12. d; 13. a; 14. c; 15. c; 16. d; 17. gametophyte, haploid; 18. sporophyte, diploid; 19. haploid, mitosis; 20. gametophyte, haploid; 21. diploid, gametophyte; 22. spore production, diploid, sporophyte.

**15.4. THE RISE OF SEED-BEARING PLANTS** [p.251]
**15.5. GYMNOSPERMS—PLANTS WITH "NAKED" SEEDS** [pp.252–253]
1. c; 2. a; 3. e; 4. b; 5. d; 6. b; 7. d; 8. b; 9. b; 10. c; 11. e; 12. b; 13. e; 14. c; 15. d; 16. e; 17. sporophyte, diploid; 18. seed; 19. sporophyte, diploid; 20. pollen sac, cone; 21. ovule, diploid; 22. female gametophyte, haploid; 23. eggs, haploid, mitosis; 24. male gametophyte, haploid; 25. megaspore, meiosis, haploid; 26. microspores, haploid, pollen grain; 27. pollen grain, female cone.

**15.6. ANGIOSPERMS—THE FLOWERING PLANTS**
[pp.254–255]
**15.7. DEFORESTATION IN THE TROPICS** [p.256]
1. d; 2. g; 3. c; 4. f; 5. b; 6. a; 7. e; 8. sporophyte, diploid; 9. female gametophyte, haploid; 10. egg, haploid, mitosis; 11. ovule, diploid, seed; 12. male gametophyte, haploid, mitosis; 13. released pollen grain, haploid; 14. embryo sporophyte plant, diploid, mitosis, diploid zygote; 15. seed, ovule 16. They act as enormous sponges to absorb, hold, and then release water slowly. Be mediating downstream flows, they prevent erosion, flooding, and sedimentation that can disrupt rivers, lakes, and reservoirs. They prevent leaching of nutrients and erosion, especially on steep slopes; 17. In these forests, litter can't accumulate because the high temperatures and heavy, frequent rainfall promote rapid decomposition of organic wastes and remains. As fast as decomposers release nutrients, trees and other plants take them up. Deep, nutrient-rich topsoils simply cannot form; 18. If shifting cultivation is practiced on small, widely spaced plots, a forest isn't necessarily damaged much.

*Self-Quiz*
1. b; 2. c; 3. a; 4. b; 5. e; 6. e; 7. c; 8. c; 9. d; 10. e.

---

# Chapter 16    Animal Evolution

*Interpreting and Misinterpreting the Past* [p.259]

**16.1. OVERVIEW OF THE ANIMAL KINGDOM**
[pp.260–261]
1. Germany; 2. dinosaurs; 3. birds; 4. *Archaeopteryx*; 5. jellyfish; 6. worms; 7. 150 million; 8. multicelled; 9. heterotrophic; 10. sexually; 11.ectoderm; 12. endoderm; 13. mesoderm; 14. invertebrates; 15. bilaterally; 16. back; 17. tube-like gut with two openings; 18. g; 19. f; 20. j; 21. c; 22. d; 23. h; 24. i; 25. e; 26. k; 27. b; 28. a.

**16.2. GETTING ALONG WELL WITHOUT ORGANS**
[pp.262–263]
**16.3. FLATWORMS—INTRODUCING ORGAN SYSTEMS** [pp.263–264]
1. no; 2. yes; 3. yes; 4. none; 5. radial; 6. bilateral; 7. none; 8. sac-like; 9. sac-like; 10. none; 11. nerve net; 12. cephalized; 13. c; 14. a; 15. c; 16. a; 17. b; 18. b; 19. a, b; 20. b.

**16.4. ANNELIDS—SEGMENTS GALORE**
[pp.264–265]
**16.5. THE EVOLUTIONARILY PLIABLE MOLLUSKS**
[pp.266–267]
1. b; 2. a; 3. a; 4. b; 5. b; 6. b; 7. b; 8. a; 9. b; 10. b; 11. b; 12. b; 13. b; 14. d; 15. a; 16. b; 17. b; 18. c.

**16.6. AMAZINGLY ABUNDANT ROUNDWORMS**
[p.267]
**16.7. ARTHROPODS—THE MOST SUCCESSFUL ANIMALS** [pp.268–271]
1. nematodes; 2. false coelom; 3. molt; 4. decomposers; 5. parasites; 6. Trichinosis; 7. elephantiasis; 8. one million; 9. insects; 10. exoskeleton; 11. jointed; 12. gills; 13. trachea; 14. metamorphosis; 15. b; 16. c; 17. b; 18. c; 19. c; 20. b; 21. a; 22. b; 23. a; 24. a; 25. T; 26. ticks; 27. five pairs (ten); 28. Copepods (Crustaceans); 29. T.

**16.8. THE PUZZLING ECHINODERMS** [p.272]
**16.9. EVOLUTIONARY TRENDS AMONG VERTEBRATES** [pp.273–274]
**16.10. MAJOR GROUPS OF JAWED FISHES**
[pp.275–276]
1. calcium carbonate; 2. brain; 3. tube feet; 4. stomachs; 5. bilateral; 6. radial; 7. notochord; 8. dorsal; 9. brain; 10. gill slits; 11. tail; 12. anus; 13. filter; 14. craniates; 15. cartilage; 16. bone; 17. jaws; 18. Vertebrates; 19. vertebral column; 20. c; 21. b; 22. e; 23. g; 24. h; 25. f; 26. f; 27. a; 28. d; 29. a, d.

**16.11. EARLY AMPHIBIOUS TETRAPODS**
[pp.276–277]
**16.12. THE RISE OF AMNIOTES** [pp.278–280]
1. Amphibians; 2. vision, hearing, balance; 3. salamanders; 4. frogs; 5. toads; 6. skin; 7. reproduction; 8. Reptiles; 9. amniotes; 10. keratin; 11. kidneys; 12. synapsids; 13. sauropsids; 14. dinosaurs; 15. 140; 16. asteroid; 17. birds; 18. c; 19. b; 20. c; 21. d.

**16.13. FROM EARLY PRIMATES TO HUMANS**
[pp.281–286]
1 prosimians; 2. tarsiers; 3. anthropoids; 4. humans; 5. hominids; 6. daytime vision; 7. bipedalism; 8. power; 9. precision; 10. teeth; 11. culture; 12. language; 13. rodents; 14. shrews; 15. 36; 16. six to seven; 17. *Australopithecus afarensis*; 18. Africa; 19. stone; 20. *Homo habilis*; 21. *Homo erectus*; 22. two; 23. *Homo sapiens*; 24. 160,000; 25. African emergence; 26. multiregional.

*Self-Quiz*
1. a; 2. b; 3. b; 4. d; 5. b; 6. b; 7. c; 8. d; 9. c; 10. d; 11. d; 12. c; 13. a; 14. c; 15. a.

*Chapter Objectives/Review Questions*
1. Cnidaria, jellyfish; 2. roundworm; 3. Arthropod, chelicerate; 4. Annelid, oligochaete; 5. flatworm, planaria; 6. Arthropod, crustacean; 7. Mollusk, gastropod; 8. Chordate, bony fish; 9. Arthropod, insect; 10. Echinoderm, sea star; 11. Chordate, mammal; 12. flatworm, tapeworm; 13. mollusk, cephalopod; 14. Annelid, polychaete; 15. Chordate, shark; 16. Chordate, amphibian.

# Chapter 17 Plants and Animals: Common Challenges

*Too Hot to Handle* [p.289]

**17.1. LEVELS OF STRUCTURAL ORGANIZATION** [pp.290–291]
1. 97°F to 100°F; 2. blood; 3. skin; 4. sweating; 5. 105°F; 6. heat stroke; 7. brain damage; 8. death; 9. Anatomy; 10. physiology; 11. structure; 12. function; 13. T; 14. a succession of stages during the formation of specialized tissues, organs, and organ systems; 15. how to keep from drying out (lack of water); 16. extracellular fluid; 17. T.

**17.2. RECURRING CHALLENGES TO SURVIVAL** [pp.292–293]
**17.3. HOMEOSTASIS IN ANIMALS** [pp.293–295]
**17.4. DOES HOMEOSTASIS OCCUR IN PLANTS?** [pp.295–296]
1. oxygen; 2. carbon dioxide; 3. diffusion; 4. carbon dioxide; 5. oxygen; 6. surface-to-volume; 7. thin (flat); 8. vascular tissues; 9. lungs; 10. capillaries; 11. immune; 12. solute–water; 13. active transport; 14. kidneys; 15. signaling molecules; 16. d(c); 17. c(a); 18. b(e); 19. e(d); 20. a(b); 21. During system acquired resistance, signaling molecules released at the injured site call for production of defensive compounds. More distant cells are also involved and they release chemicals that make the area more resistant to infection. Woody plants also use compartmentalization, the production of chemicals that wall off the infected area; 22. yellow bush lupine; 23. epidermal hairs; 24. fold; 25. moisture loss; 26. heat; 27. sun's rays; 28. night; 29. circadian.

**17.5. HOW CELLS RECEIVE AND RESPOND TO SIGNALS** [pp.297–298]
1. signal receptor; 2. transduction; 3. functional; 4. Apoptosis; 5. proteases; 6. Phagocytic; 7. cancers; 8. receptors; 9. signals.

*Self-Quiz*
1. d; 2. c; 3. d; 4. a; 5. c; 6. c; 7. c; 8. b; 9. a; 10. d.

# Chapter 18 Plant Form and Function

*Drought Versus Civilization* [p.300]

**18.1. OVERVIEW OF THE PLANT BODY** [pp.301–303]
1. Even brief drought episodes reduce photosynthesis and crop production. Plants conserve water by closing leaf stomata which stops carbon dioxide from moving in. No carbon dioxide, no sugars. A stressed plant produces fewer flowers. Then pollination may not occur. Even if its flowers are pollinated, its seeds and fruits may fall off before they ripen; 2. b (c); 3. a (e); 4. e (d); 5. c (a); 6. d (b); 7. parenchyma; 8. xylem; 9. phloem; 10. sclerenchyma; 11. collenchyma; 12. parenchyma; 13. fibers; 14. stone cells; 15. vessel element; 16. tracheid; 17. sieve tube; 18. companion cell; 19a. One; three; usually oriented in parallel; one pore or furrow; scattered throughout the stem ground tissue. b. Two; four or five; netlike array; three pores or furrows; in a ring in stem ground tissue.

**18.2. PRIMARY STRUCTURE OF SHOOTS** [pp.304–306]
1. monocot; 2. eudicot; 3. phloem; 4. xylem; 5. stomata; 6. stoma; 7. epidermis; 8. cuticle; 9. epidermis; 10. mesophyll; 11. mesophyll; 12. epidermis; 13. cuticle; 14. vein; 15. leaf; 16. apical meristem; 17. meristems; 18. cortex; 19. phloem; 20. xylem; 21. pith.

**18.3. PRIMARY STRUCTURE OF ROOTS** [pp.306–307]
1. endodermis; 2. pericycle; 3. xylem; 4. phloem; 5. cortex; 6. epidermis; 7. root hair; 8. matured; 9. elongate; 10. dividing; 11. root cap; 12. taproot; 13. fibrous; 14. lateral root.

### 18.4. SECONDARY GROWTH—THE WOODY PLANTS [pp.308–309]
1. periderm; 2. phloem; 3. heartwood; 4. sapwood; 5. bark; 6. vascular cambium; 7. early; 8. late; 9. early; 10. growth rings.

### 18.5. PLANT NUTRIENTS AND AVAILABILITY IN SOIL [pp.310–311]
1. f; 2. c; 3. i; 4. b; 5. h; 6. a; 7. j; 8. e; 9. g; 10. d.

### 18.6. HOW DO ROOTS ABSORB WATER AND MINERAL IONS? [pp.312–313]
1. mutualism; 2. Nitrogen "fixed" by bacteria; 3. root nodules; 4. scarce minerals that the fungus is better able to absorb; 5. sugars and nitrogen-containing compounds; 6. d; 7. b; 8. c; 9. a.

### 18.7. WATER TRANSPORT THROUGH PLANTS [pp.314–315]
1. stomata; 2. transpiration; 3. xylem; 4. tracheids; 5. vessel; 6. cohesion-tension; 7. transpiration; 8. tension; 9. absorbed; 10. cohesion; 11. tension; 12. xylem; 13. Hydrogen; 14. water.

### 18.8. HOW DO STEMS AND LEAVES CONSERVE WATER? [pp.316–317]
1. wilts; 2. cuticle; 3. close; 4. opens; 5. open; 6. closed; 7. do; 8. decline; 9. night.

### 18.9. HOW ORGANIC COMPOUNDS MOVE THROUGH PLANTS [pp.317–318]
1. d; 2. f; 3. b; 4. a; 5. c; 6. e.

*Self-Quiz*
1. b; 2. c; 3. a; 4. b; 5. e; 6. e; 7. c; 8. d; 9. d; 10. d; 11. c.

---

## Chapter 19  Plant Reproduction and Development

*Imperiled Sexual Partners* [p.321]

### 19.1. SEXUAL REPRODUCTION IN FLOWERING PLANTS [pp.322–325]
1. sporophyte (c); 2. flower (b); 3. meiosis (d); 4. gametophyte (f); 5. gametophyte (e); 6. fertilization (a); 7. sepal; 8. petal; 9. stamen; 10. filament; 11. anther; 12. carpel; 13. stigma; 14. style; 15. ovary; 16. ovule; 17. receptacle; 18. pollinators; 19. (humming) birds; 20. nectar; 21. beetles (flies); 22. sweet odors; 23. petals (flowers); 24. ultraviolet light; 25. stigma; 26. pollen tube; 27. 2; 28. diploid zygote; 29. $3n$ endosperm; 30. double.

### 19.2. FROM ZYGOTES TO SEEDS PACKAGED IN FRUIT [pp.325–326]
### 19.3. ASEXUAL REPRODUCTION OF FLOWERING PLANTS [p.327]
1.a. Also called seed leaves; cotyledons develop from lobes of meristematic tissue on the embryo; b. Seeds are mature ovules; c. Forms from the thickened and hardened integuments of the embryo; d. Most fruits are formed from a matured ovary but sometimes other flower parts also form fruits; 2. d; 3. a; 4. b; 5. c; 6. e; 7. c; 8. c; 9. b; 10. a; 11. b; 12. b; 13. b; 14. a; 15. b; 16. root; 17. genetically; 18. clone; 19. runners; 20. nodes; 21. cuttings; 22. tissue culture; 23. one; 24. food; 25. production (number).

### 19.4. PATTERNS OF EARLY GROWTH AND DEVELOPMENT [pp.328–329]
1. seasonal; 2. moisture; 3. seed; 4. ruptures; 5. oxygen; 6. aerobic; 7. meristematic; 8. Germination; 9. primary; 10. root; 11. coleoptile; 12. root; 13. foliage; 14. adventitious; 15. branch; 16. primary; 17. prop; 18. coat; 19. root; 20. cotyledons; 21. hypocotyl; 22. foliage; 23. cotyledons; 24. primary; 25. branch; 26. cotyledons; 27. primary; 28. nodule.

### 19.5. CELL COMMUNICATION IN PLANT DEVELOPMENT [pp.330–331]
### 19.6. ADJUSTING RATES AND DIRECTIONS OF GROWTH [pp.332–333]
1. b; 2. c; 3. d; 4. a; 5. a; 6. a; 7. e; 8. b; 9. a; 10. a; 11. a; 12. b; 13. b.

### 19.7. MEANWHILE, BACK AT THE FLOWER . . . [p.334]
### 19.8. LIFE CYCLES END, AND TURN AGAIN [p.335]
### 19.9. REGARDING THE WORLD'S MOST NUTRITIOUS PLANT [p.336]
1. d; 2. g; 3. h; 4. b; 5. a; 6. f; 7. c; 8. e; 9. b; 10. d; 11. c; 12. b; 13. a; 14. e; 15. quinoa; 16. 16; 17. lysine; 18. iron; 19. drought (frost); 20. frost (drought); 21. kwashiorkor; 22. protein-deficient.

*Self-Quiz*
1. d; 2. c; 3. b; 4. c; 5. e; 6. e; 7. d; 8. d; 9. d; 10. c.

# Chapter 20  Animal Tissues and Organ Systems

*It's All About Potential* [p.339]

## 20.1. ORGANIZATION AND CONTROL IN ANIMAL BODIES [p.340]
1. Adults retain patches of stem cells in bone marrow and other tissues. They do not culture well and they don't have many descendants. Embryonic stem cells have more potential to develop in different ways; some can give rise to all types of tissues; 2. b; 3. e; 4. d; 5. c; 6. a.

## 20.2. FOUR BASIC TYPES OF TISSUES [pp.340–345]
1. Epithelium; 2. Simple; 3. Stratified; 4. Exocrine; 5. epithelium; 6. Endocrine; 7. hormones; 8. hormone; 9. target; 10. Tight; 11. Adhering; 12. Gap; 13. Tight; 14. peptic; 15. adhering; 16. gap; 17. tight; 18. j; 19. f; 20. d; 21. b; 22. c; 23. i; 24. g; 25. h; 26. k; 27. a; 28. e; 29. b; 30. c; 31. a; 32. c; 33. b; 34. a; 35. c; 36. a; 37. c; 38. b; 39.a. Adipose; b. Cartilage; c. Blood; d. Bone; 40. smooth; 41. cardiac; 42. smooth; 43. skeletal; 44. Skeletal; 45. smooth; 46. striped; 47. Skeletal; 48. move internal organs; 49. Cardiac; 50. Nervous; 51. neurons; 52. Neuroglial; 53. neuron; 54. neurons.

## 20.3. ORGAN SYSTEMS MADE FROM TISSUES [pp.344–345]
## 20.4. SKIN—EXAMPLE OF AN ORGAN SYSTEM [pp.346–347]
1. cranial; 2. spinal; 3. thoracic; 4. abdominal; 5. pelvic; 6.a. Mesoderm; b. Endoderm; c. Ectoderm; 7. midsagittal; 8. distal; 9. proximal; 10. posterior; 11. transverse; 12. frontal; 13. anterior; 14. circulatory; 15. respiratory; 16. urinary (= excretory); 17. skeletal; 18. endocrine; 19. reproductive; 20. digestive; 21. muscular; 22. nervous; 23. integumentary; 24. lymphatic; 25. h; 26. e; 27. j; 28. g; 29. b; 30. i; 31. a; 32. f; 33. k; 34. l; 35. d; 36. c.

*Self-Quiz*
1. d; 2. b; 3. c; 4. c; 5. e; 6. c; 7. c; 8. d; 9. a; 10. d.

# Chapter 21  How Animals Move

*Pumping Up Muscles* [p.349]

## 21.1. SO WHAT IS A SKELETON? [pp.350–353]
## 21.2. HOW DO BONES AND MUSCLES INTERACT? [p.353]
1. androstenedione; 2. testosterone; 3. estrogen; 4. muscle; 5. strength; 6. controlled; 7. placebo; 8. shriveling; 9. breasts; 10. menstrual; 11. hair; 12. liver; 13. acne; 14. HDL (good); 15. b; 16. a; 17. c; 18. b; 19. a; 20. c; 21. clavicle (collar bone); 22. 12 pairs; 23. intervertebral disks and they are made of cartilage; 24. cranial; 25. facial; 26. sternum; 27. ribs; 28. vertebrae; 29. clavicle; 30. scapula; 31. humerus; 32. radius; 33. ulna; 34. carpals; 35. metacarpals; 36. phalanges; 37. pelvic girdle; 38. femur; 39. patella; 40. tibia; 41. fibula; 42. tarsals; 43. metatarsals; 44. phalanges; 45. c; 46. c; 47. a; 48. d; 49. a; 50. c; 51. b; 52. d.

## 21.3. HOW DOES SKELETAL MUSCLE CONTRACT? [pp.354–356]

## 21.4. PROPERTIES OF WHOLE MUSCLES [pp.356–358]
1. skeletal; 2. cardiac; 3. striated; 4. sarcomere; 5. myosin; 6. actin; 7. parallel; 8. ATP; 9. calcium; 10. creatine; 11. ADP; 12. aerobic respiration; 13. glucose; 14. glycogen; 15. fatty acids; 16. use glycolysis anaerobically; 17. isotonic; 18. muscle fatigue; 19. secretes toxins that spread from the site of infection; 20. vaccines have all but eradicated the disease; 21. b; 22. c; 23. a; 24. Aerobic exercises increase the number of mitochondria in both fast and slow muscle fibers and increases capillary beds that flow to them. This increases endurance. Strength training increases the mass of muscles and increases the enzymes of glycolysis.

*Self-Quiz*
1. d; 2. d; 3. b; 4. b; 5. b; 6. a; 7. d; 8. c; 9. c; 10. b.

# Chapter 22  Circulation and Respiration

*Up in Smoke* [p.360]

## 22.1. THE NATURE OF BLOOD CIRCULATION [pp.361–362]
## 22.2. CHARACTERISTICS OF HUMAN BLOOD [pp.362–363]
1. 3000; 2. immediately; 3. Cilia; 4. white blood; 5. gunk; 6. colds (asthma); 7. asthma (colds); 8. Carcinogens; 9. cancer; 10. 70 percent; 11. bad (LDL) cholesterol; 12. stickier; 13. clots; 14. b; 15. d; 16. a; 17. c; 18. 50–60 percent; 19. transport medium and solvent; 20. Erythrocytes; 21. transport oxygen; 22. 3000–6750/cc; 23. fast-acting phagocytes; 24. 1000–2700/cc; 25. immune response; 26. Platelets; 27. 250,000–300,000/cc.

## 22.3. HUMAN CARDIOVASCULAR SYSTEM [pp.364–366]
1. jugular veins; 2. superior vena cava; 3. pulmonary veins; 4. renal vein; 5. inferior vena cava; 6. iliac veins; 7. femoral vein; 8. carotid arteries; 9. aorta; 10. pulmonary arteries; 11. coronary arteries; 12. brachial artery; 13. renal artery; 14. abdominal aorta; 15. iliac arteries; 16. femoral artery; 17. superior vena cava; 18. right semilunar valve; 19. right pulmonary veins; 20. right atrium; 21. right AV valve; 22. right ventricle; 23. inferior vena cava; 24. septum; 25. aorta; 26. pulmonary trunk; 27. left semilunar valve; 28. left pulmonary veins; 29. left atrium; 30. left AV valve; 31. left ventricle; 32. endothelium; 33. pericardium; 34. myocardium; 35. c; 36. d; 37. e; 38. f; 39. a; 40. b; 41. mitochondria; 42. ATP; 43. skeletal; 44. branched; 45. ends; 46. sinoatrial (SA) node; 47. pacemaker; 48. atrioventricular (AV) node.

## 22.4. STRUCTURE AND FUNCTION OF BLOOD VESSELS [pp.367–369]

## 22.5. CARDIOVASCULAR DISORDERS [pp.370–371]
1. y; 2. y; 3. y; 4. y; 5. y; 6. y; 7. n; 8. y; 9. y; 10. y; 11. n; 12. y; 13. n; 14. n; 15. n; 16. y; 17. b; 18. d; 19. e; 20. a; 21. c; 22. spasm; 23. platelets; 24. coagulation; 25. blood flow; 26. Hypertension; 27. silent killer; 28. atherosclerosis; 29. cholesterol; 30. atherosclerotic plaque; 31. clot; 32. thrombus; 33. embolus; 34. stroke; 35. angina pectoris; 36. heart attack; 37. angioplasty.

## 22.6. THE NATURE OF RESPIRATION [pp.371–373]
## 22.7. HUMAN RESPIRATORY SYSTEM [pp.374–375]
1. surface-to-volume ratio; 2. diffusion; 3. gills; 4. lungs; 5. skin; 6. diaphragm; 7. bulk flow; 8. alveoli; 9. pulmonary; 10. glottis; 11. vocal cords; 12. epiglottis; 13. trachea; 14. oral cavity, k; 15. pleural membrane, f; 16. intercostal muscles, j; 17. diaphragm, i; 18. nasal cavity, d; 19. pharynx, c; 20. epiglottis, e; 21. larynx, b; 22. trachea, a; 23. lung, h; 24. bronchial tree (bronchi), g.

## 22.8. MOVING AIR AND TRANSPORTING GASES [pp.376–378]
## 22.9. WHEN THE LUNGS BREAK DOWN [pp.378–379]
1. inhalation; 2. exhalation; 3. thoracic (chest) cavity; 4. respiratory membrane; 5. epithelia (endothelia); 6. basement membranes; 7. oxygen; 8. carbon dioxide; 9. hemoglobin; 10. four; 11. alveolar air sacs; 12. warmer; 13. pH; 14. carbon dioxide; 15. hemoglobin; 16. bicarbonate; 17. d; 18. c; 19. b; 20. a; 21. 4 million; 22. T; 23. smokers; 24. 10–15; 25. more.

*Self-Quiz*
1. c; 2. a; 3. d; 4. a; 5. a; 6. c; 7. c; 8. b; 9. a; 10. b; 11. d.

# Chapter 23  Immunity

*The Face of AIDS* [p.382]

## 23.1. INTEGRATED RESPONSES TO THREATS [pp.383–385]
## 23.2. SURFACE BARRIERS [pp.385–387]
## 23.3. THE INNATE IMMUNE RESPONSE [pp.387–389]
1. immune; 2. antigens; 3. complement; 4. phagocytes; 5. innate; 6. invertebrate; 7. cytokines; 8. lymphocytes; 9. adaptive; 10. skin; 11. mucous membranes; 12. d; 13. c; 14. e; 15. f; 16. b; 17. a; 18. d; 19. a; 20. b; 21. a; 22. b; 23. a; 24. d; 25. d; 26. c; 27. d.

## 23.4. TAILORING RESPONSES TO SPECIFIC ANTIGENS [pp.389–391]
1. self versus nonself recognition, specificity, diversity, and memory; 2. mitotic cell division; 3. antigen-presenting; 4. phagocytosis; 5. Lysosomal; 6. MHC;

7. effector; 8. memory; 9. B; 10. antibody-mediated; 11. cell-mediated; 12. intracellular; 13. lymph nodes.

### 23.5. ANTIBODIES AND OTHER ANTIGEN RECEPTORS [pp.392–393]
### 23.6. ANTIBODY-MEDIATED IMMUNE RESPONSE [pp.394–395]
### 23.7. THE CELL-MEDIATED IMMUNE RESPONSE [pp.395–396]
1. d; 2. b; 3. e; 4. c; 5. a; 6. f; 7. variable; 8. antigens; 9. randomly; 10. bone marrow; 11. thymus gland; 12. self; 13. receptor-mediated; 14. B; 15. T; 16. dendritic; 17. clonal selection; 18. genetically identical; 19. helper T; 20. effector; 21. antibodies; 22. memory; 23. phagocytic; 24. dendritic; 25. helper; 26. cytotoxic; 27. cytokines (interleukins); 28. apoptosis; 29. macrophages; 30. natural killer.

### 23.8. DEFENSES ENHANCED OR COMPROMISED [p. 397]
### 23.9. AIDS: IMMUNITY LOST [pp.398–399]
1. e; 2. h; 3. a; 4. j; 5. b; 6. g; 7. c; 8. i; 9. f; 10. d; 11. b; 12. c; 13. c; 14. a; 15. b; 16. a.

*Self-Quiz*
1. e; 2. c; 3. a; 4. c; 5. a; 6. d; 7. d; 8. b; 9. e; 10. b.

---

# Chapter 24  Digestion, Nutrition, and Excretion

*Hips and Hunger* [p.402]

### 24.1. THE NATURE OF DIGESTIVE SYSTEMS [pp.403–404]
### 24.2. HUMAN DIGESTIVE SYSTEM [pp.404–409]
1. sixty; 2. Americans; 3. heart; 4. diabetes; 5. cancer; 6. adipose; 7. digestive; 8. incomplete; 9. mouth; 10. anus; 11. complete; 12. specialized; 13. crop; 14. gizzard; 15. intestines; 16. plants; 17. ruminants; 18. mouth, j; 19. pharynx, f; 20. esophagus, d; 21. stomach, l; 22. small intestine, e; 23. colon, g; 24. rectum, i; 25. anus, c; 26. salivary glands, b; 27. liver, k; 28. gall bladder, a; 29. pancreas, h; 30. d; 31. b; 32. c; 33. d; 34. b; 35. a; 36. b; 37. a; 38. b; 39. c.

### 24.3. HUMAN NUTRITIONAL REQUIREMENTS [pp.410–413]
### 24.4. WEIGHTY QUESTIONS, TANTALIZING ANSWERS [pp.413–414]
1. empty calories; 2. complex carbohydrates; 3. fiber; 4. essential; 5. Vegetable oils; 6. cholesterol; 7. trans; 8. heart disease; 9. stroke; 10. essential amino acids; 11. plant; 12. 2; 13. 2.5; 14. 3; 15. 6; 16. 5.5; 17. 55; 18. a. Vitamin A; b. Yellow fruits, green leafy vegetables, fortified milk, egg yolks; c. Vitamin D; d. Bone growth, calcium absorption; e. Whole grains, green vegetables, vegetable oils; f. Counters free radicals, maintains cell membranes; g. Vitamin K; h. Blood clotting, ATP formation; i. Legumes, whole grains, green leafy vegetables, meat, eggs; j. Connective tissue formation, coenzyme; k. Dark green vegetables, whole grains, yeast, lean meats, enterobacteria; l. Coenzyme in amino acid and nucleic acid formation; m. Vitamin C; n. Fruits and vegetables; 19. a. Calcium, b. Dairy products, dark green vegetables, dried legumes; c. Iodine; d. Thyroid hormone; e. Iron; f. Whole grains, green leafy vegetables, legumes, nuts, eggs, meat, molasses, dried fruit, shellfish; g. Whole grains, legumes, dairy products; h. Coenzyme role in ATP–ABP cycle; 20. 5; 21. type 2 diabetes; 22. heart; 23. breast; 24. colon; 25. twice; 26. BMI; 27. Dieting; 28. metabolic; 29. caloric (food); 30. exercising.

### 24.5. URINARY SYSTEM OF MAMMALS [p.415]
### 24.6. HOW THE KIDNEYS MAKE URINE [pp.416–418]
### 24.7. WHEN KIDNEYS BREAK DOWN [p.419]
1. kidney, d; 2. ureter, c; 3. urinary bladder, a; 4. urethra, b; 5. gut; 6. metabolism; 7. urinary; 8. respiratory; 9. sweat; 10. urea; 11. uric acid; 12. Bowman's capsule; 13. proximal tubule; 14. loop of Henle; 15. peritubular capillaries; 16. distal tubule; 17. glomerular capillaries; 18. collecting duct; 19. a; 20. c; 21. d; 22. b; 23. c; 24. c; 25.c.

*Self-Quiz*
1. c; 2. a; 3. c; 4. a; 5. c; 6. c; 7. a; 8. a; 9. d; 10. a; 11. b.

# Chapter 25  Neural Control and the Senses

*In Pursuit of Ecstasy* [p.422]

**25.1. NEURONS—THE GREAT COMMUNICATORS** [pp.423–425]
**25.2. HOW MESSAGES FLOW FROM CELL TO CELL** [pp.426–427]
**25.3. THE PATHS OF INFORMATION FLOW** [pp.428–429]

1. Ingesting Ecstasy makes you feel socially accepted, relieves anxiety, and sharpens the senses while giving you a mild high. The active ingredient in Ecstasy is MDMA, a chemical that interferes with serotonin function. MDMA causes neurons to release too much serotonin that is not normally cleared away. Serotonin molecules saturate target cell receptors with the effect of overstimulation. MDMA overdoses do not often end in death, but have in some cases; 2. dendrites (f); 3. cell body (d); 4. input (c); 5. trigger (h); 6. axon (a); 7. conducting (g); 8. axon endings (b); 9. output (e); 10. d; 11. h; 12. f; 13. b; 14. i; 15. a; 16. e; 17. c; 18. j; 19. g; 20. chemical synapse; 21. neurons; 22. neuron; 23. presynaptic; 24. neurotransmitter; 25. postsynaptic; 26. ion; 27. excite; 28. Excitatory; 29. Inhibitory; 30. acetylcholine; 31. synaptic clefts; 32. excitatory; 33. Serotonin; 34. depression; 35. Norepinephrine; 36. Dopamine; 37. GABA; 38. endorphins; 39. synaptic integration; 40. sensory; 41. motor; 42. interneurons; 43. divergent; 44. convergent; 45. reverberating; 46. eye; 47. Nerves; 48. myelin; 49. multiple sclerosis; 50. white; 51. Reflexes; 52. Sensory; 53. stretch.

**25.4. TYPES OF NERVOUS SYTEMS** [pp.430–431]
1. b; 2. c; 3. a; 4. b; 5. c; 6. a; 7. a; 8. b; 9. c.

**25.5. THE PERIPHERAL NERVOUS SYSTEM** [pp.432–433]
**25.6. THE CENTRAL NERVOUS SYSTEM** [pp.433–436]

1. d; 2. f; 3. a; 4. h; 5. b; 6. g; 7. e; 8. c; 9. medulla oblongata (c); 10. pons (e); 11. cerebellum (a); 12. hypothalamus (d); 13. thalamus (b); 14. e; 15. i; 16. d; 17. g; 18. a; 19. b; 20. h; 21. c; 22. f; 23. j.

**25.7. DRUGGING THE BRAIN** [pp.436–437]
1. b; 2. a; 3. c; 4. d; 5. a; 6. b; 7. c; 8. a; 9. d; 10. a.

**25.8. OVERVIEW OF SENSORY SYSTEMS** [p.438]
**25.9. SOMATIC SENSATIONS** [p.439]
1. stimulus; 2. Mechanoreceptors; 3. Thermoreceptors; 4. Pain; 5. Chemoreceptors; 6. Osmoreceptors; 7. Photoreceptors; 8. which; 9. frequency; 10. number; 11. programmed; 12. stronger; 13. stimulus; 14. constant; 15. sensory; 16. Perception; 17. sensation; 18. Somatic; 19. surface; 20. Meissner's; 21. Krause; 22. Ruffini; 23. Pacinian; 24. Pain; 25. somatic; 26. visceral.

**25.10. THE SPECIAL SENSES** [pp.440–445]
1. c; 2. a; 3. d; 4. c; 5. b; 6. a; 7. d; 8. d; 9. a; 10. d; 11. c; 12. a; 13. d; 14. c.

*Self-Quiz*
1. d; 2. b; 3. a; 4. c; 5. e; 6. b; 7. d; 8. a; 9. c; 10. e; 11. d; 12. a.

# Chapter 26  Endocrine Controls

*Hormones in the Balance* [p.448]

**26.1. HORMONES AND OTHER SIGNALING MOLECULES** [pp.449–451]

1. As an endocrine disrupter, this synthetic organic compound interferes with reproduction and development by mimicking or blocking the action of natural hormones; 2. e; 3. a; 4. b; 5. f; 6. d; 7. c; 8. a; 9. c; 10. a; 11. b; 12. a; 13. a; 14. b; 15. b; 16. a; 17. b; 18. a. 1, produces six releasing and inhibiting hormones; acts on different endocrine cells in the anterior pituitary lobe; produces ADH, oxytocin; ADH conserves water and oxytocin has roles in childbirth and milk secretion, the posterior pituitary lobe stores both; b. 6, ACTH, TSH, FSH, LH, GSH; prolactin and somatotropin; the first four listed stimulate other glands; prolactin and somatotropin stimulate overall growth and development; c. 6, ADH and oxytocin; stores and secretes these two hypothalamic hormones; d. 3, cortex: cortisol, aldosterone; cortisone affects glucose metabolism, aldosterone conserves sodium; e. 3, medulla: epinephrine, norepinephrine; these hormones interact, in concert with the sympathetic nervous system, to help adjust organ activities, especially during times of excitement or stress; f. 4, estrogens, progesterone; maintain primary sex organs, influence secondary sexual traits; g. 5, testosterone; develops and maintains primary sex organs, influences secondary sexual traits; h. 7, thyroxine and

triiodothyronine; roles in growth, development, metabolic control; i. 8, parathyroid hormone (PTH); increases blood level of calcium; j. 2, melatonin; affects biological clocks, overall level of activity, reproductive cycles; k. 9, thymosins; roles in white blood cell functioning; l. 10, insulin, glucagon; insulin lowers level of blood glucose, glucagon raises blood level of glucose.

### 26.2. THE HYPOTHALAMUS AND PITUITARY GLAND [pp.452–453]
1. kidney (ADH); 2. mammary (oxytocin); 3. muscles (oxytocin); 4. adrenal (ACTH); 5. thyroid (TSH); 6. testes (FSH and LH); 7. ovaries (FSH and LH); 8. mammary (PRL); 9. cells (STH or GH); 10. hypothalamus; 11. posterior; 12. anterior; 13. releasers; 14. c, e; 15. a, f; 16. b, d.

### 26.3. THYMUS, THYROID, AND PARATHYROID GLANDS [pp.453–454]

### 26.4. ADRENAL GLANDS AND STRESS RESPONSES [p.455]
### 26.5. THE PANCREAS AND GLUCOSE HOMEOSTASIS [pp.456–457]
### 26.6. HORMONES AND REPRODUCTIVE BEHAVIOR [pp.457–458]
1. negative feedback loop; 2. hypothalamus; 3. releasing hormone; 4. anterior pituitary; 5. thyroid stimulating hormone; 6. thyroid hormone (thyroxine); 7. set point; 8. c; 9. d; 10. e; 11. f; 12. g; 13. j; 14. b; 15. h; 16. i; 17. a; 18. thyroid, thyroid hormone; 19. parathyroid, parathyroid hormone; 20. adrenal cortex, cortisol; 21. pancreas, insulin; 22. pineal gland, melatonin.

*Self-Quiz*
1. b; 2. a; 3. e; 4. d; 5. b; 6. a; 7. d; 8. c; 9. b; 10. a; 11. c; 12. e; 13. d.

# Chapter 27  Reproduction and Development

*Mind-Boggling Births* [p.461]

### 27.1. METHODS OF REPRODUCTION [pp.462–463]
### 27.2. PROCESSES OF ANIMAL DEVELOPMENT [pp.464–466]
1. sexual; 2. fertilization; 3. zygote; 4. asexual; 5. budding; 6. genetically; 7. environmental; 8. parthenogenesis; 9. unfertilized eggs; 10. alleles; 11. variation; 12. offspring; 13. environment; 14. energy; 15. internal; 16. reproductive; 17. nourish; 18. yolk; 19. eggs; 20. birds; 21. embryo; 22. development; 23. yolkless; 24. e; 25. c; 26. f; 27. a; 28. d; 29. b; 30. a. mesoderm; b. ectoderm; c. endoderm; d. mesoderm; e. ectoderm; f. mesoderm; g. mesoderm; h. mesoderm; i. mesoderm; 31. c; 32. a; 33. d; 34. e; 35. b.

### 27.3. REPRODUCTIVE SYSTEM OF HUMAN MALES [pp.467–468]
1. prostate gland; 2. urinary bladder; 3. urethra; 4. erectile tissue; 5. penis; 6. testis; 7. ejaculatory duct; 8. seminal vesicle; 9. bulbourethral gland; 10. vas deferens; 11. epididymis; 12. i; 13. b; 14. g; 15. e; 16. h; 17. a; 18. d; 19. f; 20. c.

### 27.4. REPRODUCTIVE SYSTEM OF HUMAN FEMALES [pp.469–471]
1. ovary; 2. urinary bladder; 3. urethra; 4. clitoris; 5. labium minor; 6. labium major; 7. oviduct; 8. vagina; 9. cervix; 10. uterus; 11. myometrium; 12. endometrium; 13. b; 14. e; 15. c; 16. f; 17. a; 18. d.

### 27.5. HOW PREGNANCY HAPPENS [pp.472–474]
### 27.6. SEXUALLY TRANSMITTED DISEASES [pp.474–476]
1. erection; 2. clitoris; 3. coitus; 4. orgasm; 5. ejaculation; 6. vagina; 7. Sperm; 8. oviducts; 9. zona pellucida; 10. oocyte; 11. meiosis; 12. zygote; 13. j; 14. f; 15. k; 16. d; 17. h; 18. c; 19. l; 20. a; 21. e; 22. n; 23. i; 24. m; 25. b; 26. g; 27. g; 28. d; 29. a; 30. h; 31. b; 32. c; 33. e; 34. f.

### 27.7. HUMAN PRENATAL DEVELOPMENT [pp.476–483]
1. e; 2. j; 3. l; 4. h; 5. d; 6. b; 7. g; 8. k; 9. a; 10. i; 11. f; 12. c; 13. d; 14. a; 15. e; 16. c; 17. f; 18. b.

### 27.8. FROM BIRTH ONWARD [pp.483–485]
1. c; 2. b; 3. e; 4. a; 5. d; 6. sexual; 7. prenatal; 8. postnatal; 9. thirteen; 10. nineteen; 11. adulthood; 12. aging; 13. 122; 14. cancer; 15. telomerase; 16. telomeres; 17. Dolly; 18. accumulates.

*Self-Quiz*
1. d; 2. a; 3. b; 4. d; 5. a; 6. b; 7. c; 8. e; 9. d; 10. b; 11. e; 12. a; 13. d; 14. b.

# Chapter 28 Population Ecology

*The Human Touch* [p.488]

## 28.1. CHARACTERISTICS OF POPULATIONS [pp.489–490]

1. On Easter Island, a population of 15,000 humans destroyed their environment by using all of the available resources. Easter Island could no longer support a large human population. At the present time on our larger island, planet Earth, 6.3 billion people could well be in the process of destroying their environment; 2. i; 3. f; 4. l; 5. e; 6. g; 7. a; 8. d; 9. k; 10. b; 11. j; 12. c; 13. h; 14. Let X = size of the entire population. 5/25 = 20/*x*, *x* = (25)(20)/5 = 100.

## 28.2. POPULATION SIZE AND EXPONENTIAL GROWTH [pp.491–492]
## 28.3. LIMITS ON THE GROWTH OF POPULATIONS [pp.492–494]

1. immigration; 2. emigration; 3. migration; 4. zero population; 5. per capita; 6. heads; 7. 0.5; 8. 0.1; 9. reproduction; 10. time; 11. 0.4; 12. *G* = *rN*; 13. J; 14. exponential; 15. doubling; 16. biotic potential; 17.a. 0.4 per rat per month; b. 0.1 per rat per month; c. 0.3 per rat per month; 18. population growth rate; 19. environmental; 20. exponential; 21. slows; 22. dead; 23. nutrients; 24. environmental; 25. limiting factor; 26. population; 27. bacteria; 28. metabolic; 29. exponential; 30. sustainable; 31. Carrying capacity; 32. logistic growth; 33. S; 34. death; 35. birth; 36. b; 37. a; 38. a; 39. b; 40. a; 41. b; 42. b.

## 28.4. LIFE HISTORY PATTERNS [pp.495–498]

1. e; 2. c; 3. g; 4. a; 5. b; 6. f; 7. d; 8. has; 9. T; 10. T; 11. guppy; 12. faster; 13. T; 14. a genetic; 15. larger; 16. T; 17. natural selection.

## 28.5. HUMAN POPULATION GROWTH [pp.498–501]

1. Humans steadily developed the capacity to expand into new habitats and new climate zones (learning and language); humans increased the carrying capacity in their existing habitats (a shift from following migratory game herds to agriculture); human populations sidestepped limiting factors (medical practices and sanitary conditions improved as well as the ability to harness energy); 2. Density-dependent controls will kick in; 3. The total fertility rate (TFR) is the average number of children born to women of a population during their reproductive years; 4. In 1950, the worldwide TFR averaged 6.5 but by 2003, it had declined to 2.8. This still is above replacement-level fertility. Current replacement-level fertility is 2.1 for developed countries and up to 2.5 in a few developing countries; 5. d; 6. c; 7. a; 8. b; 9. b, slow growth; 10. c, rapid growth; 11. b; 12. d; 13. a; 14. c; 15. 4.6; 16. 21; 17. 25; 18. 25; 19. 1; 20. 3; 21. 3; 22. 27.

*Self-Quiz*
1. d; 2. a; 3. d; 4. b; 5. d; 6. b; 7. c; 8. a; 9. a; 10. b.

# Chapter 29 Community Structure and Biodiversity

*Fire Ants in the Pants* [p.503]

## 29.1. WHICH FACTORS SHAPE COMMUNITY STRUCTURE? [pp.504–505]

1. *Solenopsis invicta*, one of the Argentine fire ants, can be attacked and killed in its native habitat by parts of the life cycle of two flies. Both are parasitoids. Other options include importation of microbes that will infect *S. invicta* but not native ants; 2. climate; 3. food; 4. adaptive; 5. survive; 6. interact; 7. community; 8. e; 9. h; 10. g; 11. c; 12. j; 13. a; 14. k; 15. b; 16. f; 17. i; 18. d.

## 29.2. MUTUALLY BENEFICIAL INTERACTIONS [p.505]
## 29.3. COMPETITIVE INTERACTIONS [pp.506–507]

1.a. The larval stages of the moth grow only in the yucca plant; they eat only yucca seeds; b. Every plant species of the genus *Yucca* can be pollinated only by one species of the yucca moth genus; the yucca moth is the plant's only pollinator; 2. c; 3. b; 4. d; 5. b; 6. a.

## 29.4. PREDATOR–PREY INTERACTIONS [pp.508–511]

1. e; 2. a; 3. f; 4. f; 5. d; 6. f; 7. c; 8. b; 9. c; 10. e; 11. b.

## 29.5. PARASITES AND PARASITOIDS [pp.511–512]

1. Parasites; 2. nutrients; 3. predation; 4. sterility; 5. host; 6. birth; 7. death; 8. death; 9. secondary; 10. reproductive; 11. novel; 12. parasites; 13. invertebrates; 14. parasitic; 15. social; 16. T.

## 29.6. CHANGES IN COMMUNITY STRUCTURE OVER TIME [pp.513–514]

1. a; 2. c; 3. a; 4. b; 5. b; 6. a; 7. a; 8. c.

**29.7. FORCES CONTRIBUTING TO COMMUNITY INSTABILITY** [pp.515–517]
1. b; 2. d; 3. c; 4. a; 5. e.

**29.8. PATTERNS OF SPECIES DIVERSITY** [pp.518–519]
1. equator; 2. tropics; 3. sunlight; 4. resource; 5. tree; 6. populations; 7. reinforcing; 8. tree; 9. herbivores; 10. predatory; 11. parasitic; 12. reefs; 13. Island c; 14. Island c; 15. Island e.

**29.9. CONSERVATION BIOLOGY** [pp.519–523]
1. f; 2. i; 3. d; 4. h; 5. b; 6. j; 7. a; 8. k; 9. c; 10. g; 11. e.

*Self-Quiz*
1. a; 2. a; 3. d; 4. b; 5. d; 6. e; 7. e; 8. e; 9. b; 10. d.

---

# Chapter 30    Ecosystems

*Bye-Bye Bayou* [p.526]

**30.1. THE NATURE OF ECOSYSTEMS** [pp.527–529]
1. The major culprits are the rising air temperature and the burning of fossil fuels. Present wetlands will sink with a rise in sea level. Warmer water will cause massive fish kills. Heat waves and wildfires will become more intense and deaths from heat stroke will climb. More mosquitos will bring more human diseases. Severe flooding is predicted for some coastal regions; 2. e; 3. g; 4. d; 5. b; 6. i; 7. f; 8. a; 9. h; 10. c; 11. ecosystem; 12. trophic; 13. food chain; 14. food web; 15. energy; 16. four or five; 17. energy; 18. grazing; 19. carnivores; 20. detrital; 21. decomposers.

**30.2. BIOLOGICAL MAGNIFICATION IN FOOD WEBS** [p.530]
1. Biological magnification refers to a substance that degrades slowly or not at all and becomes more and more concentrated in tissues of organisms at ever-higher trophic levels of a food web. DDT is a well known and documented example. Rachel Carson, in her book, *Silent Spring,* alerted the world to the harmful effects of unregulated pesticide use. She studied a sanctuary that had been sprayed with DDT to control mosquitos.

**30.3. STUDYING ENERGY FLOW THROUGH ECOSYSTEMS** [p.531]
1. e; 2. a; 3. c; 4. d; 5. f; 6. b.

**30.4. GLOBAL CYCLING OF WATER AND NUTRIENTS** [pp.532–533]
1.a. Oxygen and hydrogen move in the form of water molecules; b. A large percentage of the nutrient is in the form of atmospheric gas such as carbon and nitrogen (mainly $CO_2$); c. Nutrients are not in gaseous forms; nutrients move from land to the seafloor and only "return" to dry land through geological uplifting, which may take millions of years; phosphorus is an example; 2. d; 3. f; 4. h; 5. i; 6. b; 7. a; 8. e; 9. g; 10. c.

**30.5. CARBON CYCLE** [pp.534–535]
**30.6. GREENHOUSE GASES, GLOBAL WARMING** [pp.536–537]
1. b; 2. a; 3. a; 4. b; 5. a; 6. b; 7. b; 8. a; 9. a; 10. b; 11. a; 12. a; 13. b.

**30.7. NITROGEN CYCLE** [pp.538–539]
1. g; 2. e; 3. i; 4. c; 5. a; 6. h; 7. d; 8. b; 9. j; 10. f.

**30.8. PHOSPHORUS CYCLE** [pp.540–541]
1. phosphorus; 2. reservoir; 3. ions; 4. ocean; 5. insoluble; 6. seafloor; 7. ions; 8. water; 9. waste; 10. decomposers; 11. hydrologic; 12. limiting; 13. human; 14. developing; 15. fertilizers; 16. algal; 17. ecosystems; 18. Phosphorus; 19. oxygen; 20. Eutrophication; 21. nutrients.

*Self-Quiz*
1. c; 2. c; 3. d; 4. b; 5. a; 6. d; 7. a; 8. d; 9. b; 10. d.

---

# Chapter 31    The Biosphere

*Surfers, Seals, and the Sea* [p.543]

**31.1. AIR CIRCULATION AND CLIMATES** [pp.544–547]
1. That winter, a massive volume of warm water from the southwestern Pacific moved east. It piled into coasts from California down through Peru, displacing currents that otherwise would have churned up tons of nutrients from the deep. This was an El Niño event; 2. Climate is affected by the complex interaction of the amount of solar radiation, the distribution of land masses and oceans, and topography; 3. f; 4. c; 5. h; 6. b; 7. a; 8. g; 9. i; 10. e; 11. d.

### 31.2. THE OCEAN, LANDFORMS, AND CLIMATES [pp.548–549]
1. heated; 2. cooled; 3. higher; 4. gravity; 5. warms; 6. currents; 7. rotation; 8. clockwise; 9. counterclockwise; 10. east; 11. summer; 12. heat; 13. warm; 14. heat; 15. Topography; 16. rain shadow; 17. wind; 18. Monsoons; 19. rainfall; 20. monsoons.

### 31.3. REALMS OF BIODIVERSITY [pp.550–556]
1. c; 2. d; 3. c; 4. e; 5. a; 6. b; 7. c; 8. d; 9. c; 10. d; 11. c; 12. a; 13. b; 14. c; 15. d; 16. c; 17. e; 18. c; 19. d; 20. c.

### 31.4. THE WATER PROVINCES [pp.557–561]
1. littoral; 2. limnetic; 3. profundal; 4. wind; 5. thermocline; 6. d; 7. b; 8. a; 9. c; 10. d; 11. e; 12. f; 13. a; 14. b; 15. a; 16. c; 17. f; 18. e; 19. a; 20. benthic (c); 21. pelagic (a); 22. neritic (b); 23. oceanic (d).

### 31.5. APPLYING KNOWLEDGE OF THE BIOSPHERE [pp.562–563]
1. f; 2. c; 3. e; 4. a; 5. g; 6. b; 7. d; 8. h.

*Self-Quiz*
1. e; 2. c; 3. d; 4. b; 5. d; 6. e; 7. b; 8. c; 9. a; 10. d.

## Chapter 32  Behavioral Ecology

*My Pheromones Made Me Do It* [p.566]

### 32.1. SO WHERE DOES BEHAVIOR START? [pp.567–571]
1. Africanized honeybees are more aggressive as a reaction to isopentyl acetate, a chemical that smells like bananas and functions as an alarm pheromone. A honeybee releases an alarm pheromone when threatened and when busily stinging something. Other bees follow the gradient of pheromone molecules diffusing through the air; 2. f; 3. d; 4. e; 5. b; 6. g; 7. a; 8. c.

### 32.2. COMMUNICATION SIGNALS [pp.572–573]
1. c; 2. a; 3. b; 4. a; 5. b; 6. d; 7. b; 8. a; 9. b; 10. a; 11. d; 12. a; 13. a.

### 32.3. MATES, OFFSPRING, AND REPRODUCTIVE SUCCESS [pp.574–575]
1. a. Hanging flies; b. Fiddler crabs; c. Sage grouse; d. Lions, sheep, elk, elephant seals, and bison; e. Midwife toads, crocodilians, Caspian terns, and grizzly bears.

### 32.4. COSTS AND BENEFITS OF SOCIAL GROUPS [pp.576–577]
1. d; 2. b; 3. a; 4. e; 5. c.

### 32.5. WHY SACRIFICE YOURSELF? [pp.578–579]
1. altruistic; 2. genes; 3. inclusive fitness; 4. relatives; 5. genetic; 6. one-half; 7. parents; 8. one-fourth; 9. self-sacrifice; 10. extended; 11. fingerprinting; 12. relatives; 13. genetically; 14. inbred; 15. unrelated; 16. behavior.

### 32.6. A LOOK AT PRIMATE SOCIAL BEHAVIOR [p.580]
### 32.7. HUMAN SOCIAL BEHAVIOR [p.581]
1. b; 2. c; 3. b; 4. b; 5. a; 6. a; 7. b; 8. b; 9. b; 10. a.

*Self-Quiz*
1. d; 2. b; 3. b; 4. b; 5. c; 6. e; 7. c; 8. b; 9. e; 10. c; 11. d.